T0250024

INTRODUCTION TO COMPUTATIONAL MODELING USING C AND OPEN-SOURCE TOOLS

José M. Garrido

Kennesaw State University
Kennesaw Georgia, USA

CRC Press
Taylor & Francis Group
Boca Raton   London   New York

CRC Press is an imprint of the
Taylor & Francis Group, an **informa** business

A CHAPMAN & HALL BOOK

Chapman & Hall/CRC
Computational Science Series

SERIES EDITOR

Horst Simon
Deputy Director
Lawrence Berkeley National Laboratory
Berkeley, California, U.S.A.

PUBLISHED TITLES

COMBINATORIAL SCIENTIFIC COMPUTING
Edited by Uwe Naumann and Olaf Schenk

CONTEMPORARY HIGH PERFORMANCE COMPUTING: FROM PETASCALE
TOWARD EXASCALE
Edited by Jeffrey S. Vetter

DATA-INTENSIVE SCIENCE
Edited by Terence Critchlow and Kerstin Kleese van Dam

PETASCALE COMPUTING: ALGORITHMS AND APPLICATIONS
Edited by David A. Bader

FUNDAMENTALS OF MULTICORE SOFTWARE DEVELOPMENT
Edited by Victor Pankratius, Ali-Reza Adl-Tabatabai, and Walter Tichy

GRID COMPUTING: TECHNIQUES AND APPLICATIONS
Barry Wilkinson

HIGH PERFORMANCE COMPUTING: PROGRAMMING AND APPLICATIONS
John Levesque with Gene Wagenbreth

HIGH PERFORMANCE VISUALIZATION:
ENABLING EXTREME-SCALE SCIENTIFIC INSIGHT
Edited by E. Wes Bethel, Hank Childs, and Charles Hansen

INTRODUCTION TO COMPUTATIONAL MODELING USING C AND
OPEN-SOURCE TOOLS
José M Garrido

INTRODUCTION TO CONCURRENCY IN PROGRAMMING LANGUAGES
Matthew J. Sottile, Timothy G. Mattson, and Craig E Rasmussen

INTRODUCTION TO ELEMENTARY COMPUTATIONAL MODELING: ESSENTIAL
CONCEPTS, PRINCIPLES, AND PROBLEM SOLVING
José M. Garrido

INTRODUCTION TO HIGH PERFORMANCE COMPUTING FOR SCIENTISTS
AND ENGINEERS
Georg Hager and Gerhard Wellein

PUBLISHED TITLES CONTINUED

MATLAB® and Simulink® are trademarks of The MathWorks, Inc. and are used with permission. The MathWorks does not warrant the accuracy of the text or exercises in this book. This book's use or discussion of MATLAB® and Simulink® software or related products does not constitute endorsement or sponsorship by The MathWorks of a particular pedagogical approach or particular use of the MATLAB® and Simulink® software.

Original art "Aztec Calendar" by Victor H. Verde.

Photograph and graphic design by Tino Garrido-Licha.

CRC Press
Taylor & Francis Group
6000 Broken Sound Parkway NW, Suite 300
Boca Raton, FL 33487-2742

© 2014 by Taylor & Francis Group, LLC
CRC Press is an imprint of Taylor & Francis Group, an Informa business

No claim to original U.S. Government works

Version Date: 20131004

International Standard Book Number-13: 978-1-4822-1678-3 (Hardback)

This book contains information obtained from authentic and highly regarded sources. Reasonable efforts have been made to publish reliable data and information, but the author and publisher cannot assume responsibility for the validity of all materials or the consequences of their use. The authors and publishers have attempted to trace the copyright holders of all material reproduced in this publication and apologize to copyright holders If permission to publish in this form has not been obtained. If any copyright material has not been acknowledged please write and let us know so we may rectify in any future reprint.

Except as permitted under U.S. Copyright Law, no part of this book may be reprinted, reproduced, transmitted, or utilized in any form by any electronic, mechanical, or other means, now known or hereafter invented, including photocopying, microfilming, and recording, or in any information storage or retrieval system, without written permission from the publishers.

For permission to photocopy or use material electronically from this work, please access www.copyright.com (http://www.copyright.com/) or contact the Copyright Clearance Center, Inc. (CCC), 222 Rosewood Drive, Danvers, MA 01923, 978-750-8400. CCC is a not-for-profit organization that provides licenses and registration for a variety of users. For organizations that have been granted a photocopy license by the CCC, a separate system of payment has been arranged.

Trademark Notice: Product or corporate names may be trademarks or registered trademarks, and are used only for identification and explanation without intent to infringe.

Library of Congress Cataloging-in-Publication Data

Garrido, Jose M.
     Introduction to computational modeling using C and open-source tools / Jose M. Garrido.
        pages cm. -- (Chapman & Hall/CRC computational science)
     Includes bibliographical references and index.
     ISBN 978-1-4822-1678-3 (hardback)
     1. Mathematical models--Data processing. 2. C (Computer program language) 3. Open source software. I. Title.

QA401.G274 2014
511.80285'5133--dc23   2013039506

Visit the Taylor & Francis Web site at
http://www.taylorandfrancis.com

and the CRC Press Web site at
http://www.crcpress.com

Contents

List of Figures

List of Tables

Preface

A *computational model* is a computer implementation of the solution to a (scientific) problem for which a mathematical representation has been formulated. These models are applied in various areas of science and engineering to solve large-scale and complex scientific problems. Developing a computational model involves formulating the mathematical representation and implementing it by applying computer science concepts, principles and methods.

Computational modeling focuses on reasoning about problems using computational thinking and multidisciplinary/interdisciplinary computing for developing computational models to solve complex problems. It is the foundation component of computational science, which is an emerging discipline that includes concepts, principles, and methods from applied mathematics and computer science.

This book presents an introduction to computational models and their implementation using the C programming language, which is still one of the most widely used programming languages. Fortran and C programming languages are the ones most suitable for high-performance computing (HPC). Although these programming languages are not new, they have evolved to become very powerful and efficient programming languages. The most recent standards are ISO/IEC 1539-1:2010 for Fortran and ISO/IEC 1999 for the C language.

MATLAB® is a registered trademark of The Mathworks, Inc.

3 Apple Hill Drive
Natick, MA 01760-2098 USA
Tel: 508 647 7000
Fax: 508-647-7001
E-mail: info@mathworks.com
Web: www.mathworks.com

The primary goal of this book is to present basic and introductory principles of computational models from the computer science perspective. The prerequisites are intermediate programming and at least Calculus I. Emphasis is on reasoning about problems, conceptualizing the problem, mathematical formulation, and the computational solution that involves computing results and visualization.

The book emphasizes analytical skill development and problem solving. The main software tools for implementing computational models are the C programming language and the Gnu Scientific Library (GSL) under Linux. The GSL is a software library of C functions that is freely available; it is a mature, stable, and well-documented library that has been well-supported since 1996. GnuPlot is an open-source versatile tool for the visualization of the data computed. Other software tools used in the book are GLPK, LP_Solve and CodeBlocks.

The material in this book is aimed at intermediate to advanced undergraduate science (and engineering) students. However, the vision in the book is to promote and introduce the principles of computational modeling as early as possible in the undergraduate curricula and to introduce the approaches of multidisciplinary and interdisciplinary computing.

This book provides a foundation for more advanced courses in scientific computing, including parallel computing using MPI, grid computing and other techniques used in high-performance computing.

The material in the book is presented in five parts. The first part is an overview of problem solving, introductory concepts and principles of computational models, and their development. This part introduces the basic modeling and techniques for designing and implementing problem solutions, independent of software and hardware tools.

The second part presents an overview of programming principles with the C programming language. The relevant topics are basic programming concepts, data definitions, programming structures with flowcharts and pseudo-code, solving problems and algorithms, arrays, pointers, basic data structures, and compiling, linking, and executing programs on Linux.

The third part applies programming principles and techniques to implement the basic computational models. It gradually introduces numerical methods and mathematical modeling principles. Simple case studies of problems that apply mathematical models are presented. Case studies are of simple linear, quadratic, geometric, polynomial, and linear systems using GSL. Computational models that use polynomial evaluation, computing roots of polynomials, interpolation, regression, and systems of linear equations are discussed. Examples and case studies demonstrate the computation and visualization of data produced by computational models.

The fourth part presents an overview of more advanced concepts needed for modeling dynamical systems. Most of the models are formulated with ordinary differential equations, and the implementation of numerical solutions is explained.

The fifth part introduces the modeling of linear optimization problems and several case studies are presented. The problem formulation to implementation of computational models with linear optimization is shown.

José M. Garrido
Kennesaw, Georgia

About the Author

José M. Garrido is professor in the Department of Computer Science, Kennesaw State University, Georgia. He holds a Ph.D. from George Mason University in Fairfax, Virginia, an M.S.C.S also from George Mason University, an M.Sc. from the University of London, and a B.S. in electrical engineering from the Universidad de Oriente, Venezuela.

Dr. Garrido's research interests are object-oriented modeling and simulation, multidisciplinary computational modeling, formal specification of real-time systems, language design and processors, and modeling systems performance. Dr. Garrido developed the Psim3, PsimJ, and PsimJ2 simulation packages for C++ and Java. He has recently developed the OOSimL, the object-oriented simulation language (with partial support from the NSF).

Dr. Garrido has published several papers on modeling and simulation, and on programming methods. He has also published seven textbooks on object-oriented simulation, operating systems, and elementary computational models.

Chapter 1

Problem Solving and Computing

1.1   Introduction

Computer problem solving attempts to derive a computer solution to a real-world problem, and a computer *program* is the implementation of the solution to the problem. A *computational model* is a computer implementation of the solution to a (scientific) problem for which a mathematical representation has been formulated. These models are applied in various areas of science and engineering to solve large-scale and complex scientific problems.

A computer program consists of data definitions and a sequence of instructions. The instructions allow the computer to manipulate the input data to carry out computations and produce desired results when the program executes. A program is normally written in an appropriate programming language.

This chapter discusses problem solving principles and presents elementary concepts and principles of problem solving, computational models, and programs.

1.2   Computer Problem Solving

Problem solving is the process of developing a computer solution to a given real-world problem. The most challenging aspect of this process is discovering the method to solve the problem. This method of solution is described by an algorithm. A general process of problem solving involves the following tasks:

1. Understanding the problem

2. Describing the problem in a clear, complete, and unambiguous form

3. Designing a solution to the problem (algorithm)

4. Developing a computer solution to the problem.

An *algorithm* is a description of the sequence of steps performed to produce the desired results, in a clear, detailed, precise, and complete manner. It describes the computations on the given data and involves a sequence of instructions or operations that are to be performed on the input data in order to produce the desired results (output data).

A program is a computer implementation of an algorithm and consists of a set of data definitions and sequences of instructions. The program is written in an appropriate programming language and it tells the computer how to transform the given data into correct results by performing a sequence of computations on the data. An algorithm is described in a semi-formal notation such as pseudo-code and flowcharts.

1.3   Elementary Concepts

A *model* is a representation of a system, a problem, or part of it. The model is simpler than, and should be equivalent to, the real system in all relevant aspects. In this sense, a model is an abstract representation of a problem. *Modeling* is the activity of building a model.

Every model has a specific purpose and goal. A model only includes the aspects of the real problem that were decided as being important, according to the initial requirements of the model. This implies that the limitations of the model have to be clearly understood and documented.

An essential modeling method is to use mathematical entities such as numbers, functions, and sets to describe properties and their relationships to problems and real-world systems. Such models are known as *mathematical models*.

A *computational model* is an implementation in a computer system of a mathematical model and usually requires high performance computational resources to execute. The computational model is used to study the behavior of a large and complex system. Developing a computational model consists of:

- Applying a formal software development process

- Applying *Computer Science* concepts, principles and methods, such as:

  - Abstraction and decomposition

  - Programming principles

  - Data structures

  - Algorithm structures

  - Concurrency and synchronization

  - Modeling and simulation

  - Multi-threading, parallel, and distributed computing for high performance (HPC)

Abstraction is a very important principle in developing computational models. This is extremely useful in dealing with large and complex problems or systems. Abstraction is the hiding of the details and leaving visible only the essential features of a particular system.

One of the critical tasks in modeling is representing the various aspects of a system at different levels of abstraction. A good abstraction captures the essential elements of a system, and purposely leaves out the rest.

Computational thinking is the ability of reasoning about a problem and formulating a computer solution. Computational thinking consists of the following elements:

- Reasoning about computer problem solving

- The ability to describe the requirements of a problem and, if possible, design a mathematical solution that can be implemented in a computer

- The solution usually requires *multi-disciplinary* and *inter-disciplinary* approaches to problem solving

- The solution normally leads to the construction of a *computational model*

Computational Science integrates concepts and principles from applied mathematics and computer science and applies them to the various scientific and engineering disciplines. Computational science is:

- An emerging multidisciplinary area

- The intersection of the more traditional sciences, engineering, applied mathematics, and computer science, and focuses on the integration of knowledge for the development of problem-solving methodologies and tools that help advance the sciences and engineering areas. This is illustrated in Figure 1.1.

- An area that has as a general goal the development of high-performance computer models.

- An area that mostly involve multi-disciplinary computational models including simulation.

When a mathematical analytical solution of the model is not possible, a numerical and graphical solution is sought and experimentation with the model is carried out by changing the parameters of the model in the computer, and studying the differences in the outcome of the experiments. Further analysis and predictions of the operation of the model can be derived or deduced from these computational experiments.

One of the goals of the general approach to problem solving is modeling the problem at hand, building or implementing the resulting solution using an appropriate tool environment (such as MATLAB® or Octave) or with some appropriate programming language, such as C.

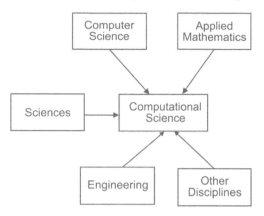

FIGURE 1.1: Computational science as an integration of several disciplines.

1.4 Developing Computational Models

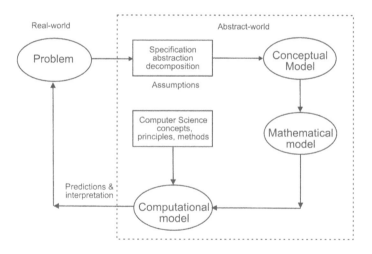

FIGURE 1.2: Development of computational models.

The process of developing computational models consists of a sequence of activities or stages that starts with the definition of modeling goals and is carried out in a possibly iterative manner. Because models are simplifications of reality there is a trade-off as to what level of detail is included in the model. If too little detail is included in the model one runs the risk of missing relevant interactions and the resultant model does not promote understanding. If too much detail is included in the model the model may become overly complicated and actually preclude the development of understanding. Figure 1.2 illustrates a simplified process for developing computational models.

Computational models are generally developed in an iterative manner. After the first version of the model is developed, the model is executed, results from the execution run are studied, the model is revised, and more iterations are carried out until an adequate level of understanding is developed. The process of developing a model involves the following general steps:

1. Definition of the *problem statement* for the computational model. This statement must provide the description of the purpose for building the model, the questions it must help to answer, and the type of expected results relevant to these questions.

2. Definition of the *model specification* to help define the conceptual model of the problem to be solved. This is a description of what is to be accomplished with the computational model to be constructed; and the assumptions (constraints), and domain laws to be followed. Ideally, the model specification should be clear, precise, complete, concise, and understandable. This description includes the list of relevant components, the interactions among the components, the relationships among the components, and the dynamic behavior of the model.

3. Definition of the *mathematical model*. This stage involves deriving a representation of the problem solution using mathematical entities and expressions and the details of the algorithms for the relationships and dynamic behavior of the model.

4. *Model implementation*. The implementation of the model can be carried out with a software environment such as MATLAB and Octave, in a simulation language, or in a general-purpose high-level programming language, such as Ada, C, C++, or Java. The simulation software to use is also an important practical decision. The main tasks in this phase are coding, debugging, and testing the software model.

5. *Verification* of the model. From different runs of the implementation of the model (or the model program), this stage compares the output results with those that would have been produced by correct implementation of the conceptual and mathematical models. This stage concentrates on attempting to document and prove the correctness of the model implementation.

6. *Validation* of the model. This stage compares the outputs of the verified model with the outputs of a real system (or a similar already developed model). This stage compares the model data and properties with the available knowledge and data about the real system. Model validation attempts to evaluate the extent to which the model promotes understanding.

A conceptual model can be considered a high-level specification of the problem

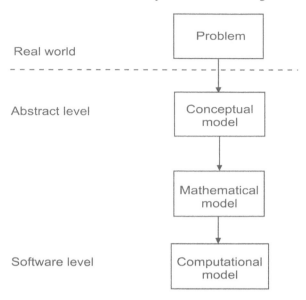

FIGURE 1.3: Model development and abstract levels.

and it is a descriptive model. It is usually described with some formal or semi-formal notation. For example, discrete-event simulation models are described with UML (the Unified Modeling Language) and/or extended simulation activity diagrams.

The conceptual model is formulated from the initial problem statement, informal user requirements, and data and knowledge gathered from analysis of previously developed models. The stages mentioned in the model development process are carried out at different levels of abstraction. Figure 1.3 illustrates the relationship between the various stages of model development and their abstraction level.

1.5   A Simple Problem: Temperature Conversion

The process of developing a computational model is illustrated in this section with an extremely simple problem: the temperature conversion problem. A basic sequence of steps is discussed for solving this problem and for developing a computational model.

1.5.1   Initial Problem Statement

American tourists visiting Europe do not usually understand the units of temperature used in weather reports. The problem is to devise some mechanism for indicating the temperature in Fahrenheit from a known temperature in Celsius.

1.5.2 Analysis and Conceptual Model

A brief analysis of the problem involves:

1. Understanding the problem. The main goal of the problem is to develop a temperature conversion facility from Celsius to Fahrenheit.

2. Finding the mathematical representation or formulas for the conversion of temperature from Celsius to Fahrenheit. Without this knowledge, we cannot derive a solution to this problem. The conversion formula is the mathematical model of the problem.

3. Knowledge of how to implement the mathematical model in a computer. We need to express the model in a particular computer tool or a programming language. The computer implementation must closely represent the model in order for it to be correct and useful.

4. Knowledge of how to test the program for correctness.

1.5.3 Mathematical Model

The mathematical representation of the solution to the problem is the formula expressing a temperature measurement F in Fahrenheit in terms of the temperature measurement C in Celsius, which is:

$$F = \frac{9}{5}C + 32 \tag{1.1}$$

Here C is a variable that represents the given temperature in degrees Celsius, and F is a derived variable, whose value depends on C.

A formal definition of a function is beyond the scope of this chapter. Informally, a *function* is a computation on elements in a set called the *domain* of the function, producing results that constitute a set called the *range* of the function. The elements in the domain are sometimes known as the input parameters. The elements in the range are called the output results.

Basically, a function defines a relationship between two (or more variables), x and y. This relation is expressed as $y = f(x)$, so y is a function of x. Normally, for every value of x, there is a corresponding value of y. Variable x is the independent variable and y is the dependent variable.

The mathematical model is the mathematical expression for the conversion of a temperature measurement in Celsius to the corresponding value in Fahrenheit. The mathematical formula expressing the conversion assigns a value to the desired temperature in the variable F, the dependent variable. The values of the variable C can change arbitrarily because it is the independent variable. The model uses real numbers to represent the temperature readings in various temperature units.

1.6   Categories of Computational Models

From the perspective of how the model changes state in time, computational models can be divided into two general categories:

1. Continuous models

2. Discrete models

A *continuous model* is one in which the changes of state in the model occur continuously with time. Often the *state variables* in the model are represented as continuous functions of time. These types of models are usually modeled as sets of difference or differential equations.

For example, a model that represents the temperature in a boiler as part of a power plant can be considered a continuous model because the state variable that represents the temperature of the boiler is implemented as a continuous function of time.

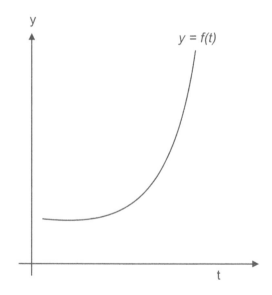

FIGURE 1.4: Continuous model.

In scientific and engineering practice, a computational model of a real physical system is often formulated as a continuous model and solved numerically by applying numerical methods implemented in a programming language. These models can also be simulated with software tools such as Simulink and Scilab which are computer programs designed for numeric computations and visualization. Figure 1.4 illustrates how the a variable changes with time.

A *discrete model* represents a system that changes its states at discrete points in time, i.e., at specific instants. The model of a simple car-wash system is a discrete-event model because an arrival event occurs, and causes a change in the state variable that represents the number of cars in the queue that are waiting to receive service from the machine (the server). This state variable and any other only change its values when an event occurs, i.e., at discrete instants. Figure 1.5 illustrates the changes in the number of cars in the queue of the model for the simple car-wash system.

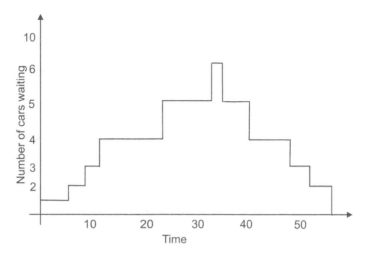

FIGURE 1.5: Discrete changes of number of cars in the queue.

Depending on the variability of some parameters, computational models can be separated into two categories:

1. Deterministic models

2. Stochastic models.

A deterministic model exhibits a completely predictable behavior. A stochastic model includes some uncertainty implemented with random variables, whose values follow a probabilistic distribution. In practice, a significant number of models are stochastic because the real systems modeled usually include properties that are inherently random.

An example of a deterministic simulation model is a model of a simple car-wash system. In this model, cars arrive at exact specified instants (but at the same instants), and all have exact specified service periods (wash periods); the behavior of the model can be completely and exactly determined.

The simple car-wash system with varying car arrivals, varying service demand from each car, is a stochastic system. In a model of this system, only the averages of these parameters are specified together with a probability distribution for the variability of these parameters. Uncertainty is included in this model because these parameter values cannot be exactly determined.

1.7 Computing the Area and Circumference of a Circle

In this section, another simple problem is formulated: the mathematical model(s) and the algorithm. This problem requires computing the area and circumference of a circle, given its radius. The mathematical model is:

$$cir = 2\pi r$$
$$area = \pi r^2$$

The high-level algorithm description in informal pseudo-code notation is:

1. Read the value of the radius of a circle, from the input device.

2. Compute the area of the circle.

3. Compute the circumference of the circle.

4. Print or display the value of the area of the circle to the output device.

5. Print or display the value of the circumference of the circle to the output device.

A more detailed algorithm description follows:

1. Read the value of the radius r of a circle, from the input device.

2. Establish the constant π with value 3.14159.

3. Compute the area of the circle.

4. Compute the circumference of the circle.

5. Print or display the value of *area* of the circle to the output device.

6. Print or display the value of *cir* of the circle to the output device.

The previous algorithm now can be implemented by a program that calculates the circumference and the area of a circle.

1.8   General Process of Software Development

For large software systems, a general software development process involves carrying out a sequence of well-defined phases or activities. The process is also known as the *software life cycle*.

The simplest approach for using the software life cycle is the *waterfall model*. This model represents the sequence of phases or activities needed to develop the software system through installation and maintenance of the software. In this model, the activity in a given phase cannot be started until the activity of the previous phase has been completed.

Figure 1.6 illustrates the sequence of phases that are performed in the waterfall software life cycle. The various phases of the software life cycle are the following:

1. *Analysis*, which results in documenting the problem description and what the problem solution is supposed to accomplish.

2. *Design*, which involves describing and documenting the detailed structure and behavior of the system model.

3. *Implementation* of the software using a programming language.

4. *Testing* and verification of the programs.

5. Installation that results in delivery, installation of the programs.

6. *Maintenance*.

There are some variations of the waterfall model of the life cycle. These include returning to the previous phase when necessary. More recent trends in system development have emphasized an iterative approach, in which previous stages can be revised and enhanced.

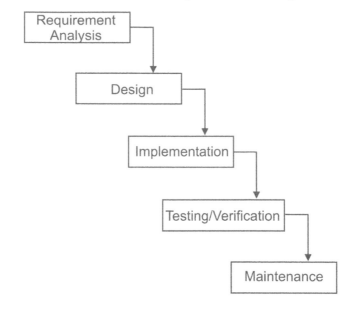

FIGURE 1.6: The waterfall model.

A more complete model of software life cycle is the *spiral model* that incorporates the construction of *prototypes* in the early stages. A prototype is an early version of the application that does not have all the final characteristics. Other development approaches involve prototyping and rapid application development (RAD).

1.9   Programming Languages

A programming language is used by programmers to write programs. This language contains a defined set of syntax and semantic rules. The syntax rules describe how to write well-defined statements. The semantic rules describe the meaning of the statements.

1.9.1   High-Level Programming Languages

A high-level programming language is a formal notation in which to write instructions to the computer in the form of a program. A programming language help programmers in the writing of programs for a large family of problems.

High-level programming languages are hardware independent and are problem-oriented (for a given family of problems). These languages allow more readable programs, and are easy to write and maintain. Examples of these languages are Pascal, C, Cobol, Fortran, Algol, Ada, Smalltalk, C++, Eiffel, and Java.

Programming languages like C++ and Java can require considerable effort to learn and master. Several newer and experimental, higher-level, object-oriented programming languages have been developed. Each one has a particular goal.

There are several integrated development environments (IDE) that facilitate the development of programs. Examples of these are: Eiffel, Netbeans, CodeBlocks, and Codelite. Other IDEs are designed for numerical and scientific problem solving that have their own programming language. Some of these computational tools are: MATLAB, Octave, Mathematica, Scilab, Stella, and Maple.

The solution to a problem is implemented by a program written in an appropriate programming language. This program is known as the *source program* and is written in a high-level programming language.

Once a source program is written, it is translated or converted into an equivalent program in *machine code*, which is the only programming language that the computer can understand. The computer can only execute instructions that are in machine code.

The program executing in the computer usually reads input data from the input device and after carrying out some computations, it writes results to the output device(s).

1.9.2   Interpreters

An interpreter is a special program that performs syntax checking of a command in a user program written in a high-level programming language and immediately executes the command. It repeats this processing for every command in the program. Examples of interpreters are the ones used for the following languages: MATLAB, Octave, PHP and PERL.

1.9.3   Compilers

A compiler is a special program that translates another program written in a programming language into an equivalent program in binary or machine code, which is the only language that the computer accepts for processing.

In addition to *compilation*, an additional step known as *linking* is required before a program can be executed. Examples of programming languages that require compilation and linking are: C, C++, Eiffel, Ada, and Fortran. Other programming languages such as Java, require compilation and interpretation.

1.9.4   Compiling and Execution of Java Programs

To compile and execute programs written in the Java programming language, two special programs are required, the compiler and the interpreter. The Java compiler checks for syntax errors in the source program and translates it into *bytecode*, which is the program in an intermediate form. The Java bytecode is not dependent on any

FIGURE 1.7: Compiling a Java source program.

particular platform or computer system. To execute this bytecode, the Java Virtual Machine (JVM), carries out the interpretation of the bytecode.

Figure 1.7 shows what is involved in compilation of a source program in Java. The Java compiler checks for syntax errors in the source program and then translates it into a program in byte-code, which is the program in an intermediate form.

FIGURE 1.8: Executing a Java program.

The Java bytecode is not dependent of any particular platform or computer system. This makes the bytecode very portable from one machine to another.

Figure 1.8 shows how to execute a program in byte-code. The Java virtual machine (JVM) carries out the interpretation of the program in byte-code.

1.9.5   Compiling and Executing C Programs

FIGURE 1.9: Compiling a C program.

Programs written in C must be compiled, linked, and loaded into memory before executing. An executable program file is produced as a result of linking. The libraries are a collection of additional code modules needed by the program. Figure 1.9 illustrates the compilation of a C program. Figure 1.10 illustrate the linkage of the program. The executable program is the final form of the program that is produced. Before a program starts to execute in the computer, it must be loaded into the memory of the computer.

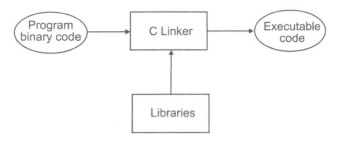

FIGURE 1.10: Linking a C program.

Summary

Computational models are used to solve large and complex problems in the various scientific and engineering disciplines. These models are implemented by computer programs coded in a particular programming language. There are several standard programming languages, such as C, C++, Eiffel, Ada, Java. Compilation is the task of translating a program from its source language to an equivalent program in machine code. Languages in scientific computing environments such as MATLAB and Octave are interpreted. Computations are carried out on input data by executing individual commands or complete programs. Application programs are programs that the user interacts with to solve particular problems. Computational models are implemented as programs.

Key Terms

computational model	mathematical model	abstraction
algorithm	conceptual model	model development
compilers	linkers	interpreters
programs	commands	instructions
programming language	Java	C
C++	Eiffel	Ada
bytecode	JVM	program execution
data definition	Source code	high-level language
simulation model	keywords	identifiers

Exercises

Exercise 1.1 Explain the differences between a computational model and a mathematical model.

Exercise 1.2 Explain the reason why the concept of abstraction is important in developing computational models.

Exercise 1.3 Investigate and write a short report on the programming languages used to implement computational models.

Exercise 1.4 What is a programming language? Why are they needed?

Exercise 1.5 Explain why there are many programming languages.

Exercise 1.6 What are the differences between compilation and interpretation in high-level programming languages?

Exercise 1.7 Explain the purpose of compilation. How many compilers are necessary for a given application? What is the difference between program compilation and program execution? Explain.

Exercise 1.8 What is the real purpose of developing a program? Can we just use a spreadsheet program such as MS Excel to solve numerical problems? Explain.

Exercise 1.9 Find and explain the differences in compiling and linking C, Java, and C++ programs.

Exercise 1.10 Explain the differences between data definitions and instructions in a program written in a high-level programming language.

Exercise 1.11 For developing small programs, is it still necessary to use a software development process? Explain. What are the main advantages in using a process for program development? What are the disadvantages?

Chapter 2

Programs

2.1 Introduction

This chapter presents an overview of the structure of a computer program, which include data definitions and basic instructions using the C programming language. Because functions are the building blocks and the fundamental components of C programs, the concepts of function definitions and function invocations are gradually explained, and complete C programs are introduced that illustrate further the role of functions. This chapter also presents concepts and principles that are used in developing computational models by implementing the corresponding mathematical models with C programs.

2.2 Programs

A *program* consists of data definitions and instructions that manipulate the data. A program is normally written in an appropriate *programming language*. It is considered part of the *software* components in a computer system. The general structure of a program consists of:

- *Data definitions*, which declare all the data to be manipulated by the instructions.

- A *sequence of instructions*, which perform the computations on the data in order to produce the desired results.

2.3   Data Definitions

The data in a program consists of one or more data items. These are manipulated or transformed by the computations (computer operations). Each data definition is specified by declaring a data item with the following:

- The *type* of the data item

- A unique *name* to identify the data item

- An optional initial *value*

The name of a data item is an *identifier* and is defined by the programmer; it must be different from any *keyword* in the programming language. The type of data defines:

- The set of possible values that the data item can take

- The set of possible operations or computations that can be applied to the data item

2.3.1   Name of Data Items

The names of the data items are used in a program for uniquely identifying the data items and are known as *identifiers*. The special text words or symbols that indicate essential parts of a programming language are known as *keywords*. These are reserved words and cannot be used for any other purpose.

For example, the problem for calculating the area of a triangle uses four data items, *a, b, c,* and *area*. The data items usually change their values when they are manipulated by the various operations. For example, the following sequence of instructions first sets the value of *y* to 34.5 then adds the value *x* and *y*; the results are assigned to *z*.

```
y = 34.5;
z = y + x;
```

The data items named *x* and *y* are known as *variables* because their values change when computations are applied on them. Those data items that do not change their values are known as *constants*. For example, *MAX_NUM, PI*, etc. These data items are given an initial value that will never change during the program execution.

When a program executes, all the data items used by the various computations are stored in the computer memory, each data item occupying a different memory location.

2.3.2 Data Types

Data types are classified into the three categories:

- Numeric

- Text

- Boolean

The numeric types are further divided into three basic types, *integer*, *float*, and *double*. The non-integer types are also known as fractional, which means that the numerical values have a fractional part.

Values of *integer* type are those that are countable to a finite value, for example, age, number of parts, number of students enrolled in a course, and so on. Values of type *float* and *double* have a decimal point; for example, cost of a part, the height of a tower, current temperature in a boiler, a time interval. These values cannot be expressed as integers. Values of type *double* provide more precision than type *float*.

Text data items are of two basic types: *char* and type *string*. Data items of type *string* consist of a sequence of characters. The values for these two types of data items are text values. The string type is the most common, such as the text value: 'Welcome!'.

The third data type is used for variables whose values can take any of two truth-values (*True* or *False*); these variables are of type *bool*. This type was first defined in the C99 standard for the C language.

2.3.3 Data Declarations in C

The data declarations are the data definitions and include the name of every variable or constant with its type. The initial values, if any, for the data items are also included in the data declaration.

In programming languages such as C, C++, and Java, a statement for the declaration of variables has the following basic syntactic structure:

```
⟨ type ⟩ ⟨ variable_name ⟩;
```

The following lines of code in the C programming language are examples of statements that declare two constants and three variables of type *int*, *float*, and *bool*.

```
const float  PI = 3.1416;
const int MAX_NUM = 100;
int count;
float weight = 57.85;
bool busy;
```

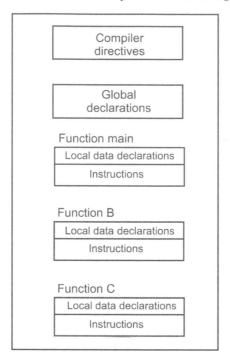

FIGURE 2.1: Structure of a C program.

2.4   Structure of a C Program

A typical program in the C language is structured as shown in Figure 2.1. It consists of several parts in the following sequence:

1. Compiler directives, which indicate to the compiler to append header files to the program that contain data and/or function declarations also known as function *prototypes*, and additional directives. A compiler directive starts with the pound symbol (#). A function prototype is only the header of a function and consists of the type of the function, the function name, and any parameters listed.

2. Global declarations, which may consist of data declarations and/or function declarations. These are global because they can be used by all functions in a program.

3. Definition of function *main*. This is the only required function in a C program; all other functions may be absent. In other words, a C program contains at lease function *main*. Execution of the program starts in this function, and normally ends in this function.

4. Definition of additional functions in the program.

Function definitions are the inclusion of programmer-defined functions that are invoked or called in the program. The other functions that can be called are the built-in functions provided by standard C *libraries* or by external libraries. In C, a library is a collection of related function definitions that may also include data declarations.

Once a function is defined, it can be called or invoked by any other function in the program. A function starts executing when it is called by another function. In a simple C program, a few functions are called in function *main*.

Before a function can be called in a C program, a function declaration is required. This declaration is also known as a function *prototype*. This is one of the reasons to have compiler directives and global declarations before the definition of function *main* and the other function definitions.

2.5 Instructions

An instruction performs a specific manipulation or computation on the data. In the source program, a computation is written with one or more appropriate language statements. In the source code, the *assignment statement* is the most fundamental statement (high-level instruction); its general form is:

⟨ *variable_name* ⟩ = ⟨ expression ⟩

The assignment *operator* is denoted by the = symbol and on the left side of this operator a variable name must always be written. On the right side of the assignment operator, an expression is written.

In the following source code, the first line is a simple assignment statement that will move the value 34.5 into variable x. The second line is a slightly more complex assignment that will perform an addition of the value of variable x and the constant 11.38. The result of the addition is assigned to variable y.

```
x = 34.5;     /* assign 34.5 to variable x */
y = x + 11.38;
```

In the previous example, the second line of source code is an assignment statement and the expression on the right side of the assignment operator is $x + 11.38$. The expression on the right side of an assignment statement may be a simple arithmetic expression or a much more complex expression.

2.6   Simple Functions

A function carries out a specific task in a program and is an internal decomposition unit because every function belongs to the program.

As mentioned previously, data declared within a function is known only to that function—the scope of the data is *local* to the function. The local data in a function has a limited lifetime; it only exists during execution of the function.

2.6.1   Function Definitions

A C program is normally *decomposed* into source files, and these are divided into functions. These are function definitions that implement the functions in the source program.

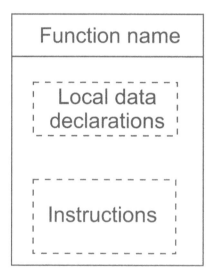

FIGURE 2.2: Structure of a C function.

Figure 2.2 illustrates the general form of a function in the C language. In the source code, the general form of a function in the C programming language is written as follows:

```
function_type function_name ( [parameters] ) {
    [ local declarations ]
    [ executable language statements ]
}
```

The relevant documentation of the function definition is described in one or more lines of comments, which begins with the characters slash-star (/*) and ends with a star-slash (*/).

The declarations that define local data in the function are optional. The instructions implement the body of the function. The following C source code shows a simple function for displaying a text message on the screen.

```
/*
   This function displays a message
   on the screen. */
void show_message () {
    printf("Computing data");
}
```

This is a very simple function and its only purpose is to display a text message on the screen. This function does not declare parameters and the type of this function is *void* to indicate that this function does not return a value.

2.6.2 Function Calls

The name of the function is used when calling or invoking the function by some other function. The function that calls another function is known as the calling function; the second function is known as the called function. When a function calls or invokes another function, the flow of control is altered and the second function starts execution immediately.

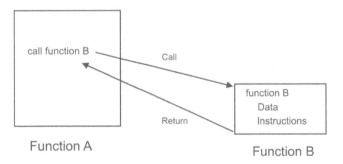

FIGURE 2.3: A function calling another function

When the called function completes execution, the flow of control is transferred back (returned) to the calling function and it continues execution from the point after it called the second function.

Figure 2.3 illustrates function *A* calling function *B*. After completing its execution, function B returns the flow of control to function A.

An example of this kind of function call is the call to function *show_message*, discussed previously. In C, the statement that calls a simple function uses the function name and an empty parentheses pair. The call to function *show_message* is written as:

```
show_message();
```

After completion, and depending on the function type, the called function may or may not return a value to the calling function.

2.7   A Simple C Program

Listing 2.1 shows a very simple but complete C program. Lines 1–6 include comments that help in documenting the program. Line 7 is a compiler directive that indicates the inclusion of the header file `stdio.h`. This contains the function prototypes for the system functions (built-in) needed for input output (I/O) of data. Line 9 starts the definition of function *main*, it indicates the type and name of the function being defined. This simple program defines only a single function: *main*. Line 10 is a local data declaration; it declares variable *x* of type *double*.

Line 11 calls the system function *printf* to display the text string within quotes, on the screen. Line 12 is a simple assignment statement; it assigns value 45.95 to variable *x*. Line 13 invokes system function *printf* to display a text string and the value of variable x. The escape character \n is a line feed control character for output; it indicates a change to a new line. Line 14 is a termination statement by returning a value zero to the operating system.

Listing 2.1: A simple program in C.

```
 1 /*
 2  Program     : welcomec.c
 3  Author      : J M Garrido, November 20, 2012.
 4  Description : Display 'Welcome C World on the screen
 5     and the value of a variable.
 6  */
 7 #include <stdio.h>
 8
 9 int main() {
10     double x;      /* variable declaration */
11     printf("Welcome to the world of C \n");
12     x = 45.95;     /* assigns a value to variable x */
13     printf("Value of x: %lf \n", x);
14     return 0;      /* execution terminates OK */
15 }
```

Figure 2.4 shows the use of the GNU C compiler tool (*gcc*) to compile and link the C program (`welcome.c`), in a Linux Terminal. Note that by default, the executable file is named `a.out`. A recommended way to use *gcc* with a more appropriate and complete name that can be specified to the C compiler is:

```
gcc -Wall -o welcome.out welcomec.c
```

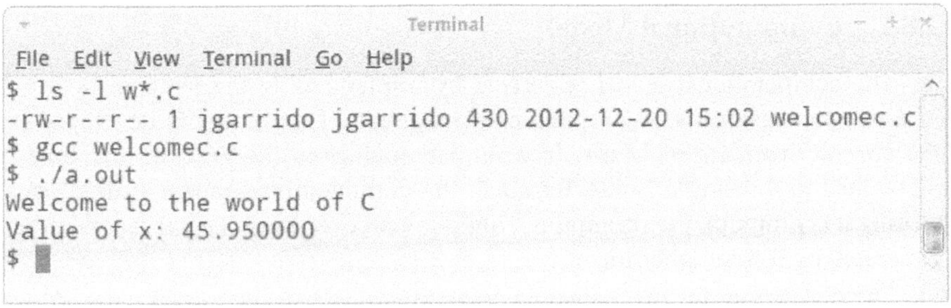

FIGURE 2.4: Compiling and executing a program in a Linux Terminal.

The compiler flag `-Wall` should always be included so the compiler will display any warning messages, in addition to the error messages. The flag `-o` is used to indicate the name *welcome.out* of the executable file generated by the compiler.

2.8 A Simple Problem: Temperature Conversion

This section revisits the temperature conversion problem, which is solved and implemented in C . The solution and implementation is derived by following a basic sequence of steps

2.8.1 Mathematical Model

The mathematical representation of the solution to the problem, the formula expressing a temperature measurement F in Fahrenheit in terms of the temperature measurement C in Celsius is:

$$F = \frac{9}{5} C + 32$$

The solution to the problem is the mathematical expression for the conversion of a temperature measurement in Celsius to the corresponding value in Fahrenheit. The mathematical formula expressing the conversion assigns a value to the desired temperature in the variable F, the dependent variable. The values of the variable C can change arbitrarily because it is the independent variable. The mathematical model uses real numbers to represent the temperature readings in various temperature units.

2.8.2 Computational Model

The computational model is derived by implementing the mathematical model in a program using the C programming language. This model is developed using the console interface of Linux. In a similar manner to the previous examples implemented in C using a Linux Terminal, the computational model is developed by writing a C program, then compiling, linking, and executing the executable program generated by the compiler/linker.

The C program for the mathematical model can be written with the *gedit* editor on Linux. This is one of many free editors available on Linux. It is simple, efficient, and easy to use. The program is another very simple C program. Listing 2.2 shows the complete source program.

Listing 2.2: Temperature conversion program in C.

```
 1 /*
 2  Program     : tempctof.c
 3  Author      : Jose M Garrido
 4  Description : Read value of temperature Celsius from
 5   console, convert to degrees Fahrenheit, and display
 6   value of this new temperature value on the output
 7   console */
 8
 9 #include <stdio.h>
10
11 int main(void) {
12      double tempc;  /* temperature in Celsius     */
13      double tempf;  /* temperature in Fahrenheit  */
14      printf("Enter value of temp in Celsius: ");
15      scanf("%lf", &tempc);
16      printf("\nValue of temp in Celsius: %lf \n", tempc);
17      tempf = tempc * (9.0/5.0) + 32.0;
18      printf("Temperature in Fahrenheit: %lf \n", tempf);
19      return 0;       /* execution terminates ok */
20 }
```

Compiling and linking the C program generates an executable file using the GNU C compiler/linker. Figure 2.5 shows a terminal window with the corresponding commands that compiles/links, and then executes the program.

Executing (running) the program computes the temperature in Fahrenheit given the value 5.0 for temperature in Celsius. The result is 41 degrees Fahrenheit.

The program can be executed several times to compute the Fahrenheit temperature starting with a given value of 10.0 for the temperature in Celsius and then repeating in increments of 5.0 degrees Celsius. The last computation is for a given value of 45.0 degrees Celsius.

Table 2.1 shows all the values of temperature in Celsius used to compute the corresponding temperature in Fahrenheit. This is a short set of results of the original problem.

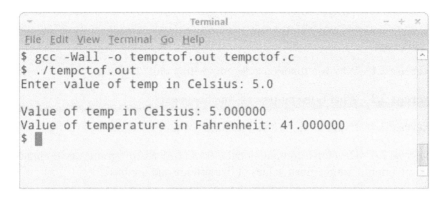

FIGURE 2.5: Compiling, linking, and executing a program.

TABLE 2.1: Celsius and Fahrenheit temperatures.

Celsius	5	10	15	20	25	30	35	40	45
Fahrenheit	41	50	59	68	77	86	95	104	113

Summary

The structure of a C computer program includes data definitions and basic instructions that manipulate the data. Functions are the building blocks and the fundamental components of C programs. Functions are defined and called in C programs. These basic programming constructs are used in developing computational models by implementing the corresponding mathematical models with C programs.

Key Terms

programs	functions	function invocation
function call	assignment statement	assignment operator
executable file	console	local declaration
terminal	main	variables
constants	compiler directive	global declaration
function prototype	header file	function definition

Exercises

Exercise 2.1 Why is a function a decomposition unit? Explain.

Exercise 2.2 What is the purpose of function *main*?

Exercise 2.3 Explain the various categories of function definitions.

Exercise 2.4 Develop a computational model (with a C program) that computes the area of a right triangle given values of the altitude and the base.

Exercise 2.5 Develop a computational model (with a C program) that computes the distance between two points in a plane: P_1 with coordinates (x_1, y_1), and P_2 with coordinates (x_2, y_2). Use the coordinate values: $(2, 3)$ and $(4, 7)$.

Exercise 2.6 Develop a computational model that computes the temperature in Celsius, given the values of the temperature in Fahrenheit.

Exercise 2.7 Develop a computational model that computes the circumference and area of a square, given the values of its sides.

Exercise 2.8 Develop a computational model (with a C program) that computes the slope of a line between two points in a plane: P_1 with coordinates (x_1, y_1), and P_2 with coordinates (x_2, y_2). Use the coordinate values: $(0, -3/2)$ and $(2, 0)$.

Chapter 3

Modular Decomposition: Functions

3.1 Introduction

A function is the main decomposition unit in a C program and it carries out a specific task in a program. When a function is called, it can receive input data from another function; the input data passed from another function is called an argument. The function can also return output data when it completes execution.

This chapter provides more details on function definitions, invocation, and decomposition. It also discusses the basic mechanisms for data transfer between two functions; several examples are included. Scientific and mathematical built-in functions are used in case studies.

3.2 Modular Decomposition

A problem is often too large and complex to deal with as a single unit. In problem solving and algorithmic design, the problem is partitioned into smaller problems that are easier to solve. The final solution consists of an assembly of these smaller solutions. The partitioning of a problem into smaller parts is known as *decomposition*. These small parts are known as *modules*, which are much easier to develop and manage.

System design usually emphasizes modular structuring, also called modular decomposition. With this approach, the solution to a problem consists of several smaller solutions corresponding to each of the subproblems. A problem is divided into smaller problems (or subproblems), and a solution is designed for each subproblem. These modules are considered building blocks for constructing larger and more complex algorithms.

3.3   Defining Functions

A C program is normally decomposed into source files, and these are divided into functions. A function is the smallest decomposition unit or module in a computational model. A function performs a related sequence of operations to accomplish a particular task. A function has to be defined before it can be used or called.

A function definition is a complete implementation of the function in the programming language. This definition includes the function header and the body of the function, which consists of local data definitions and instructions.

The function header or function prototype describes or specifies the type of data returned by the function, the name of the function, and the parameter list. This list includes the type and name of each parameter. The general form of a function definition in the C programming language is:

```
⟨ type ⟩ ⟨ function_name ⟩ ⟨ parameter_list ⟩
{
       . . [local data definitions]
       . . [instructions]
}
```

In C, the type specified in a function definition must be a primitive type, such as *int*, *float*, *double*, *char*, an aggregate type, or a programmer-defined type. The parameter list declares all the parameters of the function and must be written within parentheses.

The simplest function definition is one which has a function header that includes a *void* type, the function name, and an empty parameter list. The relevant comments for documentation of the function are described in the paragraph that starts with the the symbols '/*' and ends with '*/'.

As mentioned previously, local data declared within a function is known only to that function—the scope of the data is *local* to the function. The local data in a function has a limited lifetime; it only exists during execution of the function. The following sample C code shows a function with name *disp_mess*, which uses *void* as its type and has no parameters.

```
/*
This function displays a message
 on the screen. */
void disp_mess() {
     printf("Computing data");
}
```

3.4   Calling Functions

A function is called by another function and starts executing immediately. Upon completion, the called function returns control to the calling function and the first function continues execution.

The name of the function is used when it is called or invoked by some other function. Suppose function *main* calls function *disp_mess*. After completing execution function *disp_mess* returns control to function*main* without returning a value. Function *main* continues execution from the point after the call. The function call to function *disp_mess* is shown in following C statement.

```
disp_mess();
```

3.5   Classification of Functions

There are three groups of functions:

1. Functions that do not return any value when they are called. The previous example, function *show_message*, is a simple (or void) function because it does not return any value to the function that invoked it.

2. Functions that return a single value after completion.

3. Functions that declare one or more parameters, which are data items used in the function.

3.5.1   Simple Function Calls

Simple functions do not return a value to the calling function. An example of this kind of function is *disp_mess*, discussed previously. There is no data transfer to or from the function.

Figure 3.1 shows a function *A* calling a second function *B*. After completing its execution, the called function *B* returns the flow of control to function *A*.

3.5.2   Calling Functions that Return Data

Data transfer is possible between the calling and the called function. This data transfer may occur from the calling function to the called function, from the called function to the calling function, or in both directions.

Value-returning functions transfer data from the called function to the calling

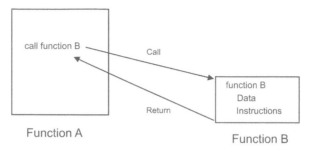

FIGURE 3.1: A function calling another function

function. In these functions, a single value is calculated or assigned to a variable and is returned to the calling function.

The called function is defined with a type, which is the type of the return value. The function *return type* may be a simple type or a programmer-defined type. After the execution of the called function completes, control is returned to the calling function with a single value. The C language statements that define the form of a value-returning function are:

```
⟨ return_type ⟩ ⟨ function_name ⟩ ⟨ parameter_list ⟩
{
      . . .
      return ⟨ return_value ⟩
}
```

The value in the return statement can be any valid expression, following the **return** keyword. The expression can include constants, variables, object references, or a combination of these.

The following example defines a function, *square*, that returns the value of the square of variable x. The computed value is returned to the calling function. In the header of the function, the type of the value returned is indicated as **float**. The code for this function definition is:

```
/*
    This function returns the square of variable x.
*/
float square () {
        float x;
        float sqx;
        x = 3.15;
        sqx = x * x;
        return sqx
}
```

The value-returning function can be called and the value returned can be used in several ways:

- Call the function in a simple assignment statement

- Call the function in an assignment statement with an arithmetic expression

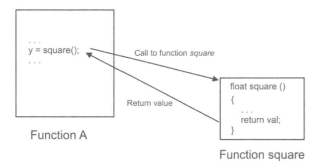

FIGURE 3.2: Function *A* calling function *square*

Figure 3.2 shows a function with name *A* that calls function *square*. The return value from function *square* is used directly by function *A* in an assignment statement that assigns the value to variable *y*.

The value returned by the called function is used by the calling function by assigning this returned value to another variable. The following example shows a function calling function *square*. The calling function assigns the value returned to variable *res*.

```
res = square();
fres = y + w * square() + 23.45;
```

The value returned is used in an assignment with an arithmetic expression after calling a function. In the example shown, after calling function *square*, the value returned is assigned to variable *res*. In the second line of C code, the function call occurs in an arithmetic expression and the final value is assigned to variable *fres*.

3.5.3 Calling Functions with Arguments

These functions include one or more data declarations in the *parameter* list. These parameter declarations are similar to local data declarations and have local scope and persistence. The function may also return a type.

The parameter list is enclosed with parenthesis and consists of one or more parameter declarations separated by commas. Each parameter declaration includes the type and name of the parameter. A function with parameters is very useful because it allows data transfer into a function when called. The data values used when calling a function are known as *arguments*. The parameters are also known as *formal parameters*, and the arguments are also known as *actual parameters*.

Every argument in the function call must be consistent with the corresponding parameter declaration in the function definition. The following example shows the definition of function *squared* that includes one parameter declaration, *p*, and returns the value of local variable, *res*.

```
double squared (double p)   {
     double res;
     res = p * p;
     return res;
}
```

Listing 3.1 shows a short but complete program in C. Line 7 is the function prototype for function *squared* and the function is called in line 14 with argument *x*. The function definition of this function has only one parameter declaration, so the function is called with one argument. It computes the square of the argument value, which is the value of variable *x* and is used as the input value in the function call. The value returned by the function call is assigned to variable *y*. In line 15, the program displays the values of *x* and *y*. The function definition of *squared* appears in line 18 to line 22.

Listing 3.1: C program to compute the square of a value.

```
 1 /*
 2  Program      : square.c
 3  Author       : Jose M Garrido
 4  Description : Compute the square of the value of a
 6     variable. */
 7 #include <stdio.h>
 8 double squared (double s); // prototype
 9 int main(void) {
10     double x;        /* variable declaration */
11     double y;
12     printf("Square of a value \n");
13     x = 45.95;       /* assigns a value to variable x */
14     y = squared(x);      /* compute square of x */
15     printf("The square of %lf is: %lf \n", x, y);
16     return 0;        /* execution terminates OK */
17 }
18 double squared (double p) {
19     double res;
20     res = p * p;
21     return res;
22 }
```

The following listing shows the Linux shell commands that compile, link, and execute the program. The output results of the execution of the program are also shown.

```
$ gcc -Wall square.c
$ ./a.out
Square of a value
The square of 45.950000 is: 2111.402500
```

In these simple cases, all arguments in a function call are used as *input* values by the function, which means that the argument values are transfered into the function. More advanced techniques of data transfer in function calls will be discussed in a subsequent chapter.

Figure 3.3 illustrates an example in which one parameter is declared in the function definition of *squared*. The function call requires an argument and a call to the function computes the square of the value in the argument; the called function returns the value computed.

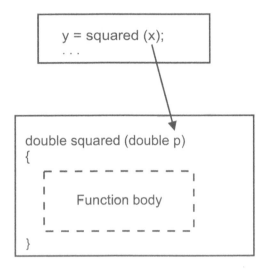

FIGURE 3.3: Calling function *squared* with an argument.

3.6 Numerical Types

Numerical values are integers or floating point values. This section briefly describes the number representation of numerical values in a computer system and consequently, and the range and precision of values.

3.6.1 Integer Numbers

In the C programming language, the types for integers are used for whole numbers and are: *int*, *short int*, and *long int*. The type *int* is the normal type for integers and the values of this type each are stored in a 4-byte (32-bit) memory block on most computer systems. This allows for integer values in the range of $-2,147,488,647$ to $2,147,483,647$.

Values of type *short int* take less memory storage, only 2 bytes (16 bits) on most computer systems and allows for a range of values from $-32,768$ to $32,767$.

Values of type *long int* may take 8 bytes or more, depending on the system. This allows for integer values in a much larger range.

The type *long long* is for very long integer values and is 128 bits long. The default mode for integer values is *signed*; the mode *unsigned* can be used for types of integer already listed.

3.6.2 Numbers with Decimal Points

The type for numbers with a decimal point are known as floating point types. In C, the types for floating-point values are: *float*, *double*, and *long double*.

Scientific notation is used to display very large and very small floating-point values. For example:

```
y = 5.77262e+12;
```

The mathematical equivalent for this value is 5.77262×10^{12} and is usually an approximate value. Scientific notation can also be used in mathematical expressions with assignments. For example:

```
x = 5.4e8 + y;
y = x * 126.5e10;
```

In the scientific notation number representation, there are two parts in a floating point numeric value: the *significand* (also known as the *mantissa*) and the *exponent*. The base of the exponent is 10. Therefore, the value of the mantissa is scaled by a power of 10. The IEEE standard for floating-point numbers specifies how a computer should store these two parts of a floating-point number. This influences not only the *range* of values of the type used, but also the *precision*.

In a 32-bit floating-point representation, the mantissa is stored in the 24-bit slot and the exponent in the remaining 8-bit slot of the complete 32-bit block allocated to the floating-point value. Each of the two slots includes a bit for the sign. This gives a precision 7 digits, and a range that depends on the value of exponent from -38 to 38. In C, the type *float* is implemented this way or in a very similar internal representation.

The type *double* is stored in an 8-byte (64 bits) block of memory. The significand part of the floating-point value is stored in a 53-bit slot, and the rest (11-bit slot) for the exponent value. This gives a precision of about 16 decimal digits and a much wider range of values.

Newer computer systems include type *long double*, which is stored in a 10-byte (80 bits) block of memory. The significand part of the floating-point value is stored in a 65-bit slot, and the rest (15 bits) for the exponent value. This gives a precision of about 19 decimal digits and a much wider range of values.

3.7   Built-in Mathematical Functions

The C programming language include libraries of built-in functions that compute the most common mathematical operations. To use the C standard mathematical library, the program must include at the top of the source program the following compiler directive:

```
#include <math.h>
```

An example of these is the library of trigonometric functions. The following language statement includes an expression that calls function *sin* applied to variable *x*, which is the *argument* written in parentheses and its value is assumed in radians. The value returned by the function is the sine of *x*.

```
y = 2.16 + sin(x);
j = 0.335;
z = x * sin(j * 0.32);
```

To compute the value of a variable to a given power, function *pow* is used for exponentiation. The function computes x^n and needs two arguments the variable and the exponent. The following simple assignment statement computes x^n, which is the value of variable *x* to the power of *n* and assigns the value computed to variable *y*.

```
y = pow(x, n);
```

To compute the square root of the value of a variable, *z* expressed mathematically as: $\sqrt{z}$, the C programming language provides the *sqrt()* function. For example, to compute the square root of *z* and assign the value to variable *q*, the C language statement is:

```
q = sqrt(z);
```

Given the following mathematical expression:

$$var = \sqrt{\sin^2 x + \cos^2 y}$$

The C assignment statement to compute the value of variable *var* uses four functions: *sqrt()*, *pow*, *sin()*, and *cos()* and is coded as:

```
var = sqrt( pow(sin(x), 2) + pow(cos(y), 2) );
```

The exponential function *exp()* computes *e* raised to the given power. The following C statement computes $q = y + xe^k$.

```
q = y + x * exp(k);
```

To compute the logarithm base e of x, denoted mathematically as $\ln x$ or $\log_e x$, C provides function *log()*. For example:

```
t = log((q-y)/x)
```

Listing 3.2 shows a complete C program that computes the area and circumference of a circle given the value of the radius. This program calls several C built-in functions. The compiler directive in line 8 provides the program with the function prototypes of functions in the standard input/output library. This is needed because the program calls functions *printf* and *scanf*. The compiler directive in line 9 provides the program with the function prototypes of the built-in mathematical library; it is necessary because the program calls function *pow* to compute the square of the radius, and accesses the constant *M_PI*, which has the value of π.

In line 15, the program reads the value of variable r from the input console (keyboard). Because variable r is used as an output argument in the call to function *scanf*, the name of the argument variable is preceded with an & symbol. In line 17 the program computes $area = \pi r^2$. In line 18, it computes $cir = 2\pi r$.

Listing 3.2: C program to compute area and circumference of a circle.

```
 1 /*
 2  Program      : areacir.c
 3  Description : Read value of the radius of a circle from
 4      input console, compute the area and circumference,
 5      display value of of these on the output console.
 6  Author       : Jose M Garrido, Nov 27 2012.
 7  */
 8 #include <stdio.h>
 9 #include <math.h>
10 int main(void) {
11      double r;     /* radius of circle */
12      double area;
13      double cir;
14      printf("Enter value of radius: ");
15      scanf("%lf", &r);
16      printf("\nValue of radius: %lf \n", r);
17      area = M_PI * pow(r, 2);
18      cir = 2.0 * M_PI * r;
19      printf("Value of area: %lf \n", area);
20      printf("Value of circumference: %lf \n", cir);
21      return 0;       /* execution terminates ok */
22 }
```

Summary

Functions are fundamental modular units in a C program. Calling functions involves several mechanisms for data transfer. Calling simple functions does not involve data transfer between the calling function and the called function. Value-returning functions return a value to the calling function. Calling functions that define one or more parameters involve values sent by the calling function and used as input in the called function.

Key Terms		
modules	functions	function definition
function call	local declaration	arguments
return value	assignment	parameters

Exercises

Exercise 3.1   Why is a function a decomposition unit? Explain.

Exercise 3.2   What is the purpose of function *main*?

Exercise 3.3   Explain variations of data transfer among functions.

Exercise 3.4   Write the C code of a function definition that includes more than two parameters.

Exercise 3.5   Write the C code that calls the function that was defined with more than two parameters.

Exercise 3.6   Develop a C program that defines two functions, one to compute the area of a triangle, the other function to compute the circumference of a triangle. The program must include the function prototypes and call these functions from *main*.

Exercise 3.7   Develop a C program that defines two functions, one to compute the area of a circle, the other function to compute the circumference of a circle. The program must include the function prototypes and call these functions from *main*.

Exercise 3.8   Develop a C program that defines two functions, one to compute the area of an ellipse, the other function to compute the circumference of an ellipse. The program must include the function prototypes and call these functions from *main*.

Exercise 3.9   Develop a computational model (with a C program) that computes the slope of a line between two points in a plane: P_1 with coordinates (x_1, y_1), and P_2 with coordinates (x_2, y_2). The program should include two functions: *main* and *slopef*. The parameters of the second function (*slopef*) are the coordinates of the points in a plane. Use the coordinate values: $(0, -3/2)$ and $(2, 0)$.

Chapter 4

More Concepts of Computational Models

4.1   Introduction

This chapter starts with a brief discussion on errors that occur in computations. Then the important concept of rate of change is discussed. Another case study of a computational model is introduced with a simple problem: the free-falling object. An implementation in C is included in order to explain the solution to the problem discussed. The chapter then presents concepts and principles on simulation and simulation models; problem and model decomposition; software tools for implementing (programming), running the model, and the visualization of results. Types of simulation models are briefly explained: continuous, discrete, deterministic, and stochastic models.

4.2   Introduction to Errors in Computing

Performing numerical computations involves considering several important concepts:

- Number representation

- The number of significant digits

- Precision and accuracy

- Errors

4.2.1   Number Representation

Numeric values use numbers that are represented in number systems. Numbers based on decimal representation use base 10. This is the common number representation used by humans. There are three other relevant number systems: binary (base 2), octal (base 8), and hexadecimal (base 16). Digital computers use the base 2, or binary, system. In a digital computer, a binary number consists of a number of binary digits or *bits*.

Any number z that is not zero ($z \neq 0$) can be written in scientific notation using decimal representation in the following manner:

$$z = \pm.d_1 d_2 d_3 \cdots d_s \times 10^p$$

Each d_i is a decimal digit (has a value from $0, 1 \ldots 9$). Assuming that $d_1 > 0$, the part of the number $d_1 d_2 d_3 \cdots d_s$ is known as the *fraction*, or *mantissa* or *significand* of z. The quantity p is known as the *exponent* and its value is a signed integer.

The number z may be written using a binary representation, which uses binary digits or *bits* and base 2.

$$z = \pm.b_1 b_2 b_3 \cdots b_r \times 2^q$$

To represent the significand using a binary representation, the number of bits is different than the number of decimal digits in the decimal representation used previously. The exponent q also has a different value than the exponent p that is used in the decimal representation.

In a computer, most computations with numbers are performed in floating point arithmetic in which the value of a real number is approximated with a finite number of digits in its mantissa, and a finite number of digits in its exponent. In this number system, a real number that is too big cannot be represented and causes an *overflow*. In a similar manner, a real number that is too small cannot be represented and causes an *underflow*.

4.2.2 Number of Significant Digits

The number of *significant digits* is the number of digits in a numeric value that defines it to be correct. In engineering and scientific computing, it is convenient and necessary to be able to estimate how many significant digits are needed in the computed result.

The number of bits in a binary number determines the precision with which the binary number represents a decimal number. A 32-bit number can represent approximately seven digits of a decimal number. A 64-bit binary number can represent 13 to 14 decimal digits.

4.2.3 Precision and Accuracy

Precision refers to how closely a numeric value used represents the value it is representing. *Accuracy* refers to how closely a number agrees with the true value of the number it is representing. Precision is governed by the number of digits being carried in the numerical calculations. Accuracy is governed by the errors in the numerical *approximation*; precision and accuracy are quantified by the errors in a numerical calculation.

4.2.4 Errors

An *error* is the difference between the true value *tv* of a number and its approximate value *av*. The *relative error* is the proportion of the error with respect to the true value.

$$error = tv - av$$

$$rel\ error = \frac{error}{tv}$$

The accuracy of a numerical calculation is quantified by the error of the calculation. Several types of errors can occur in numerical calculations.

- Errors in the parameters of the problem

- Algebraic errors in the calculations

- Iteration errors

- Approximation errors

- Roundoff errors

An *iteration error* is the error in an iterative method that approaches the exact solution of an exact problem asymptotically. Iteration errors must decrease toward zero as the iterative process progresses. The iteration error itself may be used to determine the successive approximations to the exact solution. Iteration errors can be reduced to the limit of the computing device.

An *approximation error* is the difference between the exact solution of an exact problem and the exact solution of an approximation of the exact problem. Approximation error can be reduced only by choosing a more accurate approximation of the exact problem.

A *roundoff error* is the error caused by the finite word length employed in the calculations. Roundoff error is more significant when small differences between large numbers are calculated. Most computers have either 32 bit or 64 bit word length, corresponding to approximately 7 or 13 significant decimal digits, respectively. Some computers have extended precision capability, which increases the number of bits to 128. Care must be exercised to ensure that enough significant digits are maintained in numerical calculations so that roundoff is not significant.

In many engineering and scientific calculations, 32 bit arithmetic is sufficient. However, in many other applications, 64 bit arithmetic is required. In a few special situations, 128 bit arithmetic maybe required. Such calculations are called *double precision* or *quad precision*, respectively. Many computations are evaluated using 64 bit arithmetic to minimize roundoff errors.

A floating-point number has three components: *sign*, the *exponent*, and the *significand*. The exponent is a signed integer represented in biased format (a fixed bias is added to it to make it into an unsigned number). The significand is a fixed-point number in the range $[1, 2)$. Because the binary representation of the significand always starts with one (1) and dot, this is fixed and hidden and only the fractional part of the significand is explicitly represented.

Except for integers and some fractions, all binary representations of decimal numbers are *approximations*, because of the finite number of bits used. Thus, some loss of precision in the binary representation of a decimal number is unavoidable. When binary numbers are combined in arithmetic operations such as addition, multiplication, etc., the true result is typically a longer binary number which cannot be represented exactly with the number of available bits in the binary number capability of the digital computer. Thus, the results are rounded off in the last available binary bit. This rounding off gives rise to roundoff errors, which can accumulate as the number of calculations increases.

The most common representation of numbers for computations dealing with values in a wide range is the floating-point format. The IEEE floating-point standard format (ANSI/IEEE Standard 754-1985) is used by the computer industry. Other formats will differ in their parameters and representation details, but the basic tradeoffs and algorithms remain the same.

Two of the most common floating point formats are: short (32-bit) and long (64-bit) floating-point formats. The short format has adequate range and precision for most common applications (magnitudes ranging from 1.2×10^{38} to 3.4×10^{38}). The long format is used for highly precise computations or those involving extreme variations in magnitude (from about 2.2×10^{308} to 1.8×10^{308}).

The value zero has no proper representation. For zero and other special values, the smallest and largest exponent codes (all 0s and all 1s in the biased exponent field) are not used for ordinary numbers. An all-0s word (0s in sign, exponent, and significand fields) represents +0; similarly, 0 and $\pm\infty$ have special representations, as does any nonsensical or indeterminate value, known as "not a number" (NaN).

When an arithmetic operation produces a result that is not exactly representable in the format being used, the result must be rounded to some representable value. The ANSI/IEEE standard prescribes several rounding options.

4.3   Average and Instantaneous Rate of Change

A mathematical function defines the relation between two (or more) variables. This relation is expressed as: $y = f(x)$. In this expression, variable y is a function of variable x, and x is the *independent variable* because for a given value of x, there is a corresponding value of y.

The average rate of change of a variable, y, with respect to variable, x, (the independent variable) is defined over a finite interval, Δx.

The Cartesian plane consists of two directed lines that perpendicularly intersect their respective zero points. The horizontal directed line is called the *x-axis* and the

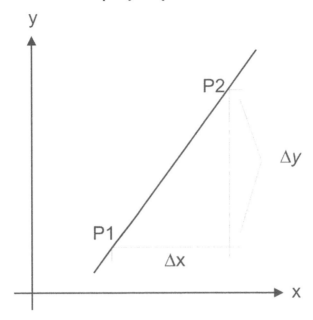

FIGURE 4.1: The slope of a line.

vertical directed line is called the *y-axis*. The point of intersection of the x-axis and the y-axis is called the *origin* and is denoted by the letter O.

The graphical interpretation of the average rate of change of a variable with respect to another is the *slope of a line* drawn in the Cartesian plane. The vertical axis is usually associated with the values of the dependent variable, y, and the horizontal axis is associated with the values of the independent variable, x.

Figure 4.1 shows a straight line on the Cartesian plane. Two points on the line, P_1 and P_2, are used to compute the slope of the line. Point P_1 is defined by two coordinate values (x_1, y_1) and point P_2 is defined by the coordinate values (x_2, y_2). The horizontal distance between the two points, Δx, is computed by the difference $x_2 - x_1$. The vertical distance between the two points is denoted by Δy and is computed by the difference $y_2 - y_1$.

The *slope* of the line is the inclination of the line and is computed by the expression $\Delta y / \Delta x$, which is the same as the average rate of change of a variable y over an interval Δx. Note that the slope of the line is constant, on any pair of points on the line.

As mentioned previously, if the dependent variable y does not have a linear relationship with the variable x, then the graph that represents the relationship between y and x is a curve instead of a straight line. The average rate of change of a variable y with respect to variable x over an interval Δx, is computed between two points, P_1 and P_2. The line that connects these two points is called a *secant* of the curve. The average rate on that interval is defined as the slope of that secant. Figure 4.2 shows a secant to the curve at points P_1 and P_2.

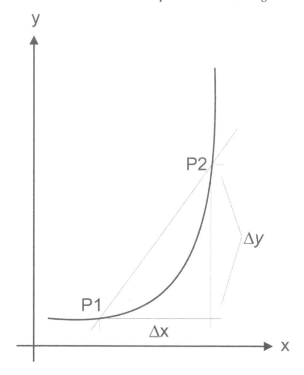

FIGURE 4.2: The slope of a secant.

The *instantaneous rate of change* of a variable, y, with respect to another variable, x, is the value of the rate of change of y at a particular value of x. This is computed as the slope of a line that is tangent to the curve at a point P.

Figure 4.3 shows a tangent of the curve at point P_1. The instantaneous rate of change at a specified point $P1$ of a curve can be approximated by calculating the slope of a secant and using a very small interval, in different words, choosing Δx very small. This can be accomplished by selecting a second point on the curve closer and closer to point $P1$ (in Figure 4.3), until the secant almost becomes a tangent to the curve at point $P1$.

Examples of rate of change are: the average velocity, $\bar{v}$, computed by $\Delta y/\Delta t$, and the average acceleration $\bar{a}$, computed by $\Delta v/\Delta t$. These are defined over a finite time interval, Δt.

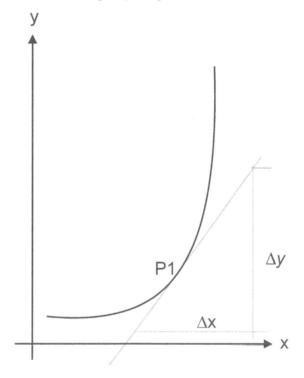

FIGURE 4.3: The slope of a tangent.

4.4   The Free-falling Object

A problem is solved by developing a computational model and executing it. The model development generally follows the computational model development process discussed previously. A C program is used to implement and run the model.

4.4.1   Initial Problem Statement

There is a need to know the vertical position and the velocity of a free-falling object as time passes. The solution to this problem is the calculation of vertical distance traveled and the velocity as the object approaches the ground. Several relevant questions related to the free-falling object need to be answered. Some of these are:

1. How does the acceleration of gravity affect the motion of the free-falling object?

2. How does the height of the free-falling object change with time, while the object is falling?

3. How does the velocity of the free-falling object change with time, while the object is falling?

4. How long does the free-falling object take to reach ground level, given the initial height, y_0? This question will not be answered here, it is left as an exercise.

4.4.2   Analysis and Conceptual Model

A brief analysis of the problem involves:

1. Understanding the problem. The main goal of the problem is to develop a model to compute the vertical positions of the object from the point where is was released and the speed accordingly with changes in time.

2. Finding the relevant concepts and principles on the problem being studied. Studying the mathematical expressions for representing the vertical distance traveled and the vertical velocity of the falling object. This knowledge is essential for developing a mathematical model of the problem.

3. Listing the limitations and assumptions about the mathematical relationships found.

4.4.2.1   Assumptions

The main assumption for this problem is that near the surface of the earth, the acceleration, due to the force of gravity, is constant with value 9.8 m/s^2, which is also 32.15 ft/s^2. The second assumption is that the object is released from rest. The third important assumption is that the frictional drag, due to resistance of the air, is not considered.

4.4.2.2   Basic Definitions

The vertical motion of an object is defined in terms of displacement (y), velocity (v), acceleration (g), and time (t).

A time change, denoted by Δt, is a finite interval of time defined by the final time instance minus the initial time instance of the interval of time: ($t_2 - t_1$). A change of displacement is denoted by Δy, and it represents the difference in the vertical positions of the object in a finite interval: ($y_2 - y_1$). In a similar manner a change of velocity is denoted by Δv, and it represents the difference in the velocities in a finite interval: ($v_2 - v_1$).

The velocity is the *rate of change* of displacement, and the acceleration is the rate of change of velocity. The average velocity, denoted by $\bar{v}$, is the average rate of change of displacement with respect to time on the interval Δt. The average acceleration, denoted by $\bar{a}$, is the average rate of change of the velocity with respect to time on the interval Δt. These are defined by the following mathematical expressions:

$$\bar{v} = \frac{\Delta y}{\Delta t} \qquad \bar{a} = \frac{\Delta v}{\Delta t}$$

4.4.3 Mathematical Model

The mathematical representation of the solution to the problem consists of the mathematical formulas expressing the the vertical displacement and the velocity of the object in terms of the time since the object was released and began free fall. Note that a general way to compute the average velocity, $\bar{v}$, is computed from the following expression:

$$\bar{v} = \frac{v_0 + v}{2},$$

with v_0 being the initial velocity and v the final velocity in that interval.

The mathematical model of the solution for a vertical motion of a free-falling object is considered next. Recall that in this model, the air resistance is ignored and the vertical acceleration is the constant $-g$. The vertical position as the object falls is expressed by the equation:

$$y = y_0 + v_0 t - \frac{gt^2}{2} \tag{4.1}$$

The velocity of the object at any time is given by the equation:

$$v_y = v_0 - gt, \tag{4.2}$$

where y is the vertical position of the object; t is the value of time; v is the vertical velocity of the object; v_0 is the initial vertical velocity of the object; and y_0 is the initial vertical position of the object. Equation 4.1 and Equation 4.2 represent the relationship among the variables: vertical position, vertical velocity, initial velocity, time instant, and initial vertical position of the object.

Note that in this model, the system state changes continuously with time. This type of state change is different compared with the example of the car-wash system. In the model of the car-wash system, the system state changes only at discrete instants of time. Another difference is that the model of the car-wash system includes an informal mathematical model, whereas the model of the free-falling object is a formal mathematical model, because it can be expressed completely by a set of mathematical equations (or expressions).

4.4.4 Computational Model

The next step is to implement the mathematical model using a C program. The computational model has an dependent part that corresponds to the mathematical expression (formula) for the vertical position, y, and the vertical velocity v_y of the object, and an independent part that allows arbitrary values given for time t. This really means that the computational model will use the equations (Equation 4.1 and Equation 4.2) defined previously.

For the implementation of the mathematical model of the free-falling object, a C program is developed that has the general structure discussed previously.

Listing 4.1 shows the source program in C. Two constants are first declared in lines 11 and 12. Note that the initial velocity, v_0, is zero. The value for the gravity acceleration is G, in meters per seconds squared, and the value of the initial height is Y_0, in meters. In C, the convention is to name the symbolic constants in upper case.

The symbolic constants and variables are declared on lines 10–14. The value of time is specified in line 15 by assigning a value of time to variable t. The vertical position for the time t of the falling object is computed in line 16. The vertical velocity at time t is computed in line 17. The output statement in line 18 displays on the screen the values of variables t and y. Line 19 displays the values of t and vel.

Listing 4.1: Free-falling object program in C.

```
 1 /*
 2  Program      : freefall.c
 3  Description : Compute the vertical position and velocity
 4       at the specified time of a free falling object.
 5       Values are in meters and seconds.
 6  Author       : Jose M Garrido, November 22, 2012.
 7  */
 8 #include <stdio.h>
 9 int main(void) {
10     const double G = 9.8;      /* acceleration of gravity */
11     const double Y_0 = 40.0; /* initial height           */
12     double y;                /* vertical position of object */
13     double vel;              /* vertical velocity */
14     double t;                /* specified time */
15     t = 1.2;                 /* value of time to compute y */
16     y = Y_0 - 0.5 * (G * t * t);
17     vel = - G * t;
18     printf("Vertical position for time: %lf is: %lf\n",
               t, y);
19     printf("Vertical velocity for time: %lf is: %lf\n",
               t, vel);
20     return 0;       /* execution terminates ok */
21 }
```

The following Linux shell command uses the GNU C compiler to compile and link the program `freefall.c` and the name of the executable program generated is `freefall.out`. The second command executes the program and the results are displayed on the output console.

```
$ gcc -Wall freefall.c -o freefall.out
$ ./freefall.out
Vertical position for time: 1.200000 is:  32.944000
Vertical velocity for time: 1.200000 is:  -11.760000
$
```

To compute the vertical position and velocity of the falling object for several values of time, the program is executed several times. Table 4.1 shows most of the values used of the height and the vertical velocity computed with the values of time shown. This table represents a simple and short set of results of the original problem.

TABLE 4.1: Values of height and vertical velocity.

t	0.0	0.5	0.7	1.0	1.2	1.8	2.2	2.5	2.8
y	40.0	38.7	37.6	35.1	32.9	24.12	16.28	9.37	1.58
v_y	0.0	−4.9	−6.86	−9.8	−11.7	−17.6	−21.5	−24.5	−27.4

4.5 Simulation: Basic Concepts

A *simulation model* can approximate some properties of a real system over a relatively long simulation interval, and show dynamic behavior of the system represented. A simulation model is a computational model, but not all computational models are simulation models.

Simulation is a set of techniques, methods, and tools for developing a simulation model of a system, and using and manipulating the simulation model to gain more knowledge about the dynamic behavior of a system. The purpose of simulation is to manipulate a simulation model to gain understanding about the behavior of the real system that the model represents.

A *simulation run* is the manipulation of a model in such a way that it operates on time or space to compress it, thus enabling one to perceive the interactions that would not otherwise be apparent because of their separation in time or space. A simulation run generally refers to a computerized execution of the model which is run over time to study the implications of the defined interactions.

An important goal of modeling and simulation is to gain a level of understanding of the interaction of the parts of a system, and of the system as a whole.

One of the great values of simulation is its ability to effect a time and space compression on the system, essentially allowing one to perceive, in a matter of minutes, interactions that would normally unfold over very lengthy time periods.

4.5.1 Simulation Models

A system is part of the real world under study that can be separated from the rest of its environment for a specific purpose. Such a system is called the *real system* because it is inherently part of the real world.

Recall that a model is an abstract representation of a real system. The model is simpler than the real system, but it should be equivalent to the real system in all relevant aspects. A simulation model is a computational model that has the following goals:

1. To study some relevant aspects of the dynamic behavior of a system by observing the operation of the system, using the sequence of events, or trace from the simulation runs

2. To estimate various performance measures

A simulation model is a computational model implemented as a set of procedures that when executed in a computer, *mimic* the behavior (in some relevant aspects) and the static structure of the real system. This type of model uses numerical methods as possibly the only way to achieve a solution. Simulation models include as much detail as necessary, that is, the representation of arbitrary complexity. The output of the model depends on its reaction to the following types of input:

- The passage of time;

- Data from the environment;

- Events (signals) from the environment.

Because there two general types of computational models: continuous and discrete, from the perspective of how the model changes state in time during a simulation, there are two general categories of simulation:

1. Continuous

2. Discrete-event

Continuous simulation is used mainly to model and study the dynamic behavior of physical systems. The continuous time response of these models is represented by ordinary differential equations. The general purpose of a simulation model is to study the dynamic behavior of a system, i.e., the state changes of the model as time advances. The state of the model is defined by the values of its attributes, which are represented by state variables. For example in a discrete-event model, the number of waiting customers to be processed by a simple barbershop is represented as a state variable (an attribute), which changes its value with time. Whenever this attribute changes value, the system changes its state. For a model to be useful, it should allow the user to:

- Manipulate the model by supplying it with a corresponding set of inputs;

- Observe its behavior or output

- Predict the behavior of the real system from the behavior of the model, under the same circumstances

The reasons for developing a simulation model and carrying out simulation runs are:

- It may be too difficult, dangerous, and/or expensive to experiment with the real system.

- The real system is non-existant; simulations are used to study the behavior of a future system (to be built).

After a simulation model has been completely developed, the model is used to study the real system and solve the original problem. After being developed, a simulation model is used for:

- Designing experiments

- Performing simulation runs and collect data

- Analysis and interpretation—drawing inferences

- Documentation

4.5.2 Variability with Time

Depending on the variability of some parameters with respect to time, simulation models can be separated into two categories:

1. Deterministic models

2. Stochastic models.

A deterministic model displays a completely predictable behavior. A stochastic model includes some uncertainty implemented with random variables, whose values follow a probabilistic distribution. In practice, a significant number of models are stochastic because the real systems being modeled usually include inherent uncertainty properties.

An example of a deterministic simulation model is a model of a simple car-wash system. In this model, cars arrive at exact specified instants (but at the same instants), and all have exact specified service periods (wash periods); the behavior of the model can be completely and exactly determined.

The simple car-wash system with varying car arrivals, varying service demand from each car, is a stochastic system. In a model for this system, only the averages of these parameters are specified together with a probability distribution for the variability of these parameters. Uncertainty is included in this model because these parameter values cannot be exactly determined.

4.5.3 Simulation Results

A simulation run is an experiment carried on the simulation model for some period of observation; the time dimension is one of the most important in simulation. Several simulation runs are usually necessary, in order to achieve some desired solution.

The results of experimenting with discrete-event simulation models, i.e., simulation runs, can be broken down into two sets of outputs:

1. Trace of all the events that occur during the simulation period and all the information about the state of the model at the instants of the events; this directly reflects the dynamic behavior of the model;

2. Performance measures — the summary results of the simulation run.

The trace allows the users to verify that the model is actually interacting in the manner according to the model's requirements. The performance measures are the outputs that are analyzed for estimates used for capacity planning or for improving the current real system.

Summary

The process of developing computational models consists of a series of phases for constructing the computational model. The various phases are: defining the problem statement and its specification, building the conceptual model, the mathematical model, the computational model, verification and validation. Simulation and simulation models are additional important concepts introduced in the study of computational models. Types of models are continuous, discrete, deterministic, and stochastic models. Computational models are implemented with the C programming language. Useful mathematical concepts discussed are: rates of change and the area under a curve. Errors that occur in computing are important to consider due to finite precision and the number of significant digits.

Key Terms

mathematical model	simulation	simulation model
computational model	continuous model	discrete model
deterministic model	stochastic model	rate of change
average change	instantaneous change	slope of line
roundoff errors	precision	significant digits
decimal numbers	binary numbers	floating-point numbers

Exercises

Exercise 4.1 Develop a computational model that computes the time the free-falling object takes to reach ground level and the vertical velocity at that time instance, given the initial height, y_0. Hint: derive a mathematical solution involving Equation 4.1 and Equation 4.2.

Exercise 4.2 Assume that the initial velocity, v_0, of the free-falling object is not zero. Develop a computational model that computes the time the object takes to reach ground level and the vertical velocity at that time instance, given the initial height, y_0. Hint: derive a mathematical solution involving Equation 4.1 and Equation 4.2.

Exercise 4.3 A large tent is to be set up for a special arts display in the main city park. The tent material is basically a circular canvas; the tent size can be adjusted by increasing or decreasing the distance from the center to the edge of the canvas, varying from 5 feet to 35 feet. The cost of the event will be based on the circular area occupied, $1.25 per square foot. Develop a computational model that takes several values of the distance from the center to the outer edge of the canvas (the radius) and provide the cost for the event.

Exercise 4.4 An account is set up that pays a guaranteed interest rate, compounded annually. The balance that the account will grow to at some point in the future is known as the future value of the starting principal. Develop a computational model for calculating the future value given values of n years, P for the starting principal, and r for the rate of return expressed as a decimal (interest rate).

Chapter 5

Algorithms and Programs

5.1 Introduction

The purpose of computer problem solving is to design a solution to a problem; an algorithm is designed and implemented in a computer program. This program is executed or run with the appropriate input data to get the desired results.

This chapter presents concepts and explanations of design structures, which are used in designing an algorithm. A complete case study is developed.

5.2 Problem Solving

The design of the solution to a problem is described by an *algorithm*, which is a sequence of steps that need to be carried out to achieve the solution to a problem. The implementation of an algorithm and the corresponding data definitions is carried out with a with a standard programming language such as C or with software tools such as MATLAB® and Octave.

Developing a computational model involves implementing a computer solution to solve some real-world problem. Design of a solution to the problem requires finding some method to solve the problem.

Analyzing the problem includes understanding the problem, identifying the given (input) data and the required results. Designing a solution to the problem consists of defining the necessary computations to be carried out on the given data to produce the final required results.

5.3   Algorithms

The algorithm is usually broken down into smaller tasks; the overall algorithm for a problem solution is decomposed into smaller algorithms, each defined to solve a subtask.

An algorithm is a clear, detailed, precise, and complete description of the sequence of steps performed to produce the desired results. An algorithm can be considered the transformation on the given data and involves a sequence of commands or operations that are to be carried out on the data in order to produce the desired results.

A computer implementation of an algorithm consists of a group of data definitions and one or more sequences of instructions to the computer for producing correct results when given appropriate input data. The implementation of an algorithm will normally be in the form of a program, which is written in an appropriate programming language and it indicates to the computer how to transform the given data into correct results.

An algorithm is often described in a semiformal notation such as pseudo-code and flowcharts.

5.4   Algorithm Description

Designing a solution to a problem consists of designing and defining an algorithm, which will be as general as possible in order to solve a family or group of similar problems. An algorithm can be described at several levels of abstraction. Starting from a very high level and general level of description of a preliminary design, to a much lower level that has more detailed description of the design.

Several notations are used to describe an algorithm. An algorithmic notation is a set of general and informal rules used to describe an algorithm. Two widely used notations are:

- Flowcharts

- Pseudo-code

A flowchart is a visual representation of the flow of data and the operations on this data. A flowchart consists of a set of symbol blocks connected by arrows. The arrows that connect the blocks show the order for describing a sequence of design or action steps. The arrows also show the flow of data.

Several simple flowchart blocks are shown in Figure 5.1. Every flowchart block has a specific symbol. A flowchart always begins with a *start* symbol, which has an

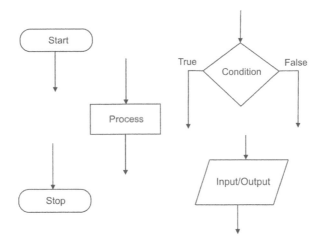

FIGURE 5.1: Simple flowchart symbols.

arrow pointing from it. A flowchart ends with a *stop* symbol, which has one arrow pointing to it.

The *process* or *transformation* symbol is the most common and general symbol, shown as a rectangular box. This symbol represents any computation or sequence of computations carried out on some data. There is one arrow pointing to it and one arrow pointing out from it.

The *selection* flowchart symbol has the shape of a vertical diamond and represents a *selection* of alternate paths in the sequence of design steps. It is shown in Figure 5.1 with a condition that is evaluated to *True* or *False*. This symbol is also known as a *decision block* because the sequence of instructions can take one of two directions in the flowchart, based on the evaluation of a condition.

The *input-output* flowchart symbol is used for a data input or output operation. There is one arrow pointing into the block and one arrow pointing out from the block.

An example of a simple flowchart with several basic symbols in shown in Figure 5.2. For larger or more complex algorithms, flowcharts are used mainly for the high-level description of the algorithms and pseudo-code is used for describing the details.

Pseudo-code is a notation that uses a few simple rules and English for describing the algorithm that defines a problem solution. It can be used to describe relatively large and complex algorithms. It is relatively easy to convert the pseudo-code description of an algorithm to a computer implementation in a high-level programming language.

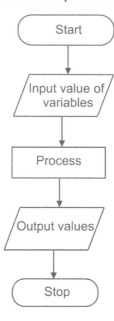

FIGURE 5.2: A simple flowchart example.

5.5   Design Structures

There are four fundamental design structures with which any algorithm can be described. These can be used with flowcharts and with pseudo-code notations. The basic design structures are:

- *Sequence*, any task can be broken down into a sequence of steps.

- *Selection*, this part of the algorithm takes a decision and selects one of several alternate paths of flow of actions. This structure is also known as alternation or conditional branch.

- *Repetition*, this part of the algorithm has a block or sequence of steps that are to be executed zero, one, or more times. This structure is also known as looping.

- *Input-output*, the values of variables are read from an input device (such as the keyword) or the values of the variables (results) are written to an output device (such as the screen)

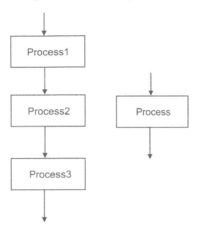

FIGURE 5.3: A flowchart with a sequence.

5.5.1 Sequence

A sequence structure consists of a group of computational steps that are to be executed one after the other, in the specified order. The symbol for a sequence can be directly represented by a *processing* block of steps in a flowchart.

A sequence can also be shown as a flow of flowchart blocks. Figure 5.3 illustrates the sequence structure for both cases. The sequence structure is the most common and basic structure used in algorithmic design.

5.5.2 Selection

With the selection structure, one of several alternate paths will be followed based on the evaluation of a condition. Figure 5.4 illustrates the selection structure. In the figure, the actions or instructions in *Processing1* are executed when the condition is true. The instructions in *Processing2* are executed when the condition is false.

A flowchart example of the selection structure is shown in Figure 5.5. The condition of the selection structure is *len* > 0 and when this condition evaluates to true, the block with the action `add 3 to k` will execute. Otherwise, the block with the action `decrement k` is executed.

5.5.3 Repetition

The repetition structure indicates that a set of action steps are to be repeated several times. Figure 5.6 shows this structure. The execution of the actions in the *Processing* block are repeated while the condition is true. This structure is also known as the *while loop*.

A variation of the repetition structure is shown in Figure 5.7. The actions in the

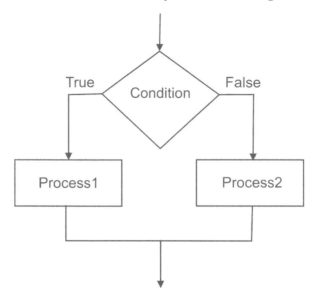

FIGURE 5.4: Selection structure in flowchart form.

Processing block are repeated until the condition becomes true. This structure is also known as the *repeat-until* loop.

5.6   Implementing Algorithms

The implementation of an algorithm is carried out by writing the code in an appropriate programming language, such as Ada, Fortran, Eiffel, C, C++, or Java .

Programming languages have well-defined syntax and semantic rules. The syntax is defined by a set of grammar rules and a vocabulary (a set of words). The legal sentences are constructed using sentences in the form of *statements*. There are two groups of words that are used to write the statements: *reserved words* and *identifiers*.

Reserved words are the keywords of the language and have a predefined purpose. These are used in most statements. Examples are: *for, end, function, while*, and *if*. Identifiers are names for variables, constants, and functions that the programmer chooses, for example, *height, temperature, pressure, number_units*, and so on.

This section discusses simple pseudo-code and computational model statements, such as the assignment and I/O statements.

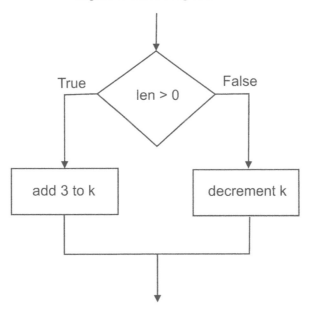

FIGURE 5.5: An example of the selection structure.

5.6.1 Assignment and Arithmetic Expressions

The *assignment* statement is used to assign a value to a variable. This variable must be written on the left-hand side of the equal sign, which is the assignment operator. The value to assign can be directly copied from another variable, from a constant value, or from the value that results after evaluating an expression.

In the following example, the first statement assigns the constant value 21.5 to variable y. The second statement assigns the value 4.5 to variable x. In the the third assignment statement in the example, variable z is assigned the result of evaluating the expression, $y + 1.5 \times x^3$. In pseudo-code, these assignment statements are simply written as follows:

```
set y = 21.5
set x = 4.5
set z = y+1.5x³
```

In C these are implemented as:

```
y = 21.5;
x = 4.5;
z = y + 10.5 * pow(x, 3);
```

5.6.2 Simple Numeric Computations

An assignment is often written with an arithmetic expression that is evaluated and the result is assigned to a variable. For example, add 15 to the value of variable

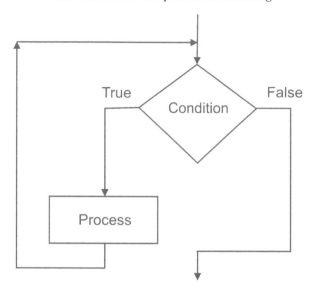

FIGURE 5.6: While loop of the repetition structure.

x, subtract the value of variable *y* from variable *z*. In pseudo-code, the *add* and the *subtract* statements can be used in simple arithmetic expressions instead of the "+" and "−" operator symbols. The two statements are written as follows:

```
add 15 to x
subtract y from z
```

In the example, in the first statement, the new value of *x* is assigned by adding 15 to the previous value of *x*. In the second assignment, the new value of variable *z* is assigned the value that results after subtracting the value of *y* from the previous value of *z*.

To add or subtract 1 in pseudo-code, the *increment* and the *decrement* statements can be used. For example, the statement increment j, adds the constant 1 to the value of variable *j*. In a similar manner, the statement decrement k subtracts the constant value 1 from the value of variable *k*.

In C, the statements in the previous examples are written as:

```
x = x + 15;
z = z - y;
```

In the previous examples, only simple arithmetic expressions were used in the assignment statements. These are addition, subtraction, multiplication, and division. More complex calculations use various numerical functions in computational models, such as square root, exponentiation, and trigonometric functions. For example, the value of the expression $\cos p + q$ assigned to variable *y* and $\sqrt{x-y}$ assigned to variable *q*. The statements in pseudo-code are:

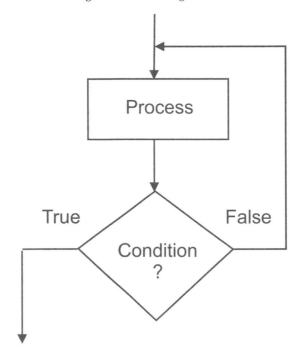

FIGURE 5.7: Repeat-until loop of the repetition structure.

$$y = \cos(p) + q$$
$$q = \sqrt{x - y}$$

In C, these assignment statements use the mathematical functions *cos* and *sqrt*.

```
y = cos(p) + q;
q = sqrt(x - y);
```

To assign the value of the mathematical expression $x^n \times y \times sin^{2m} x$ to variable z, the statement in pseudo-code is:

$$z = x^n \quad y \quad sin^{2m} x$$

In C, this assignment statement is written as follows:

```
z = pow(x, n) * y * pow(sin(x), (2*m));
```

5.6.3 Simple Input/Output

Input and output statements are used to read (input) data values from the input device (e.g., the keyboard) and write (output) data values to an output device (mainly to the computer screen). The flowchart symbol is shown in Figure 5.8.

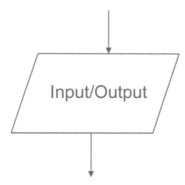

FIGURE 5.8: Flowchart data input/output symbol.

In pseudo-code, display statement that can be used for the output of a list of variables and literals; it is written with the keyword *display*.

The output statement writes the value of one or more variables to the output device. The variables do not change their values. The general form of the output statement in pseudo-code is:

```
display ( data_list )
display "value of x= ", x, "value of y = ", y
```

In C, function *printf* can display a single string and data value, so it can be used individually to display every variable and its value. The general form for this function call in C is:

```
printf (" [string], [format operators]", [variable list] );
```

The previous example outputs the value of two variables, *x* and *y*, so the output function is used two times.

```
printf ("values of x: %lf and y:   %lf  \n", x, y);
```

The input statement using pseudo-code reads a value of a variable from the input device (e.g., the keyboard). This statement is written with the keywords *read*, for input of a single data value and assign to a variable.

The following two lines of pseudo-code include the general form of the input statement and an example that uses the *read* statement to read a value of variable *y*.

```
read ( var_name )

read y
```

The input statement implies an assignment statement for the variable *y*, because the variable changes its value to the new value that is read from the input device.

The C programming languages provide several functions for console input/output. The function for input reads the value of a variable from the input device. The name of the function is *scanf*, and is used with a format operator. The general form for input in C for reading the value of a numeric floating-point variable, *var*, is shown in the following line of code.

```
scanf (" %lf", &var);
```

A string message is typically included to prompt the user for input of a data value. This is implemented with the *printf* function. The previous example for reading the value of variable *y* is written in C as follows:

```
printf ("Enter value of y: ");
scanf ("%lf", &y);
```

5.7 Computing Area and Circumference

For this problem, a computational model is developed that computes the area and circumference of a circle. The input value of the radius is read from the keyboard and the results written to the screen.

5.7.1 Specification

The specification of the problem can be described as a high-level algorithm in informal pseudo-code notation:

1. Read the value of the radius of a circle, from the input device.

2. Compute the area of the circle.

3. Compute the circumference of the circle.

4. Output or display the value of the area of the circle to the output device.

5. Output or display the value of the circumference of the circle to the output device.

5.7.2 Algorithm with the Mathematical Model

A detailed description of the algorithm and the corresponding mathematical model follows:

1. Read the value of the radius r of a circle, from the input device.

2. Establish the constant π with value 3.14159.

3. Compute the area of the circle, $area = \pi r^2$.

4. Compute the circumference of the circle $cir = 2\pi r$.

5. Print or display the value of *area* of the circle to the output device.

6. Print or display the value of *cir* of the circle to the output device.

The following lines of pseudo-code completely define the algorithm.

```
read r
π = 3.1416
area = π r²
cir = 2 π r
display "Area = ", area, " Circumference = ", cir
```

The computational model is implemented in the C programming language and is stored in file *areacir.c*, which is shown in Listing 5.1.

Listing 5.1: C program for computing the area and circumference.

```
 1 /*
 2  Program      : areacir.c
 3  Author       : Jose M Garrido, Nov 21 2012.
 4  Description : Read value of radius of circle from input
 5      compute the area and circumference, display value
 6      of these on the output console.
 7  */
 8
 9 #include <stdio.h>
10 #include <math.h>
11
12 int main(void) {
13      double r;    /* radius of circle */
14      double area;
15      double cir;
16      printf("Enter value of radius: ");
17      scanf("%lf", &r);
18      printf("\nValue of radius: %lf \n", r);
19      area = M_PI * pow(r, 2);
20      cir = 2.0 * M_PI * r;
21      printf("Value of area: %lf \n", area);
22      printf("Value of circumference: %lf \n", cir);
23      return 0;       /* execution terminates ok */
24 }
```

The following listing shows the shell Linux commands that compile and link the file `areacir.c` and execute the file `areacir.out`.

```
$ gcc -Wall areacir.c -o areacir.out
$ ./areacir.out
Enter value of radius: 3.15

Value of radius: 3.150000
Value of area: 31.172453
Value of circumference: 19.792034
```

Summary

An algorithm is a precise, detailed, and complete description of a solution to a problem. The notations to describe algorithms are flowcharts and pseudo-code. Flowcharts are a visual representation of the execution flow of the various instructions in the algorithm. Pseudo-code is an English-like notation to describe algorithms.

The design structures are sequence, selection, repetition, and input-output. These algorithmic structures are used to specify and describe any algorithm. The implementation of the algorithm is carried by programs written in the C programming language.

Key Terms			
algorithm	flowcharts	pseudo-code	variables
constants	declarations	structure	sequence
action step	selection	repetition	input/output
statements	data type	identifier	design

Exercises

Exercise 5.1 Write the algorithm and data descriptions for computing the area of a triangle. Use flowcharts.

Exercise 5.2 Write the algorithm and data descriptions for computing the area of a triangle. Use pseudo-code.

Exercise 5.3 Develop a computational model for computing the area of a triangle.

Exercise 5.4   Write the algorithm and data descriptions for computing the perimeter of a triangle. Use pseudo-code.

Exercise 5.5   Write the algorithm and data descriptions for computing the perimeter of a triangle. Use flowcharts.

Exercise 5.6   Develop a computational model with a computational model program for computing the perimeter of a triangle.

Exercise 5.7   Write the data and algorithm descriptions in flowchart and in pseudo-code to compute the conversion from a temperature reading in degrees Fahrenheit to Centigrade. The algorithm should also compute the conversion from Centigrade to Fahrenheit.

Exercise 5.8   Develop a computational model with a computational model program to compute the conversion from a temperature reading in degrees Fahrenheit to Centigrade. The program should also compute the conversion from Centigrade to Fahrenheit.

Exercise 5.9   Write an algorithm and data descriptions in flowchart and pseudo-code to compute the conversion from inches to centimeters and from centimeters to inches.

Exercise 5.10   Develop a computational model with a computational model program to compute the conversion from inches to centimeters and from centimeters to inches.

Chapter 6

Selection

6.1 Introduction

To completely describe an algorithm four design structures are used: sequence, selection, repetition, and input/output. This chapter explains the selection structure and the corresponding statements in pseudo-code and C statements for implementing computational models.

Conditions are expressions that evaluate to a truth-value (true or false). Conditions are used in the selection statements. Simple conditions are formed with relational operators for comparing two data items. Compound conditions are formed by joining two or more simple conditions with *logical* operators.

The solution to a quadratic equation is discussed as an example of applying the selection statements.

6.2 Selection Structure

The selection design structure is also known as alternation, because alternate paths are considered, based on the evaluation of a condition. This section covers describing the selection structure with flowcharts and pseudo-code, the concepts associated with conditional expressions, and implementation in the C programming language.

6.2.1 Selection Structure with Flowcharts

The selection structure is used for decision-making in the logic of a program. Figure 6.1 shows the selection design structure using a flowchart. Two possible paths for the execution flow are shown. The condition is evaluated, and one of the paths is selected. If the condition is true, then the left path is selected and *Processing 1* is performed. If the condition is false, the other path is selected and *Processing 2* is performed.

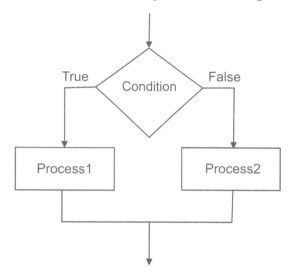

FIGURE 6.1: Flowchart of the selection structure.

6.2.2   Selection with Pseudo-Code

With pseudo-code, the selection structure is written with an **if** statement, also known as an if-then-else statement. This statement includes three sections: the condition, the then-section, and the else-section. The else-section is optional. Several keywords are used in this statement: **if, then, else,** and **endif**. The general form of the **if** statement in pseudo-code is:

```
if ⟨ condition ⟩
    then
        ⟨ statements in Processing 1 ⟩
    else
        ⟨ statements in Processing 2 ⟩
endif
```

When the condition is evaluated, only one of the two alternatives will be carried out: the one with the statements in *Processing 1* if the condition is true, or the one with the statements in *Processing 2* if the condition is not true. The pseudo-code of the selection structure in Figure 6.1 is written with an **if** statement as follows:

```
if condition is true
then
    perform instructions in Processing 1
else
    perform instructions in Processing 2
endif
```

6.2.3 Implementing Selection with the C Language

Similarly to pseudo-code, the C language includes the **if** statement that allows checking for a specified condition and executing statements if the condition is met. Note that C does not have the **then** keyword, but uses braces ({) and (}) instead of delimit blocks of code. The general form of the **if** statement in is:

```
if (⟨ condition ⟩ )
        { ⟨ statements in Processing 1 ⟩ }
else
        { ⟨ statements in Processing 2 ⟩ }
```

The first line specifies the condition to check. If the condition evaluates to **true**, the program executes the body of the first indented block of statements, otherwise the second indented block of statements after the **else** keyword is executed.

C does not require you to indent the body of an **if** statement, but it makes the code more readable and it is good programming practice.

6.2.4 Conditional Expressions

Conditional or Boolean expressions are used to form conditions. A condition consists of an expression that evaluates to a truth-value, *true* or *false*.

A conditional expression can be constructed by comparing the values of two data items and using a relational operator. The following list of relational operators can be included in a conditional expression:

1. Equal, $=$

2. Not equal, $\neq$

3. Less than, $<$

4. Less or equal to, $\leq$

5. Greater than, $>$

6. Greater or equal to, $\geq$

With pseudo-code, these relational operators are used to construct conditions as in the following examples:

$$\texttt{temp} \leq 64.75$$
$$x \geq y$$
$$p = q$$

Pseudo-code syntax also provides additional keywords that can be used for the relational operators. For example, the previous conditional expressions can also be written as follows:

`temp` **less or equal to** `64.75`

`x` **greater or equal to** `y`

`p` **equal to** `q`

With the C programming language, the relational operators are written using the following symbols: ==, for "equal," and !=, for "not equal." Note that == is the operator that tests equality, and = is the assignment operator. If = is used in the condition of an **if** statement, it will result in a logical error. The other relational operators in C are: <, >, <=, and >=.

In programming, a variable that contains a truth or logical value is often known as a *flag* because it flags the status of some condition.

6.2.5 Example with Selection

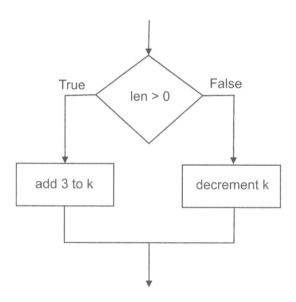

FIGURE 6.2: Example of selection structure.

In the following example, the condition to be evaluated is: *len* > 0, and a decision is taken to select which computation to carry out on variable *k*. Figure 6.2 shows the flowchart for part of the algorithm that includes this selection structure. In pseudo-code, this example is written as:

> **if** `len` **greater than** 0
>
> **then**
>
>> **add** 3 to k
>
> **else**
>
>> **decrement** k
>
> **endif**

The previous example in pseudo-code can also be written in the following manner:

```
if len > 0
then
    k = k + 3
    y = x × 35.45
else
    k = k - 1
endif
```

In C, the previous example is written as follows:

```
if (len > 0)   {
    k= k+3;
    y = x * 35.45;
}
else
    k=k-1;
```

Note that the second block of statements within the *if* statement does not have to be enclosed in braces because it consists of only one statement.

6.3   A Computational Model with Selection

The following problem involves developing a computational model that includes a quadratic equation, which is a simple mathematical model of a second-degree equation. The solution to the quadratic equation involves complex numbers.

6.3.1   Analysis and Mathematical Model

The goal of the solution to the problem is to compute the two roots of the equation. The mathematical model is defined in the general form of the quadratic equation (second-degree equation):

$$ax^2 + bx + c = 0$$

The given data for this problem are the values of the coefficients of the quadratic equation: a, b, and c. Because this mathematical model is a second degree equation, the solution consists of the value of two roots: x_1 and x_2.

6.3.2   Algorithm for General Solution

The general solution gives the value of the two roots of the quadratic equation, when the value of the coefficient a is not zero ($a \neq 0$). The values of the two roots are:

$$x_1 = \frac{-b + \sqrt{b^2 - 4ac}}{2a} \qquad x_2 = \frac{-b - \sqrt{b^2 - 4ac}}{2a}$$

The expression inside the square root, $b^2 - 4ac$, is known as the discriminant. If the discriminant is negative, the solution will involve complex roots. Figure 6.3 shows the flowchart for the general solution and the following listing is a high-level pseudo-code version of the algorithm.

```
Input the values of coefficients  a,  b, and
c
Calculate value of the discriminant
if the value of the discriminant is less than
zero
     then calculate the two complex roots
     else calculate the two real roots
endif
display the value of the roots
```

6.3.3   Detailed Algorithm

The algorithm in pseudo-code notation for the solution of the quadratic equation is:

```
read the value of a from the input device
read the value of b from the input device
read the value of c from the input device
compute the discriminant, disc = b² - 4ac
if discriminant less than zero
then
      // roots are complex
      compute x1 = (-b + √disc)/2a
      compute x2 = (-b - √disc)/2a
else
      // roots are real
      compute x1 = (-b + √disc)/2a
      compute x2 = (-b - √disc)/2a
endif
display values of the roots: x1 and x2
```

Listing 6.1 shows the C program that implements the algorithm for the solution of the quadratic equation, which is stored in the file `solquadra.c`.

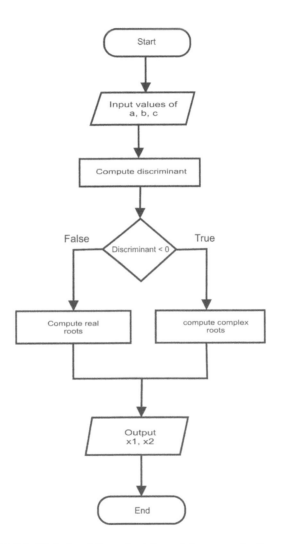

FIGURE 6.3: High-level flowchart for solving a quadratic equation.

Listing 6.1 C program that computes the roots of a quadratic equation.

```c
/*
 Program      : solquad.c
 Author       : Jose M Garrido, Nov 21 2012.
 Description : Compute the roots of a quadratic equation.
    Read the value of the coefficients: a, b, and c from
    the input console, display value of roots.
*/
#include <stdio.h>
#include <math.h>
int main(void) {
    double a;   /* coefficient a */
    double b;
    double c;
    double disc; /* discriminant */
    double x1r;  /* real part of root 1  */
    double x1i;  /* imaginary part of root 1 */
    double x2r;
    double x2i;
    printf("Enter value of cofficient a: ");
    scanf("%lf", &a);
    printf("\nValue of a: %lf \n", a);
    printf("Enter value of cofficient b: ");
    scanf("%lf", &b);
    printf("\nValue of a: %lf \n", b);
    printf("Enter value of cofficient c: ");
    scanf("%lf", &c);
    printf("\nValue of a: %lf \n", c);

    disc = pow(b, 2) - 4.0 * a * c;
    printf("disc: %lf \n", disc);
    if (disc < 0.0)   {
        /* complex roots */
        disc = -disc;
        x1r = -b/(2.0 * a);
        x1i = sqrt(disc)/(2.0 * a);
        x2r = x1r;
        x2i = -x1i;
        printf ("Complex roots \n");
        printf ("x1r: %lf x1i: %lf \n", x1r, x1i);
        printf ("x2r: %lf x2i: %lf \n", x2r, x2i);
    }
    else {
        /* real roots */
        x1r = (-b + sqrt(disc))/(2.0 * a);
        x2r = (-b - sqrt(disc))/(2.0 * a);
        printf("Real roots: \n");
        printf("x1: %lf x2: %lf \n", x1r, x2r);
    }
    return 0;      /* execution terminates ok */
}
```

The following shell commands compile, link, and execute the C program solquadra.c. The program prompts the user for the three values of the coefficients, calculates the roots, then displays the value of the roots.

```
$ gcc -Wall solquad.c -lm
$ ./a.out
Enter value of cofficient a: 1.5

Value of a: 1.500000
Enter value of cofficient b: 2.25

Value of a: 2.250000
Enter value of cofficient c: 5.5

Value of a: 5.500000
disc: -27.937500
Complex roots
x1r: -0.750000 x1i:  1.761865
x2r: -0.750000 x2i: -1.761865
```

6.4   Multi-Level Selection

There are often decisions that involve more than two alternatives. The general **if** statement with multiple paths is used to implement this structure. The case structure is a simplified version of the selection structure with multiple paths.

6.4.1   General Multi-Path Selection

The **elseif** clause is used to expand the number of alternatives. The **if** statement with *n* alternative paths has the general form:

> **if** ⟨ *condition* ⟩
>   **then**
>   ⟨ *block1* ⟩
>   **elseif** ⟨ *condition2* ⟩
>   **then**
>   ⟨ *block2* ⟩
>   **elseif** ⟨ *condition3* ⟩
>   **then**
>   ⟨ *block3* ⟩
>   **else**
>   ⟨ *blockn* ⟩
> **endif**

Each block of statements is executed when that particular path of logic is selected. This selection depends on the conditions in the multiple-path **if** statement that are evaluated from top to bottom until one of the conditions evaluates to true. The following example shows the **if** statement with several paths.

```
if y > 15.50
then increment x
elseif y > 4.5
then add 7.85 to x
elseif y > 3.85
then compute x = y*3.25
elseif y > 2.98
then x = y + z*454.7
else x = y
endif
```

In C, this example is very similar as the following listing shows.

```
if  (y > 15.50)
    x = x + 1;
elseif  (y > 4.5)
    x = x + 7.85;
elseif  (y > 3.85)
    x = y * 3.25;
elseif  (y > 2.98)
    x = y + z * 454.7;
else
    x = y;
```

6.4.2 Case Structure

The case structure is a simplified version of the selection structure with multiple paths. In pseudo-code, the **case** statement evaluates the value of a single variable or simple expression of a number or a text string and selects the appropriate path. The case statement also supports a block of multiple statements instead of a single statement in one or more of the selection options. The general case structure is:

```
case ⟨ select_var ⟩ of
   value    var_value1 : ⟨ block1 ⟩
   value    var_value2 : ⟨ block2 ⟩
      . . .
   value    var_valuen : ⟨ blockn ⟩
endcase
```

In the following example, the pressure status of a furnace is monitored and the pressure status is stored in variable *press_stat*. The following case statement first evaluates the pressure status and then assigns an appropriate text string to variable *msg*, then this variable is displayed.

```
case press_stat of
    value 'D' : msg = "Very dangerous"
    value 'A' : msg = "Alert"
    value 'W' : msg = "Warning"
    value 'N' : msg = "Normal"
    value 'B' : msg = "Below normal"
endcase
display "Pressure status: ", msg
```

The default option of the **case** statement can also be used by writing the keywords **default** or **otherwise** in the last case of the selector variable. For example, the previous example can be enhanced by including the default option in the case statement:

```
case press_stat of
    value 'E' : msg = "Very dangerous"
    value 'D' : msg = "Alert"
    value 'H' : msg = "Warning"
    value 'N' : msg = "Normal"
    value 'B' : msg = "Below normal"
    otherwise
        msg = "Pressure rising"
endcase
display "Pressure status: ", msg
```

In C, the pseudo-code **case** statement is implemented by the **switch** statement. The general form of this statement is:

```
switch { ⟨ select_var ⟩
    case   var_value1 :  ⟨ block1 ⟩
    case   var_value2 :  ⟨ block2 ⟩
       . . .
    case   var_valuen :  ⟨ blockn ⟩
    default
         ⟨ blockn ⟩
}
```

In the previous example of the pressure status of a furnace in variable *press_stat*. The following switch statement first evaluates the pressure status and then assigns an appropriate text string to variable *msg*, then this variable is displayed.

```
switch press_stat {
   case 'D' :
       strcpy(msg, "Very dangerous");
   case 'A' :
       strcpy(msg, "Alert");
   case 'W' :
       strcpy(msg, "Warning");
   case 'N' :
       strcpy(msg, "Normal");
   case 'B' :
       strcpy(msg, "Below normal");
   default :
       strcpy(msg, "Pressure rising");
}
printf ("Pressure status: %s \n", msg);
```

6.5   Complex Conditions

The logical operators **and, or**, and **not** are used to construct more complex conditional expressions from simpler conditions. The general forms of complex conditions from two simple conditions, *condexp1* and *condexp2* in pseudo-code are:

```
condexp1 and condexp2
condexp1 or condexp2
not condexp1
```

The following example includes the **not** operator:

$$\textbf{not} \quad (x \leq q)$$

The following example in pseudo-code includes the **or** logical operator:

```
if a < b  or x ≥ y
then
    ( block_1 )
else
    ( block_2 )
endif
```

The previous condition can also be written in the following manner:

a **less than** b **or** x **is greater or equal to** y

The C programming language provides several logical operators and functions for constructing complex conditions from simpler ones. The basic logical operators for scalars (individual numbers) in C are: the && (and) operator, the || (or) operator, and the ! (not) operator.

In the following example, two simple conditions are combined with the && operator (and). The result condition evaluated to *true*.

```
x = 5.25;
(2.5 < x) && (x < 21.75)
```

In C, the boolean type is available for C99 compliant compilers, as type *bool*. The header file addtypes.h defines type bool.

Summary

The selection structure is also known as alternation. It evaluates a condition then follows one of two (or more) paths. The two general selection statements **if** and **case** statements are explained in pseudo-code and in C. The first one is applied when there are two or more possible paths in the algorithm, depending on how the condition evaluates. The case statement is applied when the value of a single variable or expression is evaluated, and there are multiple possible values.

The condition in the **if** statement consists of a conditional expression, which evaluates to a truth-value (true or false). Relational operators and logical operators are used to form more complex conditional expressions.

Key Terms			
selection	alternation	condition	if statement
case statement	relational operator	logical operator	truth-value
then	else	endif	endcase
otherwise	elseif	end	switch statement

Exercises

Exercise 6.1 Develop a computational model that computes the conversion from gallons to liters and from liters to gallons. Include a flowchart, pseudo-code design and a complete implementation in C. The user inputs the string: "gallons" or "liters"; the model then computes the corresponding conversion.

Exercise 6.2 Develop a computational model to calculate the total amount to pay for movie rental. Include a flowchart, pseudo-code design and a complete implementation in C. The movie rental store charges $3.50 per day for every DVD movie. For every additional period of 24 hours, the customer must pay $0.75.

Exercise 6.3 Develop a computational model that finds and displays the largest of several numbers, which are read from the input device. Include a flowchart, pseudo-code design and a complete implementation in C.

Exercise 6.4 Develop a computational model that finds and displays the smallest of several numbers, which are read from the input device. Include a flowchart, pseudo-code design and a complete implementation in C.

Exercise 6.5 Develop computational model program that computes the gross and net pay of the employees. The input quantities are employee name, hourly rate, number of hours, percentage of tax (use 14.5%). The tax bracket is $115.00. When the number of hours is greater than 40, the (overtime) hourly rate is 40% higher. Include a flowchart, pseudo-code design and a complete implementation in C.

Exercise 6.6 Develop a computational model that computes the distance between two points. A point is defined by a pair of values (x, y). Include a flowchart, pseudo-code design and a complete implementation in C. The distance, d, between two points, $P_1(x_1, y_1)$ and $P_2(x_2, y_2)$ is defined by:

$$d = \sqrt{(x_2 - x_1)^2 + (y_2 - y_1)^2}$$

Exercise 6.7 Develop a computational model that computes the fare in a ferry transport for passengers with motor vehicles. Include a flowchart, pseudo-code design and a complete implementation in C. Passengers pay an extra fare based on the vehicle's weight. Use the following data: vehicles with weight up to 780 lb pay $80.00, up to 1100 lb pay $127.50, and up to 2200 lb pay $210.50.

Exercise 6.8 Develop a computational model that computes the average of student grades. The input data are the four letter grades for various work submitted by the students. Include a flowchart, pseudo-code design and a complete implementation in the C programming language.

Chapter 7

Repetition

7.1 Introduction

This chapter presents the repetition structure and specifying, describing, and implementing algorithms in developing computational models. This structure and the corresponding statements are discussed with flowcharts, pseudo-code, and in the C programming language. The repetition structure specifies that a block of statements be executed repeatedly based on a given condition. Basically, the statements in the process block of code are executed several times, so this structure is often called a *loop* structure. An algorithm that includes repetition has three major parts in its form:

1. the initial conditions

2. the steps that are to be repeated

3. the final results.

There are three general forms of the repetition structure: the *while* loop, the *repeat-until* loop, and the *for* loop. The first form of the repetition structure, the *while* construct, is the most flexible. The other two forms of the repetition structure can be expressed with the *while* construct.

7.2 Repetition with the While Loop

The *while* loop consists of a conditional expression and block of statements. This construct evaluates the condition before the process block of statements is executed. If the condition is true, the statements in the block are executed. This repeats while the condition evaluates to true; when the condition evaluates to false, the loop terminates.

7.2.1 While-Loop Flowchart

A flowchart with the *while* loop structure is shown in Figure 7.1. The process block consists of a sequence of actions.

The actions in the process block are performed while the condition is true. After the actions in the process block are performed, the condition is again evaluated, and the actions are again performed if the condition is still true; otherwise, the loop terminates.

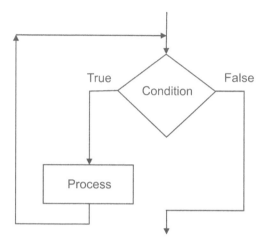

FIGURE 7.1: A flowchart with a while-loop.

The condition is tested first, and then the process block is performed. If this condition is initially false, the actions in the block are not performed.

The number of times that the loop is performed is normally a finite number. A well-defined loop will eventually terminate, unless it has been specified as a non-terminating loop. The condition is also known as the *loop condition*, and it determines when the loop terminates. A non-terminating loop is defined in special cases and will repeat the actions forever.

7.2.2 While Structure in Pseudo-Code

The form of the *while* statement includes the condition, the actions in the process block written as statements, and the keywords *while*, *do*, and *endwhile*. The block of statements is placed after the *do* keyword and before the *endwhile* keyword. The following lines of pseudo-code show the general form of the while-loop statement that is shown in the flowchart of Figure 7.1.

```
while ⟨ condition ⟩   do
    ⟨ block of statements ⟩
endwhile
```

The following example has a *while* statement with a condition that checks the value of variable j. The block of statements is performed repeatedly while the condition j <= MAX_NUM is true.

```
while j <= MAX_NUM do
    set sum = sum + 12.5
    set y = x * 2.5
    add 3 to j
endwhile
display "Value of sum: ", sum
display "Value of y: ", y
```

7.2.3 While Loop in the C Language

The following lines of code show the general form of the while-loop statement in C; it is similar to the pseudo-code statement and follows the loop definition shown in the flowchart of Figure 7.1.

```
while ⟨ condition ⟩
    ⟨ block of statements ⟩
end
```

The previous example has a *while* statement with a condition that checks the value of variable *j*. The block of statements is enclosed in braces and repeats while the condition j <= MAX_NUM is true. The following lines of code show the C implementation.

```
while ( j <= MAX_NUM) {
    sum = sum + 12.5;
    y = x * 2.5;
    j = j + 3;
{
moutput('Value of sum: ', sum)
moutput('Value of y: ', y)
```

7.2.4 Loop Counter

As mentioned previously, in the while-loop construct the condition is tested first and then the statements in the statement block are performed. If this condition is initially false, the statements are not performed.

The number of times that the loop is performed is normally a finite integer value. For this, the condition will eventually be evaluated to false, that is, the loop will terminate. This condition is often known as the *loop condition*, and it determines when the loop terminates. Only in some very special cases, the programmer can decide to write an infinite loop; this will repeat the statements in the repeat loop forever.

A *counter* variable stores the number of times (also known as iterations) that the loop executes. The counter variable is incremented every time the statements in the loop are performed. The variable must be initialized to a given value, typically to 0 or 1.

In the the following pseudo-code listing, there is a *while* statement with a counter variable with name *loop_counter*. This counter variable is used to control the number of times the block statement is performed. The counter variable is initially set to 1, and is incremented every time through the loop.

```
Max_Num = 25       // maximum number of times to execute
set loop_counter = 1   // initial value of counter
while loop_counter < Max_Num do
        increment loop_counter
        display "Value of counter: ", loop_counter
endwhile
```

The first time the statements in the block are performed, the loop counter variable *loop_counter* has a value equal to 1. The second time through the loop, variable *loop_counter* has a value equal to 2. The third time through the loop, it has a value of 3, and so on. Eventually, the counter variable will have a value equal to the value of *Max_Num* and the loop terminates.

7.2.5   Accumulator Variables

An *accumulator* variable stores partial results of repeated calculations. The initial value of an accumulator variable is normally set to zero.

For example, an algorithm that calculates the summation of numbers from input, includes an accumulator variable. The following pseudo-code statement accumulates the values of *innumber* in variable *total* and it is included in the while loop:

```
total = 0.0
while j < Max_num
    add innumber to total
endwhile
display "Total accumulated: ", total
```

After the *endwhile* statement, the value of the accumulator variable *total* is displayed using the pseudo-code statement to print the string "Total accumulated" and the value of *total*. In programming, each counter and accumulator variable serves a specific purpose. These variables should be well documented.

7.2.6   Summation of Input Numbers

The following simple problem applies the concepts and implementation of while loop and accumulator variable. The problem computes the summation of numeric values inputed from the main input device. Computing the summation should proceed while the input values are greater than zero.

The pseudo-code that describes the algorithm uses an input variable, an accumulator variable, a loop counter variable, and a conditional expression that evaluates whether the input value is greater than zero.

```
set innumber = 1.0   // number with dummy initial value
set loop_counter = 0
set sum = 0.0         // initialize accumulator variable
while innumber > 0.0   do
      display "Type number: "
      read innumer
      if innumber > 0.0
      then
           add innumber to sum
           increment loop_counter
           display "Value of counter: ", loop_counter
      endif
endwhile
display "Value of sum: ", sum
```

Listing 7.1 shows the C program that implements the summation problem. The program is stored in file summ.c.

Listing 7.1 Source C program for summation.

```
 1 /* Program    : summ.c
 2  Description : Compute the summation of numbers greater
 3      than zero. The program reads from input
 4      console, displays value of the sum of numbers.
 5  Author      : Jose M Garrido, Nov 21 2012.
 6 */
 7 #include <stdio.h>
 8 #include <math.h>
 9 int main(void) {
10      double innumber;    /* number to read and sum */
11      double msum = 0.0;
12      printf("Enter value of input number: ");
13      scanf("%lf", &innumber);
14      printf("\nValue of input number: %lf \n", innumber);
15      while ( innumber > 0.0) {
16          msum = msum + innumber;
17          printf("Enter value of input number: ");
18          scanf("%lf", &innumber);
19          printf("\nValue input number: %lf \n", innumber);
20      }
21      printf("Value of summation: %lf \n", msum);
22      return 0;      /* execution terminates ok */
23 }
```

The following output listing shows the shell commands used to compile and link the source file summ.c, and execute the object file summ.out.

```
$ gcc -Wall summ.c -o summ.out -lm
$ ./summ.out
Enter value of input number: 12.5

Value of input number: 12.500000
Enter value of input number: 10.0

Value of input number: 10.000000
Enter value of input number: 5.5

Value of input number: 5.500000
Enter value of input number: 4.25

Value of input number: 4.250000
Enter value of input number: 133.75

Value of input number: 133.750000
Enter value of input number: 0.0

Value of input number: 0.000000
Value of summation: 166.000000
```

7.3   Repeat-Until Loop

The *repeat-until* loop is a control flow structure that allows actions to be executed repeatedly based on a given condition. This construct consists of a process block of actions and the condition.

The actions within the process block are executed first, and then the condition is evaluated. If the condition is not true the actions within the process block are executed again. This repeats until the condition becomes true.

Repeat-until structures check the condition after the block is executed; this is an important difference from the while loop, which tests the condition before the actions within the block are executed. Figure 7.2 shows the flowchart for the repeat-until structure.

The pseudo-code statement of the repeat-until structure corresponds directly with the flowchart in Figure 7.2 and uses the keywords *repeat*, *until*, and *endrepeat*. The following portion of code shows the general form of the repeat-until statement.

```
repeat
     ( statements in block )
until ( condition )
endrepeat
```

The following lines of code shows the pseudo-code listing of a repeat-until statement for the previously discussed problem.

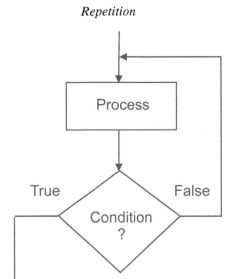

FIGURE 7.2: A flowchart with a repeat-until structure.

```
set innumber = 1.0   // dummy initial value
set l_counter = 0
set sum = 0.0        // accumulator variable
repeat
        display "Type number: "
        read innumer
        if innumber > 0.0
        then
            add innumber to sum
            increment l_counter
            display "Value of counter: ", l_counter
        endif
until innumber <= 0.0
endrepeat
display "Value of sum: ", sum
```

In C, the repeat-until loop has the general form:

```
do
{
      [ block of statements ]
}
while (condition);
```

7.4   For Loop Structure

The *for* loop structure explicitly uses a loop counter; the initial value and the
final value of the loop counter are specified. The *for* loop is most useful when the
number of times that the loop is carried out is known in advance. In pseudo-code,
the *for* statement includes the keywords: *for, to, downto, do,* and *endfor* and has the
following general form in pseudo-code:

```
for 〈 counter 〉 = 〈 initial_val 〉 to 〈
final_val 〉
do
      Block of statements
endfor
```

On every iteration, the loop counter is automatically incremented. The last time
through the loop, the loop counter reaches its final value and the loop terminates. The
for loop is similar to the *while* loop in that the condition is evaluated before carrying
out the operations in the repeat loop.

The following portion of pseudo-code uses a *for* statement for the repetition part
of the summation problem.

```
for j = 1 to num do
      set sum = sum + 12.5
      set y = x * 2.5
endfor
display "Value of sum: ", sum
display "Value of y: ", y
```

Variable *j* is the counter variable, which is automatically incremented and is used
to control the number of times the statements in a block will be performed. In C, the
previous portion of code is written as follows.

```
for ( j = 1; j <= num; j++ ) {
      sum = sum + 12.5;
      y = x * 2.5;
}
printf ("Value of sum: %lf \n", sum);
printf ("Value of y: %lf \n", y)
```

7.4.1   Summation Problem with a For Loop

Using the for-loop construct of the repetition structure, the algorithm for the
summation of input data can be defined in a relatively straightforward manner with
pseudo-code. The most significant difference from the previous design is that the

number of data inputs from the input device is included at the beginning of the algorithm.

```
set innumber = 1.0  // number with dummy initial value
set sum = 0.0       // initialize accumulator variable
display "Number of input data to read: "
read MaxNum
for loop_counter = 1 to MaxNum do
      display "Type number: "
      read innumer
      if innumber > 0.0
      then
          add innumber to sum
          display "Value of counter: ", loop_counter
      endif
endfor
display "Value of sum: ", sum
```

Listing 7.2 shows the C source program that implements the summation problem with a for-loop. The program is stored in file summfor.c.

Listing 7.2 C source program for summation with for-loop.

```
 1 /*
 2 Program     : summfor.c
 3 Author      : Jose M Garrido, Nov 21 2012.
 4 Description : Compute the summation of numbers greater
 5    than zero. The program reads from input
 6    console, displays value of the sum of numbers.
 7 */
 8 #include <stdio.h>
 9 #include <math.h>
10 int main(void) {
11      double innumber;   /* number to read and sum */
12      double msum = 0.0;
13      int max_num;   /* maximum values to read */
14      int j;          /* loop counter  */
15      printf("Enter number of input values: ");
16      scanf("%d", &max_num);
17      printf("\nValue of input number: %d \n", max_num);
18      for ( j = 1; j <= max_num; j++) {
19          printf("Enter value of input number: ");
20          scanf("%lf", &innumber);
21          if ( innumber > 0.0 )
22              msum = msum + innumber;
23      }
24      printf("Value of summation: %lf \n", msum);
25      return 0;      /* execution terminates ok */
26 }
```

7.4.2 Factorial Problem

The *factorial* operation, denoted by the symbol !, can be defined in a general and informal manner as follows:

$$y! = y(y-1)(y-2)(y-3)\ldots 1$$

For example, the factorial of 5 is:

$$5! = 5 \times 4 \times 3 \times 2 \times 1$$

7.4.2.1 Mathematical Specification of Factorial

A mathematical specification of the factorial function is as follows, for $y \geq 0$:

$$y! = \left\{ \begin{array}{ll} 1 & \text{when } y = 0 \\ y(y-1)! & \text{when } y > 0 \end{array} \right.$$

The base case in this definition is the value of 1 for the function if the argument has value zero, that is $0! = 1$. The general (recursive) case is $y! = y(y-1)!$, if the value of the argument is greater than zero. This function is not defined for negative values of the argument.

7.4.2.2 Computing Factorial

In the following C program, the factorial function is named *fact* and has one argument: the value for which the factorial is to be computed. For example, $y!$ is implemented as function *fact(y)*.

Listing 7.3. shows a C program, *factp.c* that includes function *main* that calls function *fact* to compute the factorial of a number and displays this value on the console.

Listing 7.3 C source program for computing factorial.

```
 1 /*
 2  Program     : factp.c
 3  Author      : Jose M Garrido, Nov 28 2012.
 4  Description : Compute the factorial of a number.
 5 */
 6 #include <stdio.h>
 7 #include <math.h>
 8 int fact (int y);      /* function fact prototype */
 9 int main(void) {
10      int num;
11      int r;
12      printf("Enter number: ");
13      scanf("%d", &num);
14      printf("\nValue of input number: %d \n", num);
15      r = fact(num);
16      printf(" Factorial of %d is: %d \n", num, r);
17      return 0;        /* execution terminates ok */
18 }
19 // the factorial function
20 int fact (int y) {
21      int f;
```

```
22        int n;
23        int res;
24        f = 1;
25        if ( y > 0 ) {
26             for ( n = 1; n <= y; n++)
27                  f = f * n;
28             res = f;
29        }
30        else if ( y == 0 )
31                res = 1;
32        else
33             res = -1;
34        return res;
35 }
```

Note that this implementation will always return −1 for negative values of the argument. The following shell commands compile, link, and execute the program for several values of the input number.

```
$ gcc -Wall factp.c -o factp.out
$ ./factp.out
Enter number: 5

Value of input number: 5
 Factorial of 5 is: 120

$ ./factp.out
Enter number: 8

Value of input number: 8
 Factorial of 8 is: 40320
$ ./factp.out
Enter number: 10

Value of input number: 10
 Factorial of 10 is: 3628800

$ ./factp.out
Enter number: -5

Value of input number: -5
 Factorial of -5 is: -1
```

Summary

The repetition structure is used in algorithms in order to perform repeatedly a group of action steps (instructions) in the process block. There are three types of loop structures: *while* loop, *repeat-until*, and *for* loop. In the *while* construct, the loop condition is tested first, and then the block of statements is performed if the condition is true. The loop terminates when the condition is false.

In the *repeat-until* construct, the group of statements in the block is carried out first, and then the loop condition is tested. If the loop condition is true, the loop terminates; otherwise the statements in the block are performed again.

The number of times the statements in the block are carried out depends on the condition of the loop. In the *for* loop, the number of times to repeat execution is explicitly indicated by using the initial and final values of the loop counter. Accumulator variables are also very useful with algorithms and programs that include loops. Simple examples of C programs are executed.

Key Terms			
repetition	loop	while	loop condition
do	endrepeat	block	loop termination
loop counter	endwhile	accumulator	repeat-until
for	to	downto	endfor
end	iterations	summation	factorial

Exercises

Exercise 7.1   Develop a computational model that computes the maximum value from a set of input numbers. Use a while loop in the algorithm and implement in C.

Exercise 7.2   Develop a computational model that computes the maximum value from a set of input numbers. Use a for loop in the algorithm and implement in C.

Exercise 7.3   Develop a computational model that computes the maximum value from a set of input numbers. Use a repeat-until loop in the algorithm and implement in C.

Exercise 7.4   Develop a computational model that finds the minimum value from a set of input values. Use a while loop in the algorithm and implement in C.

Exercise 7.5   Develop a computational model that finds the minimum value from a set of input values. Use a for loop in the algorithm and implement in C.

Exercise 7.6   Develop a computational model that finds the minimum value from a set of input values. Use a repeat-until loop in the algorithm and implement in C.

Exercise 7.7   Develop a computational model that computes the average of a set of input values. Use a while loop in the algorithm and implement in C.

Exercise 7.8 Develop a computational model that computes the average of a set of input values. Use a for loop in the algorithm and implement in C.

Exercise 7.9 Develop a computational model that computes the average of a set of input values. Use a repeat-until loop in the algorithm and implement in C.

Exercise 7.10 Develop a computational model that computes the student group average, maximum, and minimum grade. The computational model uses the input grade for every student. Use a while loop in the algorithm and implement in C.

Exercise 7.11 Develop a computational model that computes the student group average, maximum, and minimum grade. The computational model uses the input grade for every student. Use a for loop in the algorithm and implement in C.

Exercise 7.12 Develop a computational model that reads rainfall data in inches for yearly quarters from the last five years. The computational model is to compute the average rainfall per quarter, the average rainfall per year, and the maximum rainfall per quarter and for each year. Implement in C.

Exercise 7.13 Develop a computational model that computes the total inventory value amount and total per item. The computational model is to read item code, cost, and description for each item. The number of items to process is not known. Implement in C.

Chapter 8

Arrays

8.1 Introduction

An *array* stores multiple values of data with a single name and most programming languages include facilities that support arrays. The individual values in the array are known as *elements*. Many programs manipulate arrays and each one can store a large number of values in a single collection.

To refer to an individual element of the array, an integer value or variable known as the array *index* is used after the name of the array. The value of the index represents the relative position of the element in the array. Figure 8.1 shows an array with 13 elements. This array is a data structure with 13 cells, and each cell contains an element value. In the C programming language, the index values start with 0 (zero).

2.75	6.4	15.0	7.65	10.5	5.25	12.5	8.25	6.9	3.5	15.5	5.5	9.4
0	1	2	3	4	5	6	7	8	9	10	11	12

FIGURE 8.1: A simple array.

To use arrays in a C program the following steps are carried out:

1. Declare the array with appropriate name, size, and type

2. Assign initial values to the array elements

3. Access or manipulate the individual elements of the array

In C, an array is basically a static data structure because once the array is declared, its size cannot be changed. The size of an array is the number of elements it can store. An array can also be created at execution time using pointers.

A simple array has one dimension and is considered a row vector or a column vector. To manipulate the elements of a one-dimensional array, a single index is required. A vector is a single-dimension array and by default it is a row vector.

A *matrix* is a two-dimensional array and the typical form of a matrix is an array with data items organized into rows and columns. In this array, two indexes are used: one index to indicate the column and one index to indicate the row. Figure 8.2 shows a matrix with 13 columns and two rows. Because this matrix has only two rows, the index values for the rows are 0 and 1. The index values for the columns are from 0 up to 12.

FIGURE 8.2: A two-dimensional array.

8.2   Declaring an Array

An array is declared with an appropriate name, type, and size and the optional initial values of the various elements. The size of the array is the number of elements it can store.

8.2.1   Declaring Arrays in Pseudo-Code

In pseudo-code, the general form for declaring an array is:

define ⟨ *array_name* ⟩ **array** [⟨ *size* ⟩] **of type** ⟨ *array_type* ⟩

The following pseudo-code statement declares array *y*, with a capacity of 20 elements.

```
define y array [20] of type float
```

The size of the array can be specified using a symbolic (identifier) constant. For example, given the constant *MAX_NUM* with value 20, the array *y* can be declared in pseudo-code by the following statement:

```
define y array [MAX_NUM] of type float
```

8.2.2   Declaring Arrays in C

In the C programming language, an array is declared with its type, name, and size. For example, assume that the name of the one-dimensional array (vector) shown in Figure 8.1 is *precip*; it can be declared in C in the following manner:

```
double   rainf[13];
```

When the array is small, initial values can be given in the same statement that declares the array:

```
double   rainf[13] = {2.75 6.4 15.0 7.65 10.5 5.25 12.5 8.25 6.9
              3.5 15.5 5.5 9.4};
```

8.3   Operations on Arrays

A program can manipulate an array by accessing the individual elements of the array and performing some computation. Recall that an integer value known as the *index* is used with the name of the array to access an individual element of an array. In C, the range of index values starts with 0 and ends with the number equal to the size of the array minus 1.

8.3.1   Manipulating Array Elements in Pseudo-Code

To refer to an individual element in pseudo-code, the name of the array is used followed by the index value in rectangular brackets. The following statement assigns a value of 8.25 to element 5 of array *z*.

```
set  z[5] = 8.25
```

The index may be an integer constant, a symbolic constant (identifier), or an integer variable. The following statement assigns the value 4.5 to element 6 of array *y*, using variable *k* as the index.

```
define y array [MAX_NUM] of type float
set k = 6
set y[k] = 4.5
```

8.3.2   Manipulating Elements of an Array in C

After declaring an array in C, the individual elements can be used in the same manner simple variables are used. To reference an individual element of an array, the appropriate index value is written in brackets. The following assignment statements may be used to initialize the elements of array *rainf* that was declared previously:

```
rainf[0]  =  2.75;
rainf[1]  =   6.4;
rainf[2]  =  15.0;
rainf[3]  =  6.75;
rainf[4]  =   4.5;
rainf[5]  =  5.25;
rainf[6]  =   2.4;
rainf[7]  =  80.4;
rainf[8]  = 25.96;
rainf[9]  =  9.25;
rainf[10] =  45.5;
rainf[11] =  7.25;
rainf[12] =   8.5;
```

Accessing the elements of a vector in C is also known as indexing a vector. The values of the index of a one-dimensional array is an integer value that starts with zero, thus the first element of a simple array has an index value of zero. Accessing a particular element of an array uses the name of the array followed by the index value in brackets. The following assignment statement assigns a value of 14.5 to element 6 of array *y*.

```
y[5] = 14.5;
```

The index value used can be an integer constant, a symbolic constant (identifier), or an integer variable. The following lines of C code declare vector *height*, assign a value 4 to variable *k*, and assign the value 132.5 to element 5 (with index 4) of array *w*, using variable *k* as the index.

```
float w[5] = {12.5 8.25 7.0 21.35 6.55};
k = 4;
w[k] = 132.5;
```

In addition to accessing individual elements of an array in a simple assignment statement, more complex arithmetic operations are also supported in C. The following example involves a trigonometric operation with element 4 of vector *w*; the assignment sets the value to variable *x*. The C symbolic constant *M_PI* is the value for π.

```
x = sin(0.04*w[4]/M_PI);
```

8.4   Arrays as Arguments

The arguments in a function call can be used as input values, output values, or input-output values by the function. In the C programming language, there are two general techniques to pass arguments when calling a function:

- pass by value

- pass by reference

Using *pass by value*, a copy of the value of the argument is passed to the corresponding parameter of the function. With this technique, the value of the argument is not changed by the function. Any attempt to change the value is only local to the function.

Using *pass by reference*, an address of the argument is passed to the corresponding parameter. The function can change the value of this argument and when the function completes its execution, this argument will have a different value.

For input arguments, pass by value is used with simple variables. For output arguments, pass by reference is used. To pass simple variables by reference, *pointers* need to be used.

An array is always passed by reference and only the array name is used as the argument in the function call. The corresponding array parameter in the function definition does not include the & symbol before the parameter name. When the parameter is declared as a one-dimensional array, the size of the array need not be specified.

Because an array is passed by reference, the function being called can modify the value of the elements of the array. However, in the case that the function should not change the values of the elements of an array, the keyword *const* must be included before the parameter declaration in the function prototype and in the function definition. The following example illustrates this notion in the function prototype for function *myfunc*.

```
void myfunct ( const double x [], int xsize );
```

Two examples of very simple and useful functions that manipulate arrays are defined in the library *basic_lib*; these are: *arracol* and *linspace*. Function *arraycol* assigns equally-spaced values with a fixed increment. The function parameters in the function prototype and definition are: *arr*, *startv*, *lastval*, and *incr*. These are the array name, the starting value, the last value, and the increment value. The function returns the number of values assigned to the array. The parameters are declared of type *double* and the function prototype is:

```
int arraycol ( double arr[], double startv, double lastv,
    double incr);
```

In the following example, the third line of C code calls function *arraycol* to assign values to vector *y*, starting with value 3.15, having a final value of 20.5 and increments of 0.45.

```
int j;
. . .
j = arraycol ( y, 3.15, 20.5, 0.45);
```

The other function in library *basic_lib* is *linspace*, which assigns values to a single-dimension array. This function also assigns values that are equally spaced using the initial value, the final value, and the number of values in the vector. In the following example, a call to function *linspace* assigns vector *y* equally-spaced values starting with 3.15, a final value of 20.5, and a total of 20 values.

```
linspace(y, 3.15, 20.5, 20);
```

The following C source listing is the definition of function *linspace*; note that the first parameter is a one-dimensional array of type *double*.

```
/*  This function computes the num element values
      in the array starting from low to high.          */
void linspace (double  a[], double low, double high, int num) {
    int  i;          /* used as index of the array  */
    double diff;
    diff = (high - low)/(num-1);
    a[0] = low;
    for (i = 1; i < num; i++ ) {
        a[i] = a[i-1] + diff;
    }
    return;
}  // end linspace
```

8.5 A Simple Application with Arrays

The following example describes a program that computes height, velocity, and acceleration of a free-falling object by computing finite rates of change. The program results are used to plot the vertical position of the object with respect to time using GnuPlot.

Listing 8.1 shows the complete C program that computes the values of height in problem of the free falling object. The source program is stored in file `ffallobj3.c`. Function *linspace* is called to generate values of time starting at 0.0, ending at 2.94, using *N* different values. All the values of time used are stored in vector *tf*.

Listing 8.1 Computing the vertical position and velocity of the object.

```
1 /* ffallobj3.c
2  Compute height, velocity, and acceleration
3  of a free-falling object
4  Computing finite rates of change
5  Plot of vert position vs time with GnuPlot
6  This program uses the 'basic_lib' library
7  J Garrido, 1-3-2013
8 */
```

```
 9 #include <stdio.h>
10 #include <stdlib.h>
11 #include <math.h>
12 #include "basic_lib.h"
13 int main() {
14   const int N = 50;
15   const float g = 9.8;
16   double y0;  // initial vertical position
17   double dtf; // fixed time increment
18   int j;
19   double tf[N];  // vector of time
20   double hf[N];  // vector of computed vertical position
21   double * dhf;  // differences of hf
22   double * dvel; // differences of vel
23   double vel[N-1];
24   double accel[N-2];
25   printf("Free-falling object \n");
26   // printf("Type initial vertical pos: ");
27   // scanf("%lf", &y0);
28   y0 = 40.0;
29   printf("Time     Vertical Position \n");
30   linspace(tf, 0.0, 2.94, N); // values of time
31   dtf=tf[2]- tf[1];  // delta t
32   for (j = 0; j < N; j++) {
33       hf[j] = y0 - 0.5 * (g * pow(tf[j],2)); // height
34    printf(" %.4f %.3f \n", tf[j], hf[j]);
35   }
36   dhf=diff(hf, N);  // Differences in vertical pos
37   // Vertical velocity
38   printf("Vertical velocity \n");
39   for (j = 0; j < N-1; j++) {
40      vel[j] = dhf[j]/dtf;  // rate of chg vertical position
41   printf(" %f \n", vel[j]);
42   }
43   printf("Acceleration of object \n");
44   // differences of the vertical velocity vector
45   dvel = diff(vel, N-1);
46   for (j = 0; j < N-2; j++) {
47     accel[j] = dvel[j]/dtf;  // rate of change of velocity
48  printf(" %f \n", accel[j]);
49   }
50   return 0;
51 } // end main
```

The following shell commands compile, link, and compute the vertical position (height) of the free-falling object for every value of time stored in vector *tf*. The values of height are stored in the new vector, *hf*.

```
$ gcc -Wall ffallobj3.c -o ffallobj2a.out basic_lib.o -lm
$ ./ffallobj3.out  | tee ffallobj2a.gpl
Free-falling object
Time       Vertical position
 0.000000   40.000000
 0.196000   39.811762
 0.392000   39.247046
```

```
0.588000   38.305854
0.784000   36.988186
0.980000   35.294040
1.176000   33.223417
1.372000   30.776318
1.568000   27.952742
1.764000   24.752689
1.960000   21.176160
2.156000   17.223153
2.352000   12.893670
2.548000   8.187710
2.744000   3.105273
```

Figure 8.3 shows the plot of the change of the vertical position (height) of the free-falling object, with respect to time. Note that the graph represents a curve and not a straight line. This is because the mathematical relation between the variables hf and ht is non-linear, more specifically, it is a quadratic relationship between these two variables. This plot is produced by GNUplot using the results produced by the program execution.

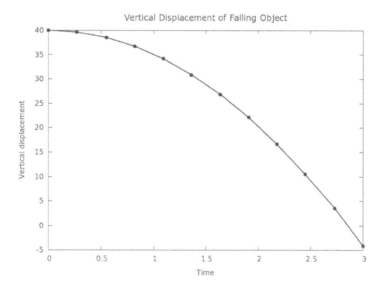

FIGURE 8.3: Plot of the values of the vertical position with time.

8.6   Arithmetic Operations with Vectors

Simple arithmetic operations can be performed that relate a vector and a number, also known as a *scalar*. These operations are: addition, subtraction, multiplication, and division. These operations are applied to all elements of the vector. For example, the following for-loop in C, adds the scalar 3.25 to all elements of vector *height*, which was previously declared with *N* elements.

```
for (j = 0; j < N; j++) {
    hf[j] = hf[j] + 3.25; // add 3.25 to all elements of hf
}
```

Several array operations can be implemented in C. The simplest array operations are: addition, subtraction, multiplication, division, and exponentiation of two vectors. Element by element operations are usually inside a loop using similar notation for multiplication, division, and exponentiation. The following lines of code in C apply several simple array operations on two small row vectors.

```
double x [6];
double y [6];
const int N = 3;
x[0] = 1.0;
x[1] = 2.0;
x[3] = 3.0;
y[0] = 4.0;
y[1] = 5.0;
y[2] = 6.0;
/* vector addition */
for (j = 0; j < N; j++)  {
    x[j] = x[j] + y[j];
    printf (" x: %lf y: %lf \n", x[j], y[j]);
}
/* vector subtraction */
for (j = 0; j < N; j++)  {
    x[j] = x[j] - y[j];
    printf (" x: %lf y: %lf \n", x[j], y[j]);
}
/* vector multiplication */
for (j = 0; j < N; j++)  {
    x[j] = x[j] * y[j];
    printf (" x: %lf y: %lf \n", x[j], y[j]);
}
```

8.7   Multi-Dimensional Arrays

A multi-dimensional array is declared with one or more dimensions. A common name for a two-dimensional array is a matrix) and this is a mathematical structure with values arranged in columns and rows. Two index values are required, one for the rows and one for the columns. Figure 8.2 shows an example of a two-dimensional array.

8.7.1 Multi-Dimensional Arrays in Pseudo-Code

To declare a two-dimensional array in pseudo-code, two index numbers are used. The first number defines the range of values for the first index (rows) and the second number defines the range of values for the second index (columns).

The following pseudo-code statements declare a two-dimensional array named *rainf* with size 30 rows and 10 columns.

```
define ROWS = 30 of type integer
define COLS = 10 of type integer
.  .  .
define rainf array [ROWS][COLS] of type float
```

To reference individual elements of a two-dimensional array, two integer numbers are used as indexes. In the following pseudo-code statements each individual element is accessed using index variables *i* and *j*. The code assigns all the elements of array *rainf* to 0.0:

```
for j = 0 to COLS - 1 do
      for i = 0 to ROWS - 1 do
            set rainf [i][j] = 0.0
      endfor
endfor
```

In these operations with two-dimensional arrays, nested loops are used: an outer loop and an inner loop. The inner loop varies the row index *i* from 0 to value *ROWS-1*, and outer loop varies the row index *j* from 0 to *COLS-1*. The assignment sets the value 0.0 to the element at row *i* and column *j*.

8.7.2 Multi-Dimensional Arrays in C

The following lines of code declare in C, vectors *x* and *w* and a two-dimensional array *z*. The element values of *x* are assigned from a starting value of 2 ending in 11 in increments of 2. For array *w*, the assignments of values starting at 1.0, a final value of 9.0, and in increments of 2.0.

The for-loop assigns values to the the elements of the double-dimension array *z* to the first row from the elements of row vector *x*, to the elements of the second row from the values in vector *w*.

```
double x[10];
double w[10];
double z[2][10];
const int N = 5;
m = arraycol(x, 2.0, 11.0, 2.0);
m = arraycol(w, 1.0, 9.0, 2.0);
for(j = 0; j < N; j++)  {
        z[0][j] = x[i];
        z[1][j] = w[i];
}
```

To access individual elements of a two dimensional array, two indexes are used. The first number is the index for rows and the second number is the index for columns. To access the value of the element in array *z* defined previously at row 1 and column 3, the syntax is: $z[1][3]$. For example:

```
zz = z[1][3] + 13.5;
```

The following lines of C code declare a two-dimensional array named *rainf* with size 3 rows and 6 columns, and assign the value 2.4 to every element of the array.

```
const int COLS = 6;
const int ROWS = 3;
double rainf[ROWS][COLS];
for (j = 1; j < COLS; j++)
   for (i = 0; i < ROWS; i ++)
           rainf [i][j] = 2.4;
```

8.7.3   Passing Multi-Dimensional Arrays

For multi-dimensional arrays, the parameter declaration in a function definition and function prototype, all index sizes except the first one must be specified. This is illustrated in the following function prototype.

```
void yfunct (double y [][200], int yrows);
```

This function prototype has a declaration with two parameters, the first one *y* is a double-dimensional array with the size of the second dimension equal to 200. The first dimension is not indicated. The second parameter is an integer that is used by the function as the value of the size of the first dimension. The function prototype of function *yfunct* may be more conveniently declared in the following form:

```
void yfunct (double y [][N], int yrows);
```

In this function prototype, N has been declared as a symbolic constant of type integer and indicates the size of the second dimension of the array. The following statements declare the symbolic constants M and N, declare a two-dimensional array *myarray*, then call function *yfunct* with two arguments.

```
const int M = 100; // number of rows
const int N = 200; // number of columns
.  .  .
double myarray [M][N];
.  .  .
   yfunct ( myarray, M );
```

8.8   Applications Using Arrays

This section discusses several simple applications of arrays; a few of these applications perform simple manipulation of arrays, other applications perform slightly more complex operations with arrays such as searching and sorting.

8.8.1   Problems with Simple Array Manipulation

The problems discussed in this section compute the average value and the maximum value in an array named *varr*. The algorithms that solve these problems examine all the elements of the array.

8.8.1.1   Average Value in an Array

To compute the average value in an array, an algorithm can be designed to first compute the summation of all the elements in the array. The accumulator variable *sum* is used to store this. Second, the algorithm computes the average value by dividing the value of *sum* by the number of elements in the array. The following listing has the pseudo-code description of the algorithm.

1. Initialize the value of the accumulator variable, *sum*, to zero.

2. For every element of the array, add its value to the accumulator variable *sum*.

3. Divide the value of the accumulator variable by the number of elements in the array, n.

The accumulator variable *sum* stores the summation of the element values in the array named *varr* with n elements. The average value, *ave*, of array *varr* using index j starting with $j = 0$ to $j = n - 1$ is expressed mathematically as:

$$ave = \frac{1}{n} \sum_{j=0}^{n-1} varr_j$$

The following listing describes the algorithm using pseudo-code. The number of values is (*num*) and the array elements are inputted from the input device.

```
description
   This algorithm computes the average of the
   elements in array varr.
   */
variables
   define sum of type float
   define ave of type float      // average value
   define j of type integer
   define num of type integer
   define array varr [N] of type float
begin
   display "Enter number of values"
   read num
   for j = 0 to num - 1 do
     display "Enter element value: "
   endfor
   //
   set sum = 0
   for j = 0 to num - 1 do
     add myarr[j] to sum
   endfor
   set ave = sum / num
   display "Average value in array: ", ave
```

Listing 8.2 shows the C function that implements the algorithm that computes the average value of the elements in the array. This code is stored in file mean.c.

Listing 8.2: C function that computes average of values in an array.

```
1 /* File: mean.c
2   This function returns the average or
3   mean value in an array with n elements. */
4
5 double mean(double x[], int n)
6 {
7     /*  Declare and initialize variables.  */
8     int k;
9     double sum=0;
10
11    /*  Determine mean values.  */
12    for (k=0; k < n; k++)
13    {
14        sum = sum + x[k];
15    }
16
17    /*  Return average value.  */
18    return ( sum/n );
19 }
```

8.8.1.2   Maximum Value in an Array

Consider a problem that deals with finding the maximum value in an array named *varr*. The algorithm with the solution to this problem also examines all the elements of the array.

The variable *max_arr* stores the maximum value found so far. The name of the index variable is *j*. The algorithm description is:

1. Read the value of the number of values, *num*, and the value of the array elements.

2. Initialize the variable *max_arr* that stores the current largest value found (so far). This initial value is the value of the first element of the array.

3. Initialize the index variable (value zero).

4. For each of the other elements of the array, compare the value of the next array element; if the value of the current element is greater than the value of *max_arr* (the largest value so far), change the value of *max_arr* to this element value, and store the index value of the element in variable *k*.

5. The index value of variable *k* is the index of the element with the largest value in the array.

The algorithm description in pseudo-code appears in the following listing.

```
description
   Find the element with the maximum
   value in the array and write its index value.  */
   variables
      define j of type integer
      define k of type integer      // index of max elelemnt
      define max_arr of type float // largest element
      define varr array [N] of type float
   begin
      display "Enter array size: "
      read num
      for j = 0 to num-1 do
         display "Enter array element: "
         read varr[j]
      endfor
      set k = 0
      set max_arr = varr[0]
      for j = 1 to num - 1 do
         if varr[j] > max_arr
         then
            set k = j
            set max_arr = varr[j]
         endif
      endfor
   display "index of max value: ", k
```

Listing 8.3 contains a C function that implements the algorithm for finding the maximum value in an array; the function is stored in the file `arrmax.c`.

Listing 8.3: A C function that finds maximum value in an array.

```
1  /* File: arrmax.c
2     This function returns the maximum
3       value in the array x with n elements.
4  */
5  double max(double x[], int n)
6  {
7       /*  Declare local variables.  */
8       int k;
9       double max_x;
10
11      /*  Find the maximum value in the array.  */
12      max_x = x[0];
13      for (k=1; k < n; k++)
14      {
15          if (x[k] > max_x)
16     max_x = x[k];
17      }
18
19      /*  Return maximum value.  */
20      return max_x;
21 }
```

8.8.2 Searching

Searching consists of looking for an array element for a particular value and involves examining some or all elements of an array. The search ends when and if an element of the array has a value equal to the requested value. Two general techniques for searching are: linear search and binary search.

8.8.2.1 Linear Search

With linear search the elements of an array are examined in a *sequential* manner starting from the first element of the array, or from some specified element. Every array element is compared with the requested or key value, and if an array element is equal to the requested value, the algorithm has found the element and the search terminates. This may occur before the algorithm has examined all the elements of the array.

The index of the element in the array that is found is the result of the search. If the requested value is not found, the algorithm indicates this with a negative result or in some other manner. The following is an algorithm description of a general linear search using a search condition of an element equal to the value of a *key*.

1. Repeat for every element of the array:

(a) Compare the current element with the requested value or key. If the value of the array element satisfies the condition, store the value of the index of the element found and terminate the search.

(b) If values are not equal, continue search.

2. If no element with value equal to the value requested is found, set the result to value −1.

The following algorithm in pseudo-code searches the array for an element with the requested value, *kval*. The algorithm outputs the index value of the element equal to the requested (or key) value *kval*. If no element with value equal to the requested value is found, the algorithm outputs a negative value.

```
description
  This algorithm defines a linear search of array varr using
  key value kvar. The result is the index value of the element
  found, or -1 */
  variables
      define kval of type float,
      define varr array [] of type float,
      define num of type integer
      define j of type integer
      define found = false of type boolean
    begin
      set j = 0
      while j < num and found not equal true do
        if varr [j] == kval
        then
            set result = j
            set found = true
        else
            increment j
        endif
      endwhile
      if found not equal true
      then
          set result = -1
      endif
    end
```

The algorithm outputs the index value of the element that satisfies the search condition, whose value is equal to the requested value *kval*. If no element is found that satisfies the search condition, the algorithm outputs a negative value.

8.8.2.2 Binary Search

Binary search is a more complex search method, compared to linear search. This search technique is very efficient compared to linear search because the number of comparisons is smaller.

The prerequisite for binary search technique is that the element values in the array to search be sorted in ascending order. The array elements to include are split into two halves or partitions of about the same size. The middle element is compared with the key (requested) value. If the element with this value is not found, the search is continues on only one partition. This partition is again split into two smaller partitions until the element is found or until no more splits are possible because the element is not found.

With a search algorithm, the efficiency of the algorithm is determined by the number of operations, which compare the element values in the array, with respect to the size of the array. The average number of comparisons with linear search for an array with N elements is $N/2$, and if the element is not found, the number of comparisons is N. With binary search, the number of comparisons is $\log_2 N$. The informal description of the algorithm is:

1. Assign the lower and upper bounds of the array to *lower* and *upper*.

2. While the lower value is less than the upper value, continue the search.

   (a) Split the array into two partitions. Compare the middle element with the key value.

   (b) If the value of the middle element is equal to the key value, terminate search and the result is the index of this element.

   (c) If the key value is less than the middle element, change the upper bound to the index of the middle element minus 1. Continue the search on the lower partition.

   (d) If the key value is greater or equal to the middle element, change the lower bound to the index of the middle element plus 1. Continue the search on the upper partition.

3. If the key value is not found in the array, the result is -1.

Summary

An array in C is a data structure that stores several values of the same type. Each of these values is known as an element. After the array has been declared the capacity of the array cannot be changed. To refer to an individual element an index is used to indicate the relative position of the element in the array.

Searching an array consists of looking for a particular element value or key. Two common search algorithms are linear search and binary search. Computing an approximation of the rate of change and the area under a curve is much more convenient using arrays, as shown in the case study discussed.

Key Terms

declaring arrays	accessing elements	array capacity
index	array element	element reference
searching	linear search	binary search
key value	algorithm efficiency	summation
accumulator		

Exercises

Exercise 8.1 Develop a computational model that computes the standard deviation of values in an array. Implement using the C programming language. The standard deviation measures the spread, or dispersion, of the values in the array with respect to the average value. The standard deviation of array X with n elements is defined as:

$$std = \sqrt{\frac{sqd}{n-1}},$$

where

$$sqd = \sum_{j=0}^{n-1} (X_j - Av)^2.$$

Exercise 8.2 Develop a computational model that finds the minimum value element in an array and returns the index value of the element found. Implement using the C programming language.

Exercise 8.3 Develop a computational model that computes the average, minimum, and maximum rainfall per year and per quarter (for the last five years) from the rainfall data provided for the last five years. Four quarters of rainfall are provided, measured in inches. Use a matrix to store these values. Implement using the C programming language.

Exercise 8.4 Develop a computational model that sorts an array using the Insertion sort technique. This sort algorithm divides the array into two parts. The first is initially empty; it is the part of the array with the elements in order. The second part of the array has the elements in the array that still need to be sorted. The algorithm takes the element from the second part and determines the position for it in the first part. To insert this element in a particular position of the first part, the elements to right of this position need to be shifted one position to the right. Implement using the C programming language.

Chapter 9

Pointers

9.1 Introduction

Programming languages such as C and C++ support pointers that allow a much more powerful and flexible manner to manipulate data structures. These data structures are mainly arrays and linked lists, which can handle large number of values in data collections.

In some cases, using pointers is much more convenient for handling variables and arrays. There are several computational tasks that can only be performed with pointers, such as dynamic memory allocation. Other computational tasks are also performed more easily with pointers.

9.2 Pointer Fundamentals

Every variable is stored in a particular memory location (allocated by the compiler and operating system) and every memory location is defined by an *address*.

A pointer variable is one that contains the location or address of memory where another variable, data value, or function is stored. A pointer is a variable whose value is the address of another variable. Pointers are one of the more versatile features of C.

Pointers are important for implementing more advanced types of data as well. For example, a linked list is a data structure that uses pointers to link individual nodes.

9.3 Pointers with C

The C programming language uses the asterisk (*) to declare a pointer variable. In C, a pointer is declared with the type of the variable it will point to. A pointer variable of type *char* can be used to point to a single character or an array of characters. A similar type can be used for a *struct*. The following are examples of simple pointer variable declarations.

```
int * jintptr;    // an integer pointer variable
int* intPtr;      // another pointer variable
double *xptr;
char* charPtr;    // declares a pointer of type char
```

To assign a value to a pointer variable the & operator is used. This operator gets the address of a variable to point at and this value can be assigned to a pointer. For example,

```
double x;
double *xptr;
x = 234.75;
xptr = &x;        // xptr now points to x
```

Figure 9.1 shows that the value of variable *x* is 234.75 and the value of pointer variable *xptr* is the address of variable *x*. Variable *xptr* points to variable *x*.

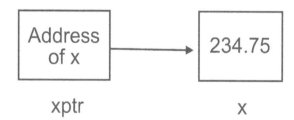

FIGURE 9.1: Pointer variable *xptr* pointing to variable *x*.

9.4 Dereferencing

Once a pointer variable has an appropriate value, which means the pointer variable is actually pointing to a variable, it can be used to dereference a variable. For example, continuing the previous lines of C code, pointer variable *xptr* can be used to change the value of the variable it currently points to. The following assignment statement changes the value of *x* using pointer *xptr*.

```
*xptr = 193.25;
```

9.5 Converting Pointer Types

In some applications it may be necessary to convert one pointer type to another. This is also known as *casting*. The general syntax for converting the types of non-pointer variables is:

```
(type) variable
```

For example, using an assignment statement, the following lines of C code change the type of a variable from *int* to *long*.

```
long xl;
int xi;
xi = 56;
xl = (long) xi;
```

For pointer variables, a similar type conversion can be used. The casting of pointer types follows an almost identical syntax, except that the asterisk (*) must be included. The general syntax is:

```
(type *) pointer_variable
```

For example, the type conversion from a pointer to an integer to a pointer to a *long* variable. The next lines of C code present the example of pointer type conversion from *int* to *long*.

```
long *xlptr;
int *xiptr;
xiptr = &xi;
xlptr = (long *) xiptr;
```

9.6 Reference Parameters

The simplest technique to get a return value from a function call is to use the function return value and this is written in an assignment statement. The main limitation is that the function call returns only one value at a time back to the calling function.

Recall that there are two general techniques to pass arguments when calling a function:

- pass by value

- pass by reference

The other technique to return data from a function is using pass by reference. The function declares reference or variable parameters that allow the return of more than one value. A reference parameter is declared as a pointer variable. The following example illustrates the definition and call of function *swap_values* using reference parameters *x* and *y*.

```
#include <stdio.h>
void swap_values (int *x, int *y); // prototype
int main()
{
        int a = 25;
        int b = 64;
        swap_values (&a, &b);
        printf ("a = %d and b = %d\n\n", a, b);
        return 0;
}
void swap_values (int *xptr, int *yptr)
{
        int temp;
        temp = *yptr;
        *yptr = *xptr;
        *xptr = temp;
}
```

Function *scanf* is a library input function that is used to get data from the console (keyboard). This function uses pointers to variables. When calling this function, the address of each variable that is read is passed to the function using the & operator. For example, the following code reads an integer value of variable *my_var* from the console:

```
int my_integer;
scanf ("%d", &my_var);
```

9.7   Pointers with Value NULL

Before assigning an actual value to a pointer variable, it is always a good practice to assign a NULL value to a pointer variable. The NULL value is a constant with a value of zero defined in the C standard libraries. A pointer that is assigned NULL is known as a null pointer and the pointer does not currently point to any variable. The following short program illustrates the use of the *NULL* value for pointer variables:

```
 1 /* This program displays the value of a pointer variable
 2      Type conversion with casting is necessary to
 3      'unsigned int' to use format '%x'
 4 */
 5 #include <stdio.h>
 6 int main ()
 7 {
 8      int *intptr;
 9      unsigned int outval;
10      intptr = NULL;
11      if (intptr == NULL) {
12          outval = (unsigned int) intptr;
13          printf("Value of intptr: %x \n", outval);
14      }
15      return 0;
16 }
```

After compiling and linking, the program is executed and it displays the following result:

```
Value of intptr: 0
```

The program includes a simple check for a null pointer using an if statement. Because a C pointer is an address which is a numeric value, there are a few basic arithmetic operations that can be performed on a pointer. These are:

- increment the value of the pointer with the ' ++ ' operator

- decrement the value of the pointer with the ' −− ' operator

- addition to the value of the pointer with the ' + ' operator

- subtraction from the value of the pointer with the ' − ' operator

The increment operator will add a value depending on the size of the type of the pointer variable. For integers that are four bytes in length, the increment operator adds 4 to the address value of the pointer variable.

The following example illustrates the use of the increment operator on a pointer variable. An array is really implemented as a pointer to the first element and it cannot directly be incremented so its address must be assigned to a pointer variable.

```
 1 #include <stdio.h>
 2 int main ()
 3 {
 4    const int ARRAY_SIZE = 5;
 5    int j;
 6    int *intptr;
 7    int myarray[] = {10, 20, 30, 40, 50};
 8    intptr = myarray;
 9    for ( j = 0; j < ARRAY_SIZE; j++) {
10        printf ("Current element: %d  value: %d \n",
```

```
                   j, *intptr);
11        intptr++;  /* point to next element of array myarray */
12     }
13     return 0;
14 }    /* end function main */
```

After compiling, linking, and executing the program, the following output is displayed:

```
$ gcc -Wall incrptr.c
$ ./a.out
Current element: 0   value: 10
Current element: 1   value: 20
Current element: 2   value: 30
Current element: 3   value: 40
Current element: 4   value: 50
```

9.8   Arrays as Pointers

In the previous section it was shown that an array can be manipulated as a pointer. This is useful when using array parameters and when an array is the return value in a function.

9.8.1   Pointer Parameters for Arrays

An array can be used as an argument in a function call. In the function definition, the corresponding parameter is declared as a pointer. Arrays as arguments are considered to be pointers by the functions receiving them. Therefore, they are always reference or variable parameters, which means that the called functions can modify the original copy of the variable. The following C program illustrates this.

```
 1 /*
 2 Program: arraymult.c
 3 This program uses an integer array as argument
 4 to call function arraymult.
 5 The function declares the parameter as an integer pointer.
 6 */
 7 #include <stdio.h>
 8 void farraymult (int *jarray, int mult); // prototype
 9 const int ARR = 7;      /* array size */
10 int main()
11 {
12     int j;
13     int marray[] = {10, 20, 30, 40, 50, 60, 70};
14     farraymult (marray, 15);
```

```
15     for (j = 0; j < ARR; j++)
16          printf("Element: %d value: %d \n ", j, marray[j]);
17     return 0;
18 }
19 void farraymult (int *parray, int pmult)
20 {
21        int j;
22        for (j = 0; j < ARR; j++)
23              parray[j] *= pmult;
24 }
```

In function *farraymult*, the first parameter is declared as an integer pointer, however, it can also be declared as an array parameter. In this case, the function declaration or prototype is:

```
void arraymult ( int parray[], int pmult );
```

Compiling, linking, and executing the program displays the following output:

```
$ gcc -Wall arrmult.c
$ ./a.out
 Element: 0 value: 150
 Element: 1 value: 300
 Element: 2 value: 450
 Element: 3 value: 600
 Element: 4 value: 750
 Element: 5 value: 900
 Element: 6 value: 1050
```

9.8.2   Functions that Return Arrays

In C, a function that returns an array specifies a pointer type as the function return type. In the following example, line 10 is the function prototype of function *farraymult* and the function definition appears in lines 22–30. The return type of the function is a pointer to an integer. Line 17 is the function call and a pointer assignment.

```
 1 /*
 2 Program: arraymult2.c
 3 This program uses an integer array as argument
 4 to call function arraymult.
 5 The function declares the parameter as an integer pointer.
 6 The function returns a new array as an integer pointer
 7 */
 8 #include <stdlib.h>
 9 #include <stdio.h>
10 int * farraymult (const int *jarray, int mult);
11 const int ARR = 7;     /* array size */
```

```
12 int main()
13 {
14     int j;
15     int marray[] = {10, 20, 30, 40, 50, 60, 70};
16     int *farray;
17     farray = farraymult (marray, 15);
18     for (j = 0; j < ARR; j++)
19         printf("Element: %d value: %d \n ", j, farray[j]);
20     return 0;
21 }
22 int * farraymult (const int *parray, int pmult)
23 {
24     int  * rarray;   /* new array */
25     int j;
26     rarray = malloc(ARR * sizeof(int));
27     for (j = 0; j < ARR; j++)
28         rarray[j] = parray[j] * pmult;
29     return rarray;
30 }
```

9.8.3   Dynamic Memory Allocation

During the execution of a program, it can request additional memory and the operating system can allocate the requested memory to the program. This technique of memory allocation is known as dynamic memory allocation. This is performed by assigning a contiguous block of memory to a pointer variable. In C, this allocation of memory is carried out with function *malloc*. Memory allocated can be resized with function *realloc* and memory can be deallocated with function *free*.

Calling function *malloc* requires one argument, the number of bytes of memory to allocate. The function returns a *void* pointer, which provides the address of the beginning of the allocated block of memory.

Calling function *realloc* function requires two arguments: the pointer to the memory block to be reallocated, and a number of *type size_t* that specifies the new size for the block. The function returns a void pointer to the newly reallocated block.

The deallocation of memory that has been allocated to a block is performed by calling function *free*. Calling this function requires only one argument, the pointer to the block to deallocate. The function does not return a value.

In the previous example, function *arraymult* creates an array of *ARR* elements by calling the library function *malloc* in line 26. The number of bytes for an integer of type *int* is 4; therefore, the total number of bytes allocated is *RR* times 4. Pointer variable *rarray* points to the block of memory allocated. In line 29, the function returns the value of pointer variable *rarray*.

9.9   Complex Data Structures

C supports two general categories of data structures that can be combined in various ways. These are arrays and structures. A structure type is a complex or composite type because it allows its components to have different types. To define a structure type, the *struct* keyword and a type name are used and enclose the data components of the structure.

9.9.1   Structure Types

The following example declares a structure type *ComplexNum* with two data components of type *double*.

```
struct ComplexNum {
    double realpart;
    double imagpart;
};
```

This definition introduces *struct ComplexNum* as a new type. This programmer-defined type can be used to declare variables of this type. In the following example, two variables: *var1* and *var2* are declared. Note that the keyword *struct* must be included when declaring these variables. Variables *var1* and *var2* are basically data structures. Using the *dot notation*, an individual component of a structure can be accessed by indicating the name of the structure, a dot, and the name of the component. In the example, a value is assigned to each individual components of the structure variable *var1*. Then the entire structure variable *var1* is copied to the structure variable *var2*.

```
struct ComplexNum var1, var2;
var1.realpart = 132.54; // assign value to component of var1
var1.imagpart = 37.32; // assign value to component of var1
var2 = var1;    // copy the whole structure var1 to var2
```

Although assignment to a structure can be performed directly from another structure of the same type with an assignment statement, direct comparison is not allowed. Instead, the individual components are compared with an *if* statement.

As mentioned previously, this type of complex data structure allows its components to have different types. The data structure is also known as a *record* and the individual components are also known as *fields*.

9.9.2   Array of Structures

An array of structure type can be declared in a similar manner to an array of a simple or primitive type. To declare an array of a structure, a previously-defined structure type is used as the type of the array.

The following example declares an array with name *cmplxarray* of 200 elements of type *struct ComplexNum*. Then values are assigned to the individual components of the first two elements of the array.

```
struct ComplexNum cmplxarray[200];
cmplxarray[0].realpart = 54.85;
cmplxarray[0].imagpart = 17.87;
cmplxarray[1].realpart = 54.85;
cmplxarray[1].imagpart = 17.87;
```

9.9.3 Pointers to Structures

To declare a pointer variable to a structure, the statement includes the structure type with an asterisk and a variable name. This is a pointer of a structure type previously defined. The following example in C illustrates declaration of pointer variable *cmplxPtr*.

```
struct ComplexNum *cmplxPtr;
```

The following program shows the definition of a structure type, declarations of structure variables, declaration of pointer variables, and use of pointer variables to structures.

In lines 7 and 8, the program declares variables *var* and *var2* of type *ComplexNum*. In line 9, it declares a pointer variable *cmplxPtr* of type *ComplexNum*. In lines 11 and 12, values are assigned to the components of variable *var*. In line 14, the address of *var* is assigned to pointer variable *cmplxPtr*. In line 15, dereferencing is used to access the value of the variable pointed at by pointer *cmplxPtr* and the value is directly assigned to variable *var2*. In lines 16 and 17, the dot notation is used to display the value of the components of *var2*. In lines 18 and 19, dereferencing is used to display the individual components of the structure pointed at by *cmplxPtr*.

```
1 #include <stdio.h>
2 struct ComplexNum {
3     double realpart;
4     double imagpart;
5 };
6 int main () {
7    struct ComplexNum var;
8    struct ComplexNum var2;
9    struct ComplexNum* cmplxPtr;
10
11   var.realpart = 132.54; // value to component of var
12   var.imagpart = 37.32;
13
14   cmplxPtr = &var;   /* cmplxPtr points now to var */
15   var2 = *cmplxPtr; /* dereferencing pointer */
16   printf("var2.real: %lf var2.imag: %lf \n", var2.realpart,
17          var2.imagpart);
18   printf("var real: %lf var imag: %lf \n",
19          (*cmplxPtr).realpart, (*cmplxPtr).imagpart);
```

```
20     return 0;
21}
```

The following C language statements declare an array of pointers to *struct ComplexNum*, then assign the address of structure variable *var* to the third element of the pointer array *cmplxPtrArray*.

```
struct ComplexNum* cmplxPtrArray[150];
cmplxPtrArray[2] = &var;
```

An alternate syntax can be used for dereferencing a pointer to a *struct*. The special characters -> after the name of the pointer variable can access any of the fields in the *struct*. This is shown in the following two statements:

```
xr = cmplxPtr->realpart;
xi = cmplxPtr->imagpart;
```

One convenient feature of the C type syntax is that it avoids the circular definition problems which come up when a pointer structure needs to refer to itself. The following definition defines a *node* in a linked list. Note that no preparatory declaration of the node pointer type is necessary.

```
struct mynode {
    int data1;
    double data2;
    struct mynode* next;   /* to the next node */
};
```

9.10   Defining Type Names

C supports the definition of new types, which basically provides new shorthand names for types. The *typedef* statement is used defining types and the syntax is:

typedef ⟨ *type* ⟩ ⟨ *type_name* ⟩

The following example defines the type name *ComplexType* as the name for the type *struct ComplexNum*.

```
typedef struct ComplexNum ComplexType;
ComplexType var;   // Declare variable var
```

The following is another example of applying *typedef*. A typical node in a linked list consists of the data components and a link component, which connects the current node to the next node. The type *NodePtr* is the type of the link component, *next*.

```
typedef struct node* NodePtr;
struct node {
    int data1;
    double data2;
    NodePtr next;
};
```

9.11   Enumerated Types

Enumerated types are arithmetic types and they are used to define variables that can only be assigned certain discrete integer values. The type is defined by enumerating the allowed values in symbolic form. The symbol *F* will have the value 0, *D* value of 1, *C* value of 2, *B* value of 3, and *A* value of 4.

```
enum Lettergrade { F, D, C, B, A };
typedef enum Lettergrade Gradet;
Gradet grade;
grade = B;
```

In the following example, the first statement defines an enumerated type with name *boolean* that includes only two possible values 0 and 1. The symbolic name *False* has a value of 0 and the symbolic name *True* has a value of 1. The second statement defines the name *Bool* for the type *enum boolean*. The third line is a declaration of variable *flag* of type *Bool*. The fourth line is an assignment statement; the value *True* is assigned to variable *flag*.

```
enum boolean {False, True};
typedef enum boolean Bool;
Bool flag;
flag = True;
```

Summary

Pointers allow a much more powerful and flexible manner to manipulate data structures. These data structures are mainly arrays and linked lists, which can handle large number of values in data collections.

There are several computational tasks that can only be performed with pointers, such as dynamic memory allocation. Other computational tasks are performed more easily with pointers.

Key Terms

declaring pointers	initializing pointers	address of a variable
dereferencing	reference parameter	dynamic memory allocation
structures	pointer to structures	type renaming
enumerated types	null pointer	type conversion

Exercises

Exercise 9.1 Develop a C program with a function that declares two pointer parameters and an integer parameter. The function must compare the characters in the two strings pointed by the first two parameters and return 0 if the first k (defined by the third parameter) characters of the strings are equal, otherwise the function returns the index value of the first character that does not match.

Exercise 9.2 Develop a C program with a function that declares two pointer parameters and an integer parameter. The function must copy the characters in the string pointed at by the first parameter to the second string. The function must copy the first k (defined by the third parameter) characters of the strings.

Exercise 9.3 Using the structure previously defined, *struct ComplexNum*, develop a C program that performs complex addition, subtraction, multiplication, and division of the complex values of the elements of array x with the complex values of the elements in array y. Include a function for each complex operation that returns a new array with the results.

Exercise 9.4 Apply the definition of the structure *invdef* and develop a C program that maintains inventory data in a warehouse in an array of structure data. The program must display the parts that have a number of units less than 5. The program must also compute the total (sales) value of the inventory.

```
struct invdef {
    int partcode;
    int numunits;
    double price;
    char pdescription[31];
};
```

Chapter 10

Linked Lists

10.1 Introduction

A linked list is a data structure that consists of a sequence of data items of the same or similar types and each data item or *node* has one or more links to another node. This data structure is dynamic in the sense that the number of data items can change. A linked list can grow and shrink during the execution of the program that is manipulating it. Recall that an **array** is also a data structure that stores a collection of data items, but the array is static because once it is created, more elements cannot be added or removed.

This chapter discusses the basic forms of simple linked lists, double-ended linked lists, and multiple linked lists. The operations possible on linked lists and higher-level data structures, such as stacks and queues, implemented with linked lists are also discussed. Abstract data types (ADTs) are discussed and defined for queues and stacks.

10.2 Nodes and Linked List

A linked list is a sequence of nodes that are connected in some fashion. A *node* is a relatively smaller data structure that contains data and one or more links that are used to connect the node to one more other nodes. In graphical form, a node may be depicted as a box, which is divided into two types of components:

- A data block that stores one or more *data components* of some specified type.

- One or more *link components* that connect the node to other nodes.

Linked lists and arrays are considered *low-level* data structures. These are used to implement *higher-level* data structures. Examples of simple higher-level data structures are *stacks* and *queues* and each one exhibits a different behavior implemented by an appropriate algorithm. More advanced and complex higher-level data structures are priority queues, trees, graphs, sets, and others.

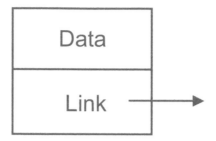

FIGURE 10.1: Structure of a node.

Figure 10.1 shows a representation of a simple node. A linked list is a data structure that consists of a chain or sequence of nodes connected in some manner. The last node on a simple linked list has a link with value *NULL*. Figure 10.2 illustrates the general form of a simple linked list. Note that head pointer *H* points to the first node (the head) of the linked list. The last node (*Node 3* in Figure 10.2) has a link that points to a black dot to indicate that the link is pointing nowhere. When comparing linked lists with arrays, the main differences observed are:

- Linked lists are dynamic in size because they can grow and shrink; arrays are static in size.

- In linked lists, nodes are linked by references and based on many nodes; whereas, an array is a large block of memory with the elements located contiguously.

- The nodes in a linked list are referenced by relationship not by position; to find a data item, always start from the first item (no direct access). Recall that access to the elements in an array is carried out using an index.

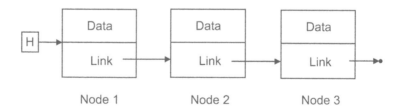

FIGURE 10.2: A simple linked list.

10.2.1 Nodes

As mentioned previously, a simple node in a linked list has a data block and a link that connects it to another node. These nodes can be located anywhere in memory and do not have to be stored contiguously in memory. The following listing shows the C code with the definition of structure for a node and the definition of the types *NodeType* and *NodePtr*. The structure for the node includes a pointer *datablock* as the data component of the node.

```
typedef struct node NodeType;
typedef NodeType * NodePtr;
struct node {
      DataT *datablock;
      NodePtr link;
};
```

A data block is another structure defined depending on the application with several variables, each of a different type. A program declares a variable of type *DataT* and assigns the values of each component of the data block. The following listing of C code illustrates this.

```
typedef struct Datablock DataT;
struct Datablock {      // data block type
      char name [31];
      int age;
      int jobcode;
};

DataT *pdata;   // declare a pointer data block
pdata = malloc(sizeof(DataT));
strcpy(pdata->name, "Joseph Hunt");
pdata->age = 45;
pdata->jobcode = 6524;
```

To create a new node with the structure previously defined for a node, a pointer of type *NodePtr* is declared, dynamic memory allocation is performed for the size of the node, and its address is assigned to the pointer. The following lines of C statements declare a pointer *newPtr* of type *NodeType*, allocate memory for a new node, assign the address of the node to *newPtr*, and assign the value of the pointer *datablock* of the node.

```
NodePtr newPtr;
newPtr = (NodePtr) malloc (sizeof(NodeType));
newPtr->datablock = pdata;
newPtr->link = NULL;
```

10.2.2 Manipulating Linked Lists

In addition to defining the data structure for the nodes, a set of operations is defined in order to create and manipulate the linked list. Some of the basic operations defined on a simple linked list are:

- Create an empty linked list

- Insert a new node at the front of the linked list

- Insert a new node at the back of the linked list

- Insert a new node at a specified position in the linked list

- Get a copy of the data in the node at the front of the linked list

- Get a copy of the data in the node at a specified position in the linked list

- Remove the node at the front of the linked list

- Remove the node at the back of the linked list

- Remove the node at a specified position in the linked list

- Traverse the list to display all the data in the nodes of the linked list

- Check whether the linked list is empty

- Check whether the linked list is full

- Find a node of the linked list that contains a specified data item

As mentioned previously, the pointer variable *H* always points to the first node of the linked list. This node is also known as the *head node* because it is the front of the linked list. If the list is empty, then the value of *H* is *NULL*. To create an empty list, all that is needed is the assignment:

```
Nodeptr H;
int numnodes;   // current number of nodes
H = NULL;
numnodes = 0;
```

To check if a list is empty, the value of the head pointer *H* needs to be compared for equality with the constant value *NULL*.

```
if ( H == NULL && numnodes == 0)
     . . .
```

A function can be defined with name *isEmpty* that checks whether the list is empty by examining the value of the head pointer *H*, which is a reference to the first node in the list. The value of this pointer has a value *NULL* when the list is empty.

To insert a new node as the first node of an *empty* linked list, then the head of the list should point to this node. Assuming that pointer variable *newPtr* points to a newly-created node, the following assignment statement inserts the node and it becomes the head of the linked list:

```
H = newPtr;
```

To insert a new node at the head of a *non-empty* linked list, the head pointer *H* is pointing at the current first node, so the value of *H* will be copied to the link component of the new node. The new value of the head pointer *H* will be value of the pointer to the new node. After creating and initializing a new node, the following statements insert the new node in the front of the linked list.

```
newPtr->link = H;
H = newPtr;
```

To remove and delete the node at the front of the linked list, the head pointer *H* will point to the second node. The node to be removed has to be destroyed after its data have been copied to another variable.

```
NodePtr cnodeptr;
cnodeptr = H;   // pointer to current head node
H = H->link;    // pointer to second node
pdata = cnodeptr->datablock; // get data
free(cnodeptr);  // de-allocate memory, destroy node
```

Traversing a linked list involves accessing every node in the list by following the links to the next node until the last node. Recall that the link of the last node is *NULL*. The following portion of code traverses a linked list to display the data of every node.

```
cnodeptr = H;    // point to first node
while ( cnodeptr != NULL ) {      // traverse list
     display_data (cnodeptr);        // display data of node
     cnodeptr = cnodeptr->link   // point to next node
}
```

A node can be inserted to the linked list at the front, at the back, or in any other place specified. For a simple linked list, insertion at the front of the list is simpler. Function *insert_front* can be defined that creates a new node and inserts the new node to the front of the linked list. Figure 10.3 shows the insertion of a new node to the front of the list.

More flexibility is obtained by having the operation to insert a node at a specified position in the linked list. For example insert a new node after current node 2. Figure 10.4 illustrates changing the links so that a new node is inserted after node 2.

10.2.3   Example of Manipulating a Linked List

Listing 10.1 is the C source code of a program that includes a global data declaration that defines the structure of a node, pointer types of the node structure, and various functions that implement the operations discussed on a linked list. Function

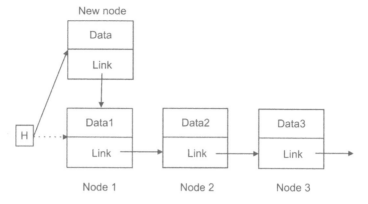

FIGURE 10.3: A new node inserted in the front of a linked list.

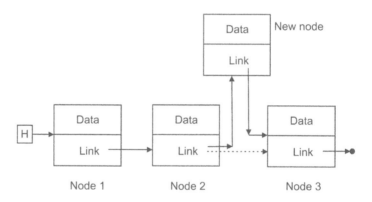

FIGURE 10.4: A new node inserted after node 2.

main, which appears on lines 20 to 62 calls these functions to create and manipulate a linked list. The C source code of this program is stored in file datablistp.c. The C files that define the linked list structures and the function definitions are stored in linked.h and linked.c.

Listing 10.1: C program for creating and manipulating a linked list.

```
 1 /* Program: datablistp.c
 2    This program shows how to create nodes and various ways
 3    to link them in a linked list using data blocks.
 4    The linked list is implemented in library 'linked'.
 5    J. M. Garrido Feb. 2013
 6    */
 7 #include <stdio.h>
 8 #include <stdlib.h>
 9 #include <string.h>
10 #include "basic_lib.h"
11 #include "linked.h"
```

```
12 struct Datablock {    // data block type
13     char name [31];
14     int age;
15     int jobcode;
16 };
17 DataT* make_dblock(char *nname, int nage, int njobcode);
18 void display_data (DataT *pdata);
19 void traverse_display();
20 int main() {
21     DataT *dblock;
22     create_list(); // create empty linked list
23     printf("Inserting data to front of list\n");
24     dblock = make_dblock("John Doe", 38, 1523);
25     insert_front(dblock);
26     dblock = make_dblock("Joseph Hunt", 59, 1783);
27     insert_front(dblock);
28     dblock = make_dblock("William Dash", 49, 2581);
29     insert_front(dblock);
30     traverse_display(); // display all blocks in list
31     printf("Removing front: ");
32     dblock = remove_front();
33     display_data(dblock);
34     free(dblock);
35     traverse_display();
36     dblock = make_dblock("David Winemaker", 48, 1854);
37     printf("Inserting to pos 2: ");
38     display_data(dblock);
39     insert_node(dblock, 2);
40     traverse_display();
41     dblock = make_dblock("Mary Washington", 54, 1457);
42     printf("Inserting 3: ");
43     display_data(dblock);
44     insert_node(dblock, 3);
45     traverse_display();
46     printf("Inserting front: ");
47     dblock = make_dblock("Ann Smith", 44, 1568);
48     display_data(dblock);
49     insert_front(dblock);
50     traverse_display();
51     printf("Remove 2:   ");
52     dblock = remove_node(2);
53     display_data(dblock);
54     traverse_display();
55     free(dblock);
56     printf("Removing last node \n");
57     dblock = remove_last();
58     display_data(dblock);
59     traverse_display();
60     free(dblock);
61     return 0;
62 }
63 DataT* make_dblock(char *pname, int page, int pjobcode) {
64     DataT * ldatab;
65     ldatab = malloc(sizeof(DataT));
66     strcpy(ldatab->name, pname);
67     ldatab->age = page;
68     ldatab->jobcode = pjobcode;
```

```
69   return ldatab;
70 };
71 // display data block
72 void display_data(DataT *pblock) {
73    printf("Data: %s %d %d \n", pblock->name,
74          pblock->age, pblock->jobcode);
75 }
76 void traverse_display () {
77    printf("Traverse and display data in list\n");
78    int i;
79    DataT *lblock;
80    if (numnodes <= 0 )
81       return;
82    lblock =  get_front();   // get first data block
83    display_data(lblock);
84    for (i=2; i <= numnodes; i++) {
85       lblock = get_next(); // get next data block
86       display_data(lblock);
87    }
88 }
```

The following listing includes the Linux shell commands to compile, link, and execute the program datablistp.c, and get the results produced.

```
$ gcc -Wall datablistp.c basic_lib.o linked.o
$ ./a.out
List all nodes in linked list
Data from node: William Dash 49 2581
Data from node: Joseph Hunt 59 1783
Data from node: John Doe 38 1523
Data removed front node: William Dash 49 2581
insert nodes at pos 2, 3, and front
List all nodes in linked list
Data from node: Ann Smith 44 1568
Data from node: Joseph Hunt 59 1783
Data from node: John Doe 38 1523
Data from node: David Winemaker 48 1854
Data from node: Mary Washington 54 1457
Data removed node 2: Joseph Hunt 59 1783
List all nodes in linked list
Data from node: Ann Smith 44 1568
Data from node: John Doe 38 1523
Data from node: David Winemaker 48 1854
Data from node: Mary Washington 54 1457
Removing last node
List all nodes in linked list
Data from node: Ann Smith 44 1568
Data from node: John Doe 38 1523
Data from node: David Winemaker 48 1854
```

10.2.4   Linked List as an Abstract Data Type

An abstract data type (ADT) is a concept used in software development that involves modularity, abstraction, and information hiding. An ADT is a collection of data definitions and well-defined operations that allows programmers to apply the ADT without being concerned about the lower-level implementations. The programmers' perceptions are at a high-level of abstraction and they should mainly understand the function specifications (function prototypes). The lower-level details, which are the function implementations (function definitions), will normally be hidden from the programmers.

In the case of a linked list, the possible operations on a linked list in the form of function prototypes are visible to all users so they can manipulate the linked list. In C, function prototypes are normally stored in header files. The various function implementations are stored in a separate file.

The program in this section that defines, creates, and manipulates a linked list is slightly more complex that the one in the previous section. The data definitions and the function prototypes of a linked list are stored in the header file linked2.h. The functions definitions (implementations) of a linked list are stored in file linked2.c. These two files (linked2.h and linked2.c) are examples of what is often referred to as a library of linked list definitions and function implementations for linked list.

The program that uses these definitions is stored in file datablistp2.c, which is a relatively short program because it is using a function library with predefined functions of a linked list. Note that the program creates two linked lists using the ADT defined in files linked2.h and linked2.c.

Listing 10.2: C program using linked list ADT library.

```
 1 /* Program: datablistp2.c
 2    This program illustrates using linked list ADTs.
 3    The linked list is implemented in library 'linked2'.
 4    J. M. Garrido Feb. 2013
 5    */
 6 #include <stdio.h>
 7 #include <stdlib.h>
 8 #include <string.h>
 9 #include "basic_lib.h"
10 #include "linked2.h"
11 // typedef struct Datablock DataT; in 'linked.h'
12 struct Datablock {    // data block type
13     char name [31];
14     int age;
15     int jobcode;
16 };
17 DataT* make_dblock(char *nname, int nage, int njobcode);
18 void display_data (DataT *pdata);
19 void traverse_display(listT *plist);
20 int main() {
21    DataT *dblock;
22    listT alist;  // a linked list
23    listT blist;  // another linked list
24    listT *alistp = &alist;
25    listT *blistp = &blist;
26    create_list(alistp, 25, "ListA"); // create empty list
27    create_list(blistp, 35, "ListB"); // create empty list
28    printf("Inserting data to front of %s \n", alistp->lname);
```

```
29    dblock = make_dblock("John Doe", 38, 1523);
30    insert_front(alistp, dblock);
31    dblock = make_dblock("Joseph Hunt", 59, 1783);
32    insert_front(blistp, dblock);
33    dblock = make_dblock("William Dash", 49, 2581);
34    insert_front(alistp, dblock);
35    printf("Number of nodes in %s: %d \n",
              alistp->lname, alistp->numnodes);
36    printf("Number of nodes in %s: %d \n", blistp->lname,
              blistp->numnodes);
37    traverse_display(alistp); // display all blocks in alist
38    traverse_display(blistp); // display all blocks in blist
39    printf("Removing front %s: ", alistp->lname);
40    dblock = remove_front(alistp);
41    display_data(dblock);
42    free(dblock);
43    traverse_display(alistp);
44    dblock = make_dblock("David Winemaker", 48, 1854);
45    printf("Inserting to pos 2 %s: ", alistp->lname);
46    display_data(dblock);
47    insert_front(blistp, dblock);
48    insert_node(alistp, dblock, 2);
49    traverse_display(alistp);
50    traverse_display(blistp);
51    dblock = make_dblock("Mary Washington", 54, 1457);
52    printf("Inserting 3 %s: ", alistp->lname);
53    display_data(dblock);
54    insert_node(alistp, dblock, 3);
55    traverse_display(alistp);
56    printf("Inserting front %s: ", alistp->lname);
57    dblock = make_dblock("Ann Smith", 44, 1568);
58    display_data(dblock);
59    insert_front(alistp, dblock);
60    traverse_display(alistp);
61    printf("Remove 2 %s:  ", alistp->lname);
62    dblock = remove_node(alistp, 2);
63    display_data(dblock);
64    traverse_display(alistp);
65    free(dblock);
66    printf("Removing last node %s ", alistp->lname);
67    dblock = remove_last(alistp);
68    display_data(dblock);
69    traverse_display(alistp);
70    printf("Number of nodes in %s: %d \n", blistp->lname,
              blistp->numnodes);
71    traverse_display(blistp);
72    free(dblock);
73    return 0;
74 }
75 DataT* make_dblock(char *pname, int page, int pjobcode) {
76   DataT * ldatab;
77   ldatab = malloc(sizeof(DataT));
78   strcpy(ldatab->name, pname);
79   ldatab->age = page;
80   ldatab->jobcode = pjobcode;
81   display_data(ldatab);
82   return ldatab;
```

```
83 };
84 // display data block
85 void display_data(DataT *pblock) {
86     printf("Data: %s %d %d \n", pblock->name,
87            pblock->age, pblock->jobcode);
88 }
89 void traverse_display (listT *plist) {
90     printf("Traverse and display data: %s \n", plist->lname);
91     int i;
92     DataT *lblock;
93     if (plist->numnodes <= 0 )
94        return;
95     lblock =  get_front(plist);   // get first data block
96     display_data(lblock);
97     for (i=2; i <= plist->numnodes; i++) {
98         lblock = get_next(plist); // get next data block
99         display_data(lblock);
100    }
101 }
```

In line 12 of the source program, a structure is defined as the data block that the program uses to store data. The required name of the structure is `Datablock` and the form of the structure may vary. Each of these data blocks will be stored in a node of the linked list by the appropriate functions in the linked list library. The name of the structure is used to define type `DataT` in the library. In lines 22 and 23, the program declares two linked lists; in lines 24 and 25 it declares and initializes pointers to these linked lists.

The following listing includes the Linux shell commands to compile, link, and execute the program `datablistp2.c`, and the results are produced. Note that the pre-compiled file `linked2.o` is linked with the program to produce the executable file.

```
$ gcc -Wall datablistp2.c linked2.o basic_lib.o
$ ./a.out
Inserting data to front of ListA
Data: John Doe 38 1523
Data: Joseph Hunt 59 1783
Data: William Dash 49 2581
Number of nodes in ListA: 2
Number of nodes in ListB: 1
Traverse and display data in: ListA
Data: William Dash 49 2581
Data: John Doe 38 1523
Traverse and display data in: ListB
Data: Joseph Hunt 59 1783
Removing front ListA: Data: William Dash 49 2581
Traverse and display data in: ListA
Data: John Doe 38 1523
Data: David Winemaker 48 1854
Inserting to pos 2 ListA: Data: David Winemaker 48 1854
Traverse and display data in: ListA
Data: John Doe 38 1523
Traverse and display data in: ListB
```

```
Data: David Winemaker 48 1854
Data: Joseph Hunt 59 1783
Data: Mary Washington 54 1457
Inserting 3 ListA: Data: Mary Washington 54 1457
Traverse and display data in: ListA
Data: John Doe 38 1523
Inserting front ListA: Data: Ann Smith 44 1568
Data: Ann Smith 44 1568
Traverse and display data in: ListA
Data: Ann Smith 44 1568
Data: John Doe 38 1523
Remove 2 ListA:  Data: John Doe 38 1523
Traverse and display data in: ListA
Data: Ann Smith 44 1568
Removing last node ListA Data: Ann Smith 44 1568
Traverse and display data in: ListA
Number of nodes in ListB: 2
Traverse and display data in: ListB
Data: David Winemaker 48 1854
Data: Joseph Hunt 59 1783
```

10.3   Linked List with Two Ends

The linked lists discussed previously have only one end, which points to the first node, and this pointer is also known as the head of the linked list. In addition to the head node, providing a pointer to the last node gives the linked list more flexibility. With two ends, a linked list has two pointers: a pointer to the first node H, the *head* or front of the list, and a pointer to the last node T, the *tail* of the linked list. Figure 10.5 illustrates a linked list with a head pointer H and a tail pointer T.

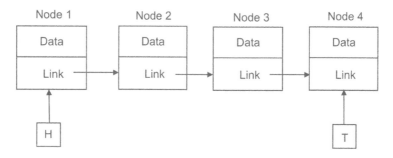

FIGURE 10.5: A linked list with two ends.

With a pointer to the last node (the *tail*), in addition to the pointer to the first node (the *head*), the linked list now provides the ability to directly add a new node to the back of the linked list without traversing it from the front. In a similar manner, the last node of a linked list can be removed without traversing it from the front.

To create an empty linked list with two ends, two pointers have to be declared and initialized, the head and the tail of the list. The structure of a node used previously does not change for a linked list with two ends.

```
NodePtr H = NULL;    // head of list
NodePtr T = NULL;    // tail of list
int numnodes = 0;
```

Linked lists with two ends are very useful and convenient for implementing higher-level data structures such as *queues*.

10.4 Double-Linked Lists

Linked lists that have nodes with only one link, a pointer to the next node, can only traverse the linked list in one direction, starting at the front and toward the tail of the list. To further enhance the flexibility of a linked list, a second link is included in the definition of the nodes. This second link is a pointer to the previous node. These linked lists are also known as *doubly linked lists*. Figure 10.6 illustrates the general form of a linked list with nodes that have two links: a pointer to the next node and a pointer to the previous node.

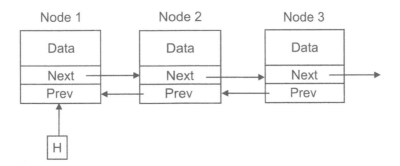

FIGURE 10.6: A linked list with two links per node.

The following listing of C statements defines a node type with two links, *next* that points to the next node in the linked list and *prev* that points to the previous node in the linked list. This is an example of a node structure with more than one link components, in addition to the data block.

```
    typedef struct node NodeType;
    typedef NodeType * NodePtr;
    struct node {
        DataT datablock;
        NodePtr next;    // pointer to next node
        NodePtr prev;    // pointer to previous node
    };
```

The linked list library *linked3* defines double-linked lists with two ends. It is stored in files linked3.h and linked3.c. The following listing shows the C code in the header file of this library.

```
// double-linked list with two ends
typedef struct Datablock DataT;
typedef struct node NodeType;
typedef NodeType * NodePtrT;
struct node {          // node type
    DataT * datablock;
    NodeType * next; // pointer to next node
 NodeType * prev; // pointer to previous node
};
//
typedef struct llist listT;
struct llist {
    NodePtrT H;   // head of list
NodePtrT T;   // tail of list
    int numnodes;
    int maxnodes;
    NodePtrT current;
    char lname[30];
};
//
void create_list(listT *plist, int maxn, char *pname);
void insert_front (listT *plist, DataT *cdata);
DataT* remove_front(listT *plist);
void traverse_display (listT *plist);
Bool empty_list (listT *plist);
Bool full_list (listT *plist);
void insert_node(listT *plist, DataT *ndata, int pos);
void insert_last(listT *plist, DataT *ndata);
DataT* remove_node(listT *plist, int pos);
DataT* get_data(listT *plist, int pos);
DataT* get_front(listT *plist);    // get data from first node
DataT* get_next(listT *plist);     // copy data from next node
DataT* remove_last(listT *plist); // remove last node
int listSize(listT *plist);
```

10.5   Higher-level Data Structures

Higher-level data structures are those used directly in problem solving and can be implemented by lower-level data structures. This section discusses and describes the structure and operations of two simple and widely-known higher-level data structures: *queues* and *stacks*. These can be implemented with low-level data structures such as arrays and linked lists.

10.5.1   Stacks

A stack is a higher-level dynamic data structure that stores a collection of data items in a data block. The data in a stack may be a data block with variables of a simple type, pointers to data objects, or a combination of these. Each node in a stack includes a data block and one or more links.

A stack has only one end: the *top* of the stack. The main characteristics of a stack are:

- Data blocks can only be inserted at the top of the stack (TOS)

- Data blocks can only be removed from the top of the stack

- Data blocks are removed in reverse order from that in which they are inserted into the stack. A stack is also known as a last in and first out (LIFO) data structure.

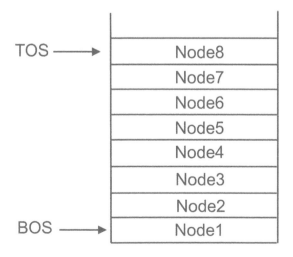

FIGURE 10.7: A stack as a dynamic data structure.

Figure 10.7 illustrates the form of a stack. It shows the top of the stack as the insertion point and the removal point. The operations that manipulate a stack ADT are:

- *create_stack*, create an empty stack.

- *sEmpty*, returns true if the stack is empty; otherwise returns false.

- *sFull*, returns true if the stack is full; otherwise returns false.

- *top*, returns a copy of the data block at the top of the stack without removing the node from the stack.

- *pop*, removes the node from the top of the stack.

- *push*, inserts a new node to the top of the stack.

- *stackSize*, returns the number of nodes currently in the stack.

The most direct way to implement a stack is with a single-list linked list in which insertions and deletions are performed at the front of the linked list. The linked list library *linked2* already implemented is used to implement a stack library, which is stored in files `stack.h` and `stack.c`. Listing 10.3 shows the C source code of a program that uses the stack ADT in the library to create and manipulate data in a stack.

Listing 10.3: C program using an stack ADT library.

```
 1 /* Program: datastack.c
 2    This program shows stack manipulation using data blocks.
 3    The stack ADT library is implemented in 'stack'.
 4    J. M. Garrido Feb. 16, 2013.
 5    */
 6 #include <stdio.h>
 7 #include <stdlib.h>
 8 #include <string.h>
 9 #include "basic_lib.h"
10 #include "stack.h"
11 // typedef struct Datablock DataT; in 'linked.h'
12 struct Datablock {     // data block type
13       char name [31];
14       int age;
15       int jobcode;
16 };
17 DataT* make_dblock(char *nname, int nage, int njobcode);
18 void display_data(DataT *pdata);
19 int main() {
20     int nums;
21     DataT *dblock;
22     stackType mstacka;
23     stackType * stackp = &mstacka;   // stack pointer
24     create_stack(stackp, 35, "stackA"); // create empty stack
25     printf("Push data into stack: ");
26     dblock = make_dblock("John Doe", 38, 1523);
```

```
27      display_data(dblock);
28      push(stackp, dblock);
29      dblock = make_dblock("Joseph Hunt", 59, 1783);
30      printf("Push data into stack: ");
31      display_data(dblock);
32      push(stackp, dblock);
33      dblock = make_dblock("William Dash", 49, 2581);
34      printf("Push data into stackA: ");
35      display_data(dblock);
36      push(stackp, dblock);
37      printf("Pop from stack: ");
38      dblock = pop(stackp);
39      display_data(dblock);
40      free(dblock);
41      dblock = make_dblock("David Winemaker", 48, 1854);
42      printf("Push into stack: ");
43      display_data(dblock);
44      push(stackp, dblock);
45      nums = stackSize(stackp);
46      printf("Size of %s: %d \n", stackp->lname, nums);
47      printf("Pop from stack: ");
48      dblock = pop(stackp);
49      display_data(dblock);
50      free(dblock);
51      return 0;
52  }
53  DataT* make_dblock(char *pname, int page, int pjobcode) {
54    DataT * ldatab;
55    ldatab = malloc(sizeof(DataT));
56    strcpy(ldatab->name, pname);
57    ldatab->age = page;
58    ldatab->jobcode = pjobcode;
59    return ldatab;
60  }
61  // display data block
62  void display_data(DataT *pblock) {
63      printf("Data: %s %d %d \n", pblock->name,
64            pblock->age, pblock->jobcode);
65  }
```

The following listing shows the Linux shell commands that compile, link, and execute the program. The results produced are also shown.

```
$ gcc -Wall datastack.c basic_lib.o stack.o linked2.o
$ ./a.out
Push data into stack: Data: John Doe 38 1523
Push data into stack: Data: Joseph Hunt 59 1783
Push data into stackA: Data: William Dash 49 2581
Pop from stack: Data: William Dash 49 2581
Push into stack: Data: David Winemaker 48 1854
Size of stackA: 3
Pop from stack: Data: David Winemaker 48 1854
```

10.5.2 Queues

A queue is a dynamic data structure that stores a collection of data items or nodes and that has two ends: the *head* and the *tail*. The rules for basic behavior of a queue are:

- Data items can only be inserted at the tail of the queue

- Data items can only be removed from the head of the queue

- Data items must be removed in the same order as they were inserted into the queue. This data structure is also known as a first in and first out (FIFO) data structure.

Figure 10.8 illustrates the form of a queue. It shows the insertion point at the tail and the removal point at the head of the queue. The operations that manipulate a queue are:

- *isEmpty*, returns true if the queue is empty; otherwise returns false.

- *isFull*, returns true if the queue is full; otherwise returns false.

- *copyHead*, returns a copy of the data object at the head of the queue without removing the object from the queue.

- *removeHead*, removes the head item from the queue

- *insertTail*, inserts a new data item into the tail of the queue.

- *queueSize*, returns the number of data items currently in the queue.

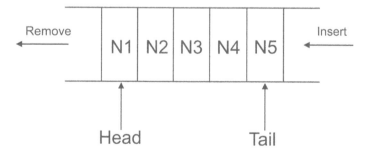

FIGURE 10.8: A queue as a dynamic data structure.

Queues can be implemented with single-linked lists, but are more efficiently implemented with double-linked lists with two ends. The linked list library *linked3* defines double-linked lists with two ends; it is stored in files `linked3.h` and `linked3.c` and is used mainly to implement queues.

Listing 10.4 shows the C source code of a program that creates a queue from a queue ADT library, which is stored in files queue.h and queue.c. The program is stored in file dataqueue.c.

Listing 10.4: C program using a queue ADT library.

```
 1 /* Program: dataqueue.c
 2    This program shows stack manipulation using data blocks.
 3    The stack is implemented in library 'stack'.
 4    J. M. Garrido Feb. 14, 2013.
 5    */
 6 #include <stdio.h>
 7 #include <stdlib.h>
 8 #include <string.h>
 9 #include "basic_lib.h"
10 #include "queue.h"
11 // typedef struct Datablock DataT; in 'linked.h'
12 struct Datablock {      // data block type
13       char name [31];
14       int age;
15       int jobcode;
16 };
17 DataT* make_dblock(char *nname, int nage, int njobcode);
18 void display_data(DataT *pdata);
19 int main() {
20     int nums;
21     DataT *dblock;
22     queueT mqueuea;
23     queueT * queuep = &mqueuea;   // queue pointer
24     create_queue(queuep, 35, "queueA"); // create empty queue
25     printf("Enqueue data: ");
26     dblock = make_dblock("John Doe", 38, 1523);
27     display_data(dblock);
28     enqueue(queuep, dblock);
29     dblock = make_dblock("Joseph Hunt", 59, 1783);
30     printf("Enqueue data: ");
31     display_data(dblock);
32     enqueue(queuep, dblock);
33     dblock = make_dblock("William Dash", 49, 2581);
34     printf("Enqueue data: ");
35     display_data(dblock);
36     enqueue(queuep, dblock);
37     nums = queueSize(queuep);
38     printf("Queue size: %s %d \n", queuep->lname, nums);
39     printf("Copy Head data from queue: %s \n", queuep->lname);
40     dblock = head(queuep);
41     display_data(dblock);
42     dblock = make_dblock("David Winemaker", 48, 1854);
43     printf("Enqueue: ");
44     display_data(dblock);
45     enqueue(queuep, dblock);
46     nums = queueSize(queuep);
47     printf("Size of queue: %d \n", nums);
48     printf("Dequeue from queue: ");
49     dblock = dequeue(queuep);
50     display_data(dblock);
51     free(dblock);
```

```
52    return 0;
53 }
54 DataT* make_dblock(char *pname, int page, int pjobcode) {
55    DataT * ldatab;
56    ldatab = malloc(sizeof(DataT));
57    strcpy(ldatab->name, pname);
58    ldatab->age = page;
59    ldatab->jobcode = pjobcode;
60    return ldatab;
61 }
62 // display data block
63 void display_data(DataT *pblock) {
64    printf("Data: %s %d %d \n", pblock->name,
65           pblock->age, pblock->jobcode);
66 }
```

The following listing shows the Linux shell commands that compile, link, and execute the program. The results produced by the program execution are also shown.

```
$ gcc -Wall dataqueue.c basic_lib.o queue.o linked3.o
$ ./a.out
Enqueue data: Data: John Doe 38 1523
Enqueue data: Data: Joseph Hunt 59 1783
Enqueue data: Data: William Dash 49 2581
Queue size: queueA 3
Copy Head data from queue: queueA
Data: John Doe 38 1523
Enqueue: Data: David Winemaker 48 1854
Size of queue: 4
Dequeue from queue: Data: John Doe 38 1523
```

Summary

Linked lists are dynamic data structures for storing a sequence of nodes. The data is part of the structure known as a node. After the list has been declared and created, it can grow or shrink by adding or removing nodes. Lists can have one or two ends. Each node may have one or more links. A link is a pointer to another node of the linked list. Abstract data types (ADTs) of higher-level data structures are implemented with linked lists. Queues and stacks are examples of higher-level data structures and can be implemented with arrays or by linked lists.

Key Terms		
linked lists	nodes	links
dynamic structure	low-level structures	high-level structures
next	previous	queues
stacks	data block	abstract data type

Exercises

Exercise 10.1   Design and implement in C a linked list in which each node represents a flight stop in a route to a destination. The data in each node is the airport code (an integer number). The airline can add or delete intermediate flight stops. It can also calculate and display the number of stops from the starting airport to the destination airport.

Exercise 10.2   The *linked2* library for linked lists defines nodes with only one link and with only one end. Develop a *stack2* library that uses double-linked lists with two ends. Develop a complete C program that creates and manipulates three stacks.

Exercise 10.3   The *linked2* library for linked lists defines nodes with only one link and with only one end. Add a function that searches for a node with a given name and returns a copy of the data block. Develop a complete C program that uses the modified library.

Exercise 10.4   Develop a library that defines an array of *N* linked lists. This can be used if the linked list is used to implement a priority queue with *N* different priorities. A node will have an additional component, which is the priority defined by an integer variable. Develop a complete C program.

Chapter 11

Text Data

11.1 Introduction

In computational models, most of the data defined and manipulated is numerical in nature. This chapter presents text data also known as text *strings*. There is a significant number of string operators and functions that manipulate strings.

11.2 C Strings

An individual text symbol is also known as a text character. In C, a text character is of type *char* and a simple string is basically a sequence of text characters. A string is implemented as a one-dimensional array with the last character being a special character with value '\0' to mark the end of the string.

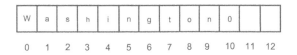

FIGURE 11.1: A C string.

An individual text character is denoted by a text symbol enclosed in single quotes, and this can be assigned to a variable of type *char*. For example:

```
char chvar;
chvar = 'g';
```

As mentioned previously, a string is declared as an array of type *char*. The size of the array must be sufficiently large to store the text string and the terminating '\0' character.

Figure 11.1 shows a string *Washington* by showing an array of characters (individual symbols of type *char*) of total size 13 although the array only has 11, including the special '\0' character that indicates the end of the string. The following statement declares string variable *inst* and and initializes it with string literal.

```
char inst[13] = "Washington";
```

Because a string is an array of characters, each individual character can be referenced using an index value. For example, the seventh character—the one with index value 6—is accessed by indicating 6 in brackets with the string variable *inst*. The result of this assignment statement is the character t.

```
char chvar;
chvar = inst[6];
```

11.3   String Input and Output

The C programming language includes a standard library of functions that performs common string operations, including input/output of strings. The two string functions to input a string from the console and display string respectively are functions *gets()* and *puts()*. Input and output of strings can also be performed with functions *scanf* and *printf*. Reading a string variable with function *scanf* can only take a word because when white space is encountered, function *scanf* terminates.

Listing 11.1 shows a C program with string input and output. This program is stored in file *string_io.c*.

Listing 11.1 C program with string input and output.

```
 1 /* Program: string_io.c
 2   This program illustrates basic string I/O in C.
 3 */
 4 #include <stdio.h>
 5 int main() {
 6   char myarraya[20];
 7   char myarrayb[20];
 8   char inst[20] = "BW State University";
 9   printf("Inst: %s \n", inst);
10   printf("Type string max 19 characters \n");
11   gets(myarraya);
12   printf("String a: ");
13   puts(myarraya);
14   printf("\nType another string (max 19 chars)\n");
15   scanf("%s", myarrayb);   /* no blank spaces */
16   printf("String b: %s\n", myarrayb);
17   return 0;
18 }
```

Lines 6 and 7 declare two string variables *myarraya* and *myarrayb*, each of size 20, that will store up to 19 text characters and the special '\0' character at the end to delimit the string. Line 8 declares and initializes string variable inst, also of size 20. Line 9 displays the value of string variable *inst*. For this, it uses the %s format. Line 11 reads a string including blank spaces from the console using function *gets*. Line 13 displays the string on the console screen using function *puts*. Line 15 reads the value of string variable *myarrayb* from the console using function *scanf*. Line 16 displays the string on the console screen using function *printf*. The following listing shows the commands to compile, link, and execute the program.

```
$ gcc -Wall string_io.c
$ ./a.out
Inst: BW State University
Type string max 19 characters
Kennesaw State U.
String a: Kennesaw State U.

Type another string (max 19 chars)
Columbus State U.
String b: Columbus
```

11.4   String Operations

There are several string functions available in the C standard function library. The function prototypes for the string functions are stored in header file *string.h*, which must be appended at the top of the C source file that uses string manipulation.

One of the most useful library functions is *strcpy* which copies the characters of one string to another. The order of the arguments of function *strcpy()* is the second string is assigned to the first one. Another useful string function is *strlen* which returns the number of characters in the specified C string excluding the trailing \0.

The following function list includes the function prototypes and a brief description of the most useful string manipulation functions.

```
char *strcpy (char *dest, char *src)
```
    Copy source string *src* into the destination *dest* string.

```
char *strncpy(char *string1, char *string2, int n)
```
    Copy the first *n* characters of string *string2* to string *string1* .

```
 int strcmp(char *string1, char *string2)
```
    Compare *string1* and *string2* to determine alphabetic order.

```
int strncmp(char *string1, char *string2, int n)
```
    Compare the first *n* characters of two strings.

```
int strlen(char *string)
```
Determine the length of a string.

```
char *strcat(char *dest, const char *src)
```
Concatenate the source string *src* to the destination string *dest*.

```
char *strncat(char *dest, const char *src, int n)
```
Concatenate *n* characters from the source string *src* to the destination string *dest*.

```
char *strchr(char *string, int c)
```
Find first occurrence of character *c* in *string*.

```
char *strrchr(char *string, int c)
```
Find last occurrence of character *c* in *string*.

```
char *strstr(char *string2, char *string1)
```
Find first occurrence of string *string1* in *string2*.

```
char *strtok(char *s, const char *delim)
```
Parse the string *s* into tokens using *delim* as delimiter.

11.5 Using the String Functions

The size of the string array can be found by calling function *strlen*. In the following example, the integer variable *slen* will have a value of 19. columns.

```
char inst[20] = "BW State University";
int slen;
slen = strlen(inst);
```

String concatenation joins two strings. In the following example, the string variable *world* is appended to the end of string variable *hello*.

```
char hello[20];
char world[20];
strcpy(hello, "Hello, there ");
strcpy(world, "World");
strcat(hello, world);  /* world appended at end of hello */
```

A substring can be accessed by specifying the pointer relative to the beginning of the string that contains the substring. In the following example, a substring of a string *mlist* that starts at position *pos* and that contains *n* characters is to be copied to destination string *dest*.

```
char mylist[25];
char dest[25];
char *mylistp;          /* string pointer */
int pos = 6;            /* substring starts at character 6 */
int n = 8;              /* length of substring */

mylistp = mylist;       /* points to beginning of mylist */
mylistp += pos - 1;     /* point to substring */
strncpy(dest, mylistp, n); /* copy substring */
```

Listing 11.2 shows the C source code of a program that illustrates the general use of several string functions. The program is stored in file: compstr.c.

Listing 11.2 C program with string operations.

```
 1 /* Program compstr.c
 2    This program performs calls to various
 3    string functions */
 4 #include <string.h>
 5 #include <stdio.h>
 6 int main() {
 7    char mystringa[25];
 8    char mystringb[45];
 9    char thirdstring[20];
10    int strflag;
11    int numchars;
12    strcpy(mystringa, "This is a C-string");
13    strcpy(mystringb, "Another C-string ");
14    strflag = strcmp( mystringa, mystringb);
15    if (strflag != 0)
16       printf("Result: strings not equal \n");
17    numchars = strlen(mystringb);
18    printf("Number chars in second string: %d \n", numchars);
19    printf("\nType third string: ");
20    // scanf("%s", thirdstring);
21    gets(thirdstring);
22    printf("Third string: %s \n", thirdstring);
23    numchars = strlen(thirdstring);
24    printf("Number chars in third string: %d \n", numchars);
25    strcat(mystringb, thirdstring);
26    printf("Strings joined: %s \n", mystringb);
27    numchars = strlen(mystringb);
28    printf("Number chars in joined string: %d \n", numchars);
29    return 0;
```

The following listing shows the commands to compile, link, and execute the previous program. Then a sample execution of the program is also shown.

```
$ gcc -Wall compstr.c
$ ./a.out
Result: strings not equal
Number of chars in second string: 17

Type third string: Problem Solving
Third string: Problem Solving
Number of chars in third string: 15
Strings joined: Another C-string Problem Solving
Number of chars in joined string: 32
```

Summary

Strings are sequences of text characters. They are implemented as one-dimensional arrays of type *char*. To refer to an individual character in a string, an integer value, known as the index, is used to indicate the relative position of the character in the string. The C programming language provides the programmer with a standard library of string functions. Several string functions have been shown in the examples presented.

Key Terms		
strings	text character	index
char arrays	char pointers	C standard string library
string pointers	string functions	substring

Exercises

Exercise 11.1   Develop a C program that takes (as input) a string with a sequence of characters and complement the character sequence into a new string that it returns. This should be performed in such a way that any character ′a′ is converted to a character ′t′, any character ′g′ is converted to character ′c′, any character ′t′ to character ′a′, and any character ′c′ to character ′g′.

Exercise 11.2   Develop a C function that converts the first character in a string to upper case, but only if the first character is a letter.

Exercise 11.3 Develop a C function that reverses the characters in a string and returns a new string.

Exercise 11.4 Develop a C function that finds the number of times a specified character occurs in a string.

Exercise 11.5 Develop a C function that finds the number of times a string (substring) occurs in a second string.

Exercise 11.6 Develop a C function that gets a substring of a string and returns the new string.

Exercise 11.7 Develop a C function that defines an array of strings. The program must read from the console the values of each string in the array, then display the string in the entire array.

Chapter 12

Computational Models with Arithmetic Growth

12.1 Introduction

Given a real-world problem, a mathematical model is defined and formulated then one or more techniques are used to implement this model in a computer to derive the corresponding computational model. This chapter presents an introduction to simple mathematical models that exhibit arithmetic growth and describes the overall behavior of simple mathematical and computational models. This includes definitions and explanations of several important concepts related to modeling.

Some of these concepts have been briefly introduced in preceding chapters. The derivation of difference and functional equations is explained; their use in mathematical modeling is illustrated with a few examples. A complete computational model implemented with the C programming language is presented and discussed.

12.2 Mathematical Modeling

The three types of methods that are used for modeling are:

1. Graphical

2. Numerical

3. Analytical

Graphical methods apply visualization of the data to help understand the data. Various types of graphs can be used; the most common one is the line graph.

Numerical methods directly manipulate the data of the problem to compute various quantities of interest, such as the average change of the population size in a year.

Analytical methods use various forms of relations and equations to allow computation of the various quantities of interest. For example, an equation can be derived that defines how to compute the population size for any given year. Each method has its advantages and limitations. The three methods complement each other and are normally used in modeling.

12.2.1 Difference Equations

A data list that contains values ordered in some manner is known as a *sequence*. Typically, a sequence is used to represent ordered values of a property of interest in some real problem. Each of these values corresponds to a recorded measure at a specific point in time. In the example of population change over a period of five years, the ordered list will contain the value of the population size for every year. This type of data is discrete data and the expression for the ordered list can be written as:

$$\langle p_1, p_2, p_3, p_4, p_5 \rangle$$

In this expression, p_1 is the value of the population of year 1, p_2 is the value of the population for year 2, p_5 is the value of the population for year 5, and so on. In this case, the length of the list is 5 because it has only five values, or terms.

Another example is the study of changes in electric energy price in a year in Georgia, given the average monthly price. Table 12.1 shows the values of average retail price of electricity for the state of Georgia.[1] The data given corresponds to the price of electric power that has been recorded every month for the last 12 months. This is another example of *discrete data*. The data list is expressed mathematically as:

$$\langle e_1, e_2, e_3, e_4, e_5, e_6, e_7, e_8, e_9, e_{10}, e_{11}, e_{12} \rangle$$

Given the data that has been recorded about a problem and that has been recorded in a list, simple assumptions can be made. One basic assumption is that the quantities in the list increase at a constant rate. This means the increment is assumed to be fixed. If the property is denoted by x, the increment is denoted by Δx, and the value of a term measured at a particular point in time is equal to the value of the preceding term and the increment added to it. This can be expressed as:

$$x_n = x_{n-1} + \Delta x \tag{12.1}$$

Another assumption is that the values of x are actually always increasing, and not decreasing. This means that the increment is greater than zero, denoted by $\Delta x \geq 0$. At any point in time, the increment of x can be computed as the difference of two consecutive measures of x and has the value given by the expression:

$$\Delta x = x_n - x_{n-1} \tag{12.2}$$

These last two mathematical expressions, Equation 12.1 and Equation 12.2, are

---

[1] U.S. Energy Information Administration — Independent Statistics and Analysis. http://www.eia.gov/

known as *difference equations* and are fundamental for the formulation of simple mathematical models. We can now derive a simple mathematical model for the monthly average price of electric energy, given the collection of monthly recorded energy price in cents per kW-h of the last 12 months. This model is formulated as:

$$e_n = e_{n-1} + \Delta e$$

The initial value of energy price, prior to the first month of consumption, is denoted by e_0, and it normally corresponds to the energy price of a month from the previous year.

TABLE 12.1: Average price of electricity (cents per kW-h) in 2010.

Month	Jan	Feb	Mar	Apr	May	Jun	Jul	Aug
Price	10.22	10.36	10.49	10.60	10.68	10.80	10.88	10.94

Month	Sep	Oct	Nov	Dec
Price	11.05	11.15	11.26	11.40

12.2.2 Functional Equations

A functional equation has the general form: $y = f(x)$, where x is the *independent variable*, because for every value of x, the function gives a corresponding value for y. In this case, y is a function of x.

An equation that gives the value of x_n at a particular point in time denoted by n, without using the previous value, x_{n-1}, is known as a *functional equation*. In this case the functional equation can also be expressed as: $x = f(n)$. From the data given about a problem and from the difference equation(s), a functional equation can be derived. Using analytical methods, the following mathematical expression can be derived and is an example of a functional equation.

$$x_n = (n-1)\Delta x + x_1 \tag{12.3}$$

This equation gives the value of the element x_n as a function of n. In other words, the value x_n can be computed given the value of n. The value of Δx has already been computed. The initial value of variable x is denoted by x_1 and is given in the problem by the first element in the sequence x.

12.3   Models with Arithmetic Growth

Arithmetic growth models are the simplest type of mathematical models. These models have a linear relationship in the variable because the values of the variable increase by equal amounts over equal time intervals. Using x as the variable, the increase is represented by Δx, and the difference equation defined in Equation 12.1:

$$x_n = x_{n-1} + \Delta x$$

Examples are time dependent models, in which a selected property is represented by a variable that changes over time.

From the given data of a real problem, to decide if the model of the problem will exhibit arithmetic growth, the given data must be processed by computing the differences of all *consecutive* values in the data. The differences thus consist of another list of values. If the differences are all equal, then the model exhibits arithmetic growth, and therefore it is a *linear model*.

Equation 12.1 and the simplifying assumption of constant growth can be applied to a wide variety of real problems, such as: population growth, monthly price changes of electric energy, yearly oil consumption, and spread of disease.

Using graphical methods, a line chart or a bar chart can be constructed to produce a visual representation of the changes in time of the variable x. Using numerical methods, given the initial value x_0 and once the increase Δx in the property x has been derived, successive values of x can be calculated using Equation 12.1. As mentioned before, with analytical methods, a functional equation can be derived that would allow the direct calculation of variable x_n at any of the n points in time that are included in the data list given. This equation is defined in Equation 12.3:

$$x_n = x_1 + \Delta x (n - 1)$$

12.4   Using the C Programming Language

The external C library *basic_lib* includes two functions that are needed to implement computational models with arithmetic growth. These functions are: *diff* and *linspace*. The first function, *diff* computes the differences of a vector (sequence) of data values given.

Calling the *diff* function requires two arguments. The first argument is the specified vector, the second argument is the size of the vector. The function produces another vector that has the values of the differences of the values in the given vector.

```
double e[10] = {1.25, 2.15, 4.55, 3.2, 1.05, 2.45, 3.85, 1.15, 2.75,
     3.55};
double *de;       /* vector with the computed differences */
int n;
n =10;    /* size of vector e */
de = diff(e, n);
```

The differences computed in vector *e* are:

```
0.9000 2.4000 -1.3500 -2.1500 1.4000 1.4000 -2.7000
     1.6000 0.8000
```

Listing 12.1 shows the C program that computes the differences and the approximate price of electricity for 12 months. The array for the monthly price is denoted by *e*, and the array for months is denoted by *m*. The array of differences is denoted by *de* and is computed in line 27 of the program. The data is taken from Table 12.1.

In the problem under study, the number of measurements is the total number of months, denoted by *n* and has a value of 12. To compute the average value of the increments in price of electric energy in the year we can use the general expression for calculating average:

$$\Delta e = \frac{1}{n} \sum_{i=1}^{i=n} de_i$$

Computing the average is implemented in C, with the *mean(v)* library function that computes the average of the values in a vector. Using *d* to denote the average increment Δe, the following statement in C illustrates this and is included in line 31 of the C program.

```
d = mean(de, N-1)
```

Listing 12.1: C program for computing the differences in price of electricity.

```
1 /*
2  Program:   priceelect.c
3  C program for computing monthly price for electric energy
4  J. M. Garrido, December 2012.
5 */
6 #include <stdio.h>
7 #include "basic_lib.h"
8
9 int main() {
10    const int N = 12;    /* 12 months */
11    double e[] = {10.22, 10.36, 10.49, 10.60, 10.68, 10.80, 10.88,
             10.94, 11.05, 11.15, 11.26, 11.40};
12    double c[N];    /* calculated price using average */
13    double *de;
14    int m[N];    /* months */
```

```
15    double deltax;        /* average difference */
16    int j;
17
18    printf("Montly price of electricity\n");
19    m[0] = 1;
20    for (j = 1; j < N; j++)
21      m[j] = m[j-1] + 1;
22    /* Array Monthly price for electric energy */
23    for (j = 0; j < N; j++)
24      printf("%lf ", e[j]);
25    /* differences in sequence e */
26    printf("\nDifferences of the given data\n");
27    de = diff (e, N);
28    for (j = 0; j < N-1; j++)
29      printf("%f ", de[j]);
30    /* average of increments */
31    deltax = mean(de, N-1);
32    printf("\nAverage difference: %lf \n", deltax);
33    /* Calculating price of electric energy */
34    c[0] = e[0];
35    for (j = 1; j < N; j++)
36      c[j] = deltax * j + e[0];
37    printf("\nCalculated prices of elect: \n");
38    for (j = 0; j < N; j++)
39      printf("%f ", c[j]);
40    printf("\n");
41
42    printf("\nData for Plotting\n");
43    for (j = 0; j < N; j++)
44      printf("%d %lf %lf \n", m[j], e[j], c[j]);
45    return 0;
46 }
```

With the value of average increment of price of electricity, the functional equation of the model can be applied to compute the price of electric energy for any month. The new array, *c*, is defined and contains all the values computed using the functional equation and the average value of the increments. This is shown in lines 34–36 of the C program.

The following listing shows the commands and results after compiling, linking and executing the C program.

```
$ gcc -Wall priceelect.c basic_lib.o
$ ./a.out > priceelect.1st
Monthly price of electricity
10.220 10.360 10.490 10.600 10.680 10.800 10.880 10.940 11.050
      11.150 11.260 11.400
Differences of the given data
0.140 0.130 0.110 0.080 0.120 0.080 0.060 0.110 0.100 0.110 0.140
Average difference: 0.107273

Calculated prices of elect:
10.220 10.327273 10.434545 10.541818 10.649091 10.756364 10.863636
      10.970909 11.078182 11.185455 11.292727 11.400
```

12.5   Producing the Charts of the Model

Two lists of values for the price of electric energy are available as arrays. The first one is the data given with the problem and is denoted by *e*, the second list was computed in the C program using the functional equation for all 12 months, and the list is denoted by *c*. The array with data representing the months of the year is denoted by *m*.

Gnuplot is used to produce charts with these three data lists or arrays. Two line charts are generated on the same plot, one line chart of array *e* with array *m*, the second chart using array *c* and array *m*. The Gnuplot commands that create and draw the charts are stored in the script file: priceelect.cgp and are shown as follows:

```
set title "Plot of Montly Price of Electricity vs time"
set xlabel "Time (secs)"
set ylabel "Monthly price of electricity"
set size 1.0, 1.0
set samples 12
plot "priceelect.gpl" u 1:2 with linespoints, "priceelect.gpl" u 1:3
    with linespoints
```

Figure 12.1 shows the line chart with the original data given in Table 12.1 and the line with computed values of the monthly price of electricity.

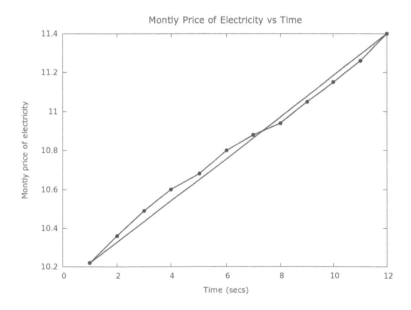

FIGURE 12.1: Given and computed values of monthly price of electric energy.

12.6   Validation of a Model

Validation of a model is the analysis that compares the values computed with the model with the actual values given. For example, starting with the first value of the monthly consumption of electric energy, the model is used to compute the rest of the monthly values of consumption. These values can then be compared to the values given.

If the corresponding values are close enough, the model is considered a reasonable *approximation* to the real system.

Summary

Several basic concepts of mathematical models are presented. A computational model is a computer implementation of a mathematical model. Making simplifying assumptions and using abstraction are important steps in formulating a mathematical model. This involves a transition from the real world to the abstract world. Simple mathematical techniques such as difference equations and functional equations are used. With arithmetic growth models, the values of the differences of the data is constant. The difference and functional equations for these models are linear.

Key Terms		
arithmetic growth	mathematical methods	graphical methods
numerical methods	analytical methods	discrete data
data list	ordered list	sequence
differences	average difference	linear models
difference equations	functional equations	model validation

Exercises

Exercise 12.1   Construct a line chart of the data list in Table 12.1 and discuss how the price of electric energy changes in a specified period.

Exercise 12.2   Construct a bar chart of the data list in Table 12.1 and discuss how the price of electric energy changes in a specified period.

Exercise 12.3 Formulate a mathematical model based on a difference equation of the data list in Table 12.1.

Exercise 12.4 Develop a C program that uses the data in Table 12.1 to compute the average increase in price of electric energy per month from Equation 12.1 and/or Equation 12.2. Start with the second month, calculate the price for the rest of the months. Discuss the difference between the data in the table and the corresponding values calculated.

Exercise 12.5 Develop a C program that uses the data from a different problem to compute the differences and decide whether the model of the problem exhibits arithmetic growth.

Chapter 13

Computational Models with Quadratic Growth

13.1 Introduction

A computational model with quadratic growth is one in which the differences in the data are not constant but growing in some regular manner. Recall that with arithmetic growth these differences are constant. This chapter presents an introduction to computational models in which the differences in the data follow a pattern of arithmetic growth.

This chapter explains the computation of difference and functional equations; their use in the mathematical model is illustrated with a few examples. A complete computational model implemented with the C programming language is presented and discussed.

13.2 Differences of the Data

The data values in a quadratic growth model do not increase (or decrease) by a constant amount. With *quadratic growth*, the differences of the data change linearly. The differences of the differences, known as the *second differences*, are constant.

The data in a problem is used to set up an ordered list of values, or sequence. This type of list is denoted as follows:

$$\langle p_1, p_2, p_3, p_4, p_5 \rangle$$

The following is an example of a data list given in a generic problem and Figure 13.1 shows the graph of these values.

$$S = \langle 6.5, 10.5, 17.5, 27.5, 40.5, 56.5, 75.5, 97.5, 122.5 \rangle$$

The differences of these values is another sequence, D, that can be derived and

171

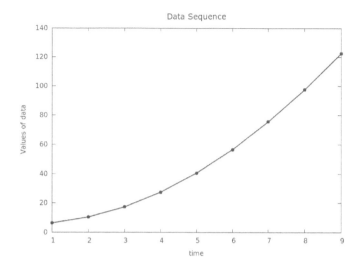

FIGURE 13.1: Plot of data in the sequence.

represents the *differences* of the values in sequence *S*. The values in *D* are the increases (or decreases) of the values in the first sequence, *S*.

$$D = \langle 4.0, 7.0, 10.0, 13.0, 16.0, 19.0, 22.0, 25.0 \rangle$$

The increases of the values of the differences, *D*, appear to change linearly and follow an arithmetic growth pattern; this is an important property of quadratic growth models. Figure 13.2 shows the graph with the data of the first differences.

The C program that declares the arrays for the data sequences in the problem, their differences, their second differences is stored in file: `differences.c`.

Listing 13.1 shows the source code of this C program. Note that the first differences are computed in line 28 and the *second differences* are computed in line 32. As mentioned in the previous chapter, the differences are computed by calling function `diff()`, which is in the *basic_lib* library.

Listing 13.1: Program that computes the first and second differences.

```
 1 /*
 2  Program:   differences.c
 3  C program for computing first and second differences of
 4  problem data
 5  J. M. Garrido, January 2013.
 6 */
 7 #include <stdio.h>
 8 #include "basic_lib.h"
 9
10 int main() {
11     const int N = 9;    /* 9 data points */
12     int x[N];
```

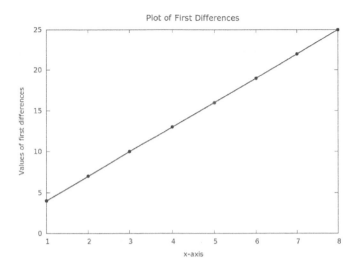

FIGURE 13.2: Plot of first differences

```
13    int j;
14    /* Array with original data sequence */
15    double s[] = {6.5, 10.5, 17.5, 27.5, 40.5, 56.5, 75.5, 97.5,
              122.5};
16    double *d;    /* first differences  */
17    double *d2;   /* second differences */
18
19    x[0] = 1;
20    for (j = 1; j < N; j++)
21       x[j] = x[j-1] + 1;
22    /*  Array with problem data  */
23    printf("Data sequence \n");
24    for (j = 0; j < N; j++)
25       printf("%lf ", s[j]);
26    /* Compute first differences in sequence e */
27    printf("\nFirst Differences of the given data\n");
28    d = diff(s, N);
29    for (j = 0; j < N-1; j++)
30       printf("%f ", d[j]);
31    /* compute differences of the differences */
32    d2 = diff(d, N-1);
33    printf("\nSecond differences\n");
34    for (j = 0; j < N-2; j++)
35       printf("%f ", d2[j]);
36    printf("\n\nData for Plotting\n");
37    for (j = 0; j < N; j++)
38       printf("%d %lf \n", x[j], s[j]);
39    printf("\nData of differences for plotting\n");
40    for (j = 0; j < N-1; j++)
41       printf("%d %lf \n", x[j], d[j]);
42    return 0;
43 }
```

The following listing shows the shell commands on Linux to compile, link, and execute the C program. The output produced by the execution of the program shows the data given by the problem, the first differences, and the second differences.

```
$ gcc -Wall differences.c basic_lib.o
$ ./a.out
Data sequence
6.500 10.500 17.500 27.500 40.500 56.500 75.500 97.500 122.500
First Differences of the given data
4.000 7.000 10.000 13.000 16.000 19.000 22.000 25.000
Second differences
3.000 3.000 3.000 3.000 3.000 3.000 3.000
```

In the example, the second differences are constant, all have value 3.0. This is another important property of quadratic growth models.

13.3   Difference Equations

An ordered data list or sequence is used to represent ordered values of a property of interest in some real problem. Each of these values can correspond to a recorded measure at a specific point in time. The expression for the values in a sequence, S, with n values can be written as:

$$S = \langle s_1, s_2, s_3, s_4, s_5 \ldots s_n \rangle$$

In a similar manner, the expression for the values in the differences, D, with m values and with $m = n - 1$, can be written as:

$$D = \langle d_1, d_2, d_3, d_4, d_5 \ldots d_m \rangle$$

Because the values in the sequence of second differences are all the same, the value is denoted by dd, and computed simply as $dd = d_n - d_{n-1}$.

To formulate the difference equation, the value of a term, s_{n+1} in the sequence is computed with the value of the preceding term s_n plus the first term of the differences d_1, plus the single value dd of the second differences added to it times $n - 1$. This equation, which is the difference equation for quadratic growth models, can be expressed as:

$$s_{n+1} = s_n + d_1 + dd\,(n-1) \tag{13.1}$$

13.4   Functional Equations

An equation that gives the value of a term x_n without using consecutive values (the previous value, x_{n-1} or the next value, x_{n+1}), is known as a *functional equation*. The difference equation for quadratic growth models, Equation 13.1 can be rewritten as:

$$s_n = s_{n-1} + d_1 + dd\,(n-2) \qquad (13.2)$$

Equation 13.2 can be manipulated by substituting s_{n-1} for its difference equation, and continuing this procedure until s_1. In this manner, a functional equation can be derived. The following mathematical expression is a general functional equation for quadratic growth models.

$$s_n = s_1 + d_1\,(n-1) + dd\,(n-2)n/2 \qquad (13.3)$$

Equation 13.3 gives the value s_n as a function of n for a quadratic growth model. The value of the first term of the original sequence is denoted by s_1, the value of the first term of the differences is denoted by d_1, and the single value, dd, is the second differences.

13.5   Examples of Quadratic Models

Equation 13.1 and Equation 13.3 represent mathematical models of quadratic growth models. These can be applied to a wide variety of real problems, such as: computer networks, airline routes, roads and highways, and telephone networks. In these models, the first differences increase in a linear manner (arithmetic growth), as the two examples discussed in previous sections of this chapter. Other models with quadratic growth involve addition of ordered values from several sequences that exhibit arithmetic growth.

13.5.1   Growth of Number of Patients

Statistical data maintained by a county with several hospitals include the number of patients every year. Table 13.1 gives the data of the number of patients in the years from 1995 to 2002, and their increases. Figure 13.3 shows the graph for the number of patients in the hospital by year from 1995 through 2002. It can be observed that the number of patients from 1995 through 2002 does not follow a straight line; it is not linear.

Table 13.1 shows that the number of patients increases every year by a constant

TABLE 13.1: Number of patients for years 1995–2002.

Year	1995	1996	1997	1998	1999	2000
Patients	5,500	8,500	13,500	20,500	29,500	40,500
Increase	0	3,000	5,000	7,000	9,000	11,000

Year	2001	2002
Patients	53,5000	68,500
Increase	13,000	15,000

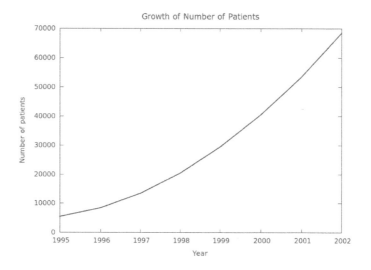

FIGURE 13.3: Number of patients for 1995–2002.

number. The differences in the number of patients from year to year increase in a regular pattern, in a linear manner. This implies that the increases of the number of patients follow an arithmetic growth.

Computing the first and second differences of the data for this problem is similar to the one already discussed.

13.5.2 Growth of Computer Networks

The following example represents a simple network that connects computers directly to each other and a number of links is necessary for the direct connection between computers. To connect two computers, 1 single link is needed. To connect 3 computers, 3 links are needed. To connect 4 computers, 6 links are needed.

It can be noted that as the number of computers increases, the number of links increases in some pattern. To connect 5 computers, 4 new links are needed to connect the new computer to the 4 computers already connected. This gives a total of 10 links. To connect 6 computers, 5 new links are needed to connect the new computer to the 5 computers that are already connected, a total of 15 links.

Let L_n denote the number of links needed to connect n computers. The difference equation for the number of links can be expressed as:

$$L_n = L_{n-1} + (n-1)$$

This equation has the same form as the general difference equation for quadratic growth, Equation 13.2. The parameters are set as: $d_1 = 0$ and $dd = 1$.

Using the expression for the difference equation for L_n, the following C program is used to construct the data sequence for L for n varying from 1 to 50. The source code is stored in the file links.c and is shown in Listing 13.2.

By executing the program, all the terms in the sequence L are computed and the results are used to produce a chart using GnuPlot. Figure 13.4 shows the graph of the number of links needed to connect n computers.

Listing 13.2: C source program that computes the number of links.

```
1  /*
2  Program:    links.c
3  C program computes the number of links needed
4  to connect n computers
5  J. M. Garrido, January 2013.
6  */
7  #include <stdio.h>
8  int main() {
9     const int M = 50;  // limit on number of computers
10    int j;
11    int n[M];        // number of computers
12    int llinks[M];   // links
13    printf("Links of computer network\n");
14
15    n[0] = 1;
16    for (j = 1; j < M; j++)
17         n[j] = n[j-1] + 1;
18    llinks[0] = 0;
19    // compute the links as n varies from 2 to m
20    for (j=1; j < M; j++)
21         llinks[j] = llinks[j-1] + (j-1);
22    // data for plot
23    printf("Number of computers   number of links\n");
24    for (j=0; j < M; j++)
25         printf(" %d   %d \n", n[j], llinks[j]);
26    return 0;
27 }
```

From the general functional equation, Equation 13.3, the functional equation for the network problem discussed can be expressed as:

$$L_{n+1} = L_1 + (n+1)n/2$$

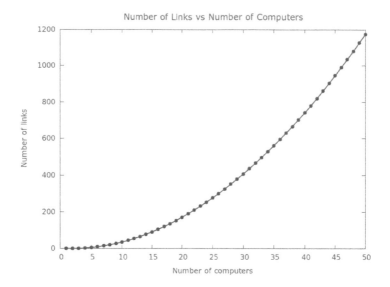

FIGURE 13.4: Number of links to connect n computers

Notice that L_1 is always zero ($L_1 = 0$) because no link is necessary when there is only one computer ($n = 1$).

$$L_n = n(n-1)/2$$

13.5.3   Models with Sums of Arithmetic Growth

A variety of models have data about a property that follows an arithmetic growth pattern. The summation or running totals of the data of this property is also important. The following example will illustrate this concept.

A county maintains data about the cable installations for multi-purpose services, such as TV, phones, Internet access and others. The data include new cable installations (in thousands) per year and the total number of cable installations per year. This data is shown in Table 13.2.

TABLE 13.2: Number of cable installations for years 1995–2002.

Year	1995	1996	1997	1998	1999	2000	2001	2002
New	1.5	1.9	2.3	2.7	3.1	3.5	3.9	4.3
Sum	1.5	3.4	5.7	8.4	11.5	15.0	18.9	25.2

For this type of data, the general principle is that the new cable installations follow an arithmetic growth pattern, and the summations of the cable installations follow a quadratic growth pattern. Two sequences are needed: *a* for the new cable installations, and *b* for the sums of cable installations.

$$a = \langle a_1, a_2, a_3, a_4, a_5 \ldots a_n \rangle \quad b = \langle b_1, b_2, b_3, b_4, b_5 \ldots b_n \rangle$$

To develop a difference equation and a functional equation of the sums of cable installations, a relation of the two sequences, *a* and *b*, needs to be expressed.

It can be observed from the data in Table 13.2 that the data in sequence *a* follows an arithmetic growth pattern. The difference equation for *a* can be expressed as:

$$a_n = a_{n-1} + \Delta a, \quad \Delta a = 400, \quad n = 2 \ldots m$$

The functional equation for *a* can be expressed as:

$$a_n = 1500 + 400\,(n-1)$$

It can also be observed from the data in Table 13.2 that the data in sequence *b* is related to the data in sequence *a*. The difference equation for sequence *b* is expressed as:

$$b_n = b_{n-1} + a_n, \quad n = 2 \ldots m$$

Substituting a_n for the expression in its functional equation and using the general functional equation for quadratic growth, Equation 13.3:

$$b_n = b_1 + d_1\,(n-1) + dd\,(n-2)n/2$$

Finally, the functional equation for the sequence *b* is expressed as:

$$b_n = 1500 + 1900n + 400(n-1)n/2$$

Summary

This chapter presented some basic concepts of quadratic growth in mathematical models. Simple mathematical techniques such as difference equations and functional equations are used in the study of models with quadratic growth. In these models the increases follow an arithmetic growth pattern; the second differences are constants. The most important difference with models with arithmetic growth is that the increases cannot be represented by a straight line, as with arithmetic models. The functional equation of quadratic growth is basically a second degree equation, also known as a quadratic equation.

	Key Terms	
quadratic growth	non-linear representation	differences
second differences	network problems	summation
quadratic equation	roots	coefficients

Exercises

Exercise 13.1   On a typical spring day, the temperature varies according to the data recorded in Table 13.3. Develop a C program to compute first and second differences. Formulate the difference equations and functional equations for the temperature. Is this discrete or continuous data? Discuss.

TABLE 13.3: Temperature changes in 12-hour period.

Time	7	8	9	10	11	12	1	2	3	4	5	6
Temp. (F^0)	51	56	60	65	73	78	85	86	84	81	80	70

Exercise 13.2   Develop a C program to compute first and second differences. Produce line chart of the data list in Table 13.1. Discuss how the number of patients changes in a specified period.

Exercise 13.3   Formulate a mathematical model based on a difference equation of a modified computer network problem similar to one discussed in this chapter. There are three servers and several client computers connected via communication links. All connections between pairs of clients require two links. The connection between servers also requires two links. Use the concepts and principles explained in this chapter to derive an equation for the number of links.

Exercise 13.4 Formulate a mathematical model based on a functional equation of a modified computer network problem similar to one discussed in this chapter. There are three servers and several client computers connected via communication links. All connections between pairs of clients require two links. The connection between servers also requires two links. Use the concepts and principles explained in this chapter to derive an equation for the number of links.

Exercise 13.5 Formulate a mathematical model based on a functional equation of a modified computer network problem similar to one discussed in this chapter. There are K servers and several client computers connected via communication links. All connections between pairs of clients require two links. The connection of a server to the other $K-1$ servers requires a single link. Use the concepts and principles explained in this chapter to derive an equation for the number of links.

Chapter 14

Models with Geometric Growth

14.1 Introduction

This chapter presents an introduction to computational models in which the data follow a pattern of geometric growth. In such models, the data exhibit growth in such a way that in equal intervals of time, the data increase by an equal percentage or factor.

The difference and functional equations in models with geometric growth are explained; their use in mathematical modeling is illustrated with a few examples. Several computational models implemented with C and the *basic_lib* library are presented and discussed.

14.2 Basic Concepts

The data in the sequence represent some relevant property of the model and are expressed as a variable s. An individual value of variable s is known as a *term* in the sequence and is denoted by s_n. A sequence with m data values or terms, is written as follows:

$$\langle s_1, s_2, s_3, s_4, s_5, \ldots, s_m \rangle$$

In models with geometric growth, the data increase (or decrease) by an equal percentage or *growth factor* in equal intervals of time. The difference equation that represents the pattern of geometric growth has the general form:

$$s_{n+1} = c\, s_n \tag{14.1}$$

In Equation 14.1, the parameter c is constant and represents the *growth factor*, and n identifies an individual value such that $n \leq m$. With geometric growth, the data increases or decreases by a fixed factor in equal intervals.

14.2.1   Increasing Data

The data in a sequence will successively increase in value when the value of the growth factor is greater than 1. For example, consider a data sequence that exhibits geometric growth with a growth factor of 1.45 and a starting value of 50.0. The sequence with 8 terms is:

$$\langle 50.0, 72.5, 105.125, 152.43, 221.02, 320.48, 464.70, 673.82 \rangle$$

Figure 14.1 shows a graph of the data with geometric growth. Note that the data increases rapidly.

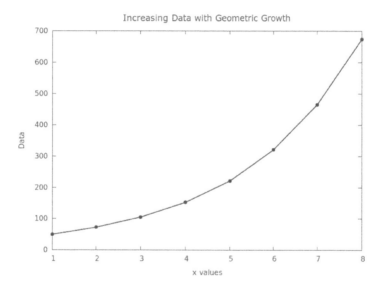

FIGURE 14.1: Data with geometric growth.

14.2.2   Decreasing Data

The data in a sequence will successively decrease in value when the value of the growth factor is less than 1. For example, consider a data sequence that exhibits geometric growth with a growth factor of 0.65 and a starting value of 850.0. The sequence with 10 terms is:

$$\langle 850.0, 552.5, 359.125, 233.43, 151.73, 98.62, 64.10, 41.66, 27.08, 17.60 \rangle$$

Figure 14.2 shows a graph of the data with geometric growth. Note that the data decreases rapidly because the growth factor is less than 1.0.

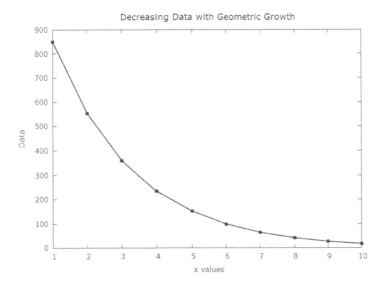

FIGURE 14.2: Data decreasing with geometric growth.

14.2.3   Case Study 1

The population of a small town is recorded every year; the increases per year are shown in Table 14.1, which gives the data about the population during the years from 1995 to 2003. The table also shows the population growth factor.

TABLE 14.1: Population of a small town during 1995–2003 (in thousands).

Year	1995	1996	1997	1998	1999	2000	2001	2002	2003
Pop.	81	90	130	175	206	255	288	394	520
Fac.	-	1.111	1.444	1.346	1.177	1.237	1.129	1.368	1.319

Note that although the growth factors are not equal, the data can be considered to grow in a geometric pattern. The values of the growth factor shown in the table are sufficiently close and the average growth factor calculated is 1.2667.

The following computational tasks are performed: (1) create the data lists (arrays) of the sequence s with the values of the original data in Table 14.1; (2) compute the average growth factor from the data in the table; (3) compute the values of a second data list y using 1.267 as the average growth factor and the difference equation $y_{n+1} = 1.267 \, y_n$, and (4) plot the graphs.

The C program that performs these tasks is stored in file: popstown.c. When the program executes, the following listing is produced.

```
Program to compute average growth factor
Given data of population per year
1995 81.000000
1996 90.000000
1997 130.000000
1998 175.000000
1999 206.000000
2000 255.000000
2001 288.000000
2002 394.000000
2003 520.000000

Factors in array s:
1.111111 1.444444 1.346154 1.177143 1.237864 1.129412 1.368056
     1.319797

Mean factor: 1.266748
1995 81.000000
1996 102.606554
1997 129.976603
1998 164.647547
1999 208.566881
2000 264.201591
2001 334.676725
2002 423.950931
2003 537.038814
```

Figure 14.3 shows a graph with two curves; one with the population in the town by year from 1995 through 2003 taken directly from Table 14.1. The other curve shown in the graph of Figure 14.3 is the computed data applying Equation 14.1 with 1.267 as the growth factor, using the difference equation $s_{n+1} = 1.267 \, s_n$.

Listing 14.1 shows the C source code of the program file: popstown.c. The factors of the population data are computed in line 24 and average growth factor is computed in line 28 of the program. The computed data is calculated in lines 31–32.

Listing 14.1: C program that computes the population average growth factor.

```
1  /* program: popstown.c
2    This program computes the growth factor of the population per
          year.
3    Uses 'basic_lib'
4    J Garrido 11-2-2012
5  */
6  #include <stdio.h>
7  #include <math.h>
8  #include "basic_lib.h"
9
10 int main() {
11   const int N = 9;
12   double s[] = {81.0, 90.0, 130.0, 175.0, 206.0, 255.0, 288.0, 394.0,
          520.0};
13   int x[] = {1995, 1996, 1997, 1998, 1999, 2000, 2001, 2002, 2003};
14   double cs[N]; /* computed values */
15   double * f;    // pointer for array of factors
16   double meanv;
```

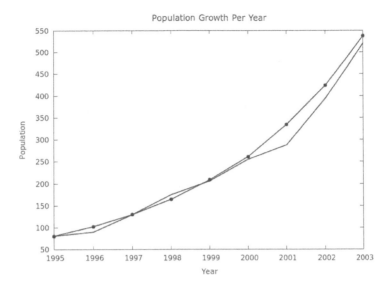

FIGURE 14.3: Population of a small town for 1995–2003.

```
17  int n;
18  int j;
19  printf("Program to compute average growth factor\n\n");
20  n = N;
21  printf("Given data of population per year\n");
22  for (j = 0; j < N; j++)
23      printf("%d %f \n", x[j], s[j]);
24  f = factor(s, n);
25  printf("\nFactors in array s: \n");
26  for (j = 0; j < N-1; j++)
27      printf("%f ", f[j]);
28  meanv = mean(f, n-1);
29  printf("\n\nMean factor: %f \n", meanv);
30  cs[0] = s[0];
31  for (j=0; j < N-1; j++)
32      cs[j+1] = meanv * cs[j];
33  for (j = 0 ; j < N; j++)
34      printf("%d %f \n", x[j], cs[j]);
35  return 0;
36  }  // end main
```

14.2.4 Case Study 2

Consider part of a water treatment process in which every application of solvents removes 65% of impurities from the water, to make it more acceptable for human consumption. This treatment has to be performed several times until the level of purity of the water is adequate for human consumption. Assume that when the water has less than 0.6 parts per gallon of impurities, it is adequate for human consumption.

In this problem, the data given is the contents of impurities in parts per gallon of water. The initial data is 405 parts per gallon of impurities and the growth factor is 0.35.

The C program shown below declares the data lists (arrays) of the sequence s with the data of the contents of impurities in parts per gallon of water The output data produced by executing the program and GnuPlot is used to plot the graphs. The program is stored in file: `watertr.c`. When the program executes, the following output listing is produced.

```
Program to compute growth of impurities in water
Initial impurity in water: 405.000000
Number of applications: 10
Factor: 0.350000
Impurities after each application
1.000000 405.000000
2.000000 141.750000
3.000000 49.612500
4.000000 17.364375
5.000000 6.077531
6.000000 2.127136
7.000000 0.744498
8.000000 0.260574
9.000000 0.091201
10.000000 0.031920
```

Figure 14.4 shows the graph of the impurities in parts per gallon of water for several applications of solvents. Listing 14.2 shows the C program that compute the impurities in parts per gallon of water, and draws the graphs.

Listing 14.2: C source program that computes the impurities of water.

```
 1 /* program: watertr.c
 2    This program computes the impurities in water given the growth
          factor.
 3    Uses 'basic_lib'
 4    J Garrido 11-2-2012
 5 */
 6 #include <stdio.h>
 7 #include <math.h>
 8 #include "basic_lib.h"
 9
10 int main() {
11     const int N = 10; /* Number of applications */
12     double s[N];
13     double x[N];
14     double f = 0.35;      /* constant factor */
15     int n;
16     int j;
17     double xf;
18     s[0] = 405.0; /* initial impurity of water */
19     printf("Program to compute growth of impurities in water\n\n");
```

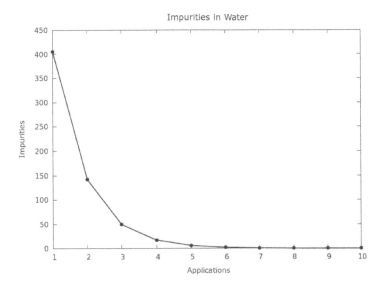

FIGURE 14.4: Impurities in water (parts/gallon).

```
20    xf = N;
21    n = arraycol(x, 1.0, xf, 1.0);
22    printf("Initial impurity in water: %f\n", s[0]);
23    printf("Number of applications: %d\n", N);
24    printf("\n\nFactor: %f \n", f);
25
26    for (j=0; j < N-1; j++)
27        s[j+1] = f * s[j];
28    printf("Impurities after each application \n");
29    for (j = 0 ; j < N; j++)
30        printf("%f %f \n", x[j], s[j]);
31    return 0;
32 }  // end main
```

14.3 Functional Equations

From the difference equation for models with geometric growth, Equation 14.1, the first few terms of sequence *s* can be written as:

$$s_2 = cs_1$$
$$s_3 = cs_2 = c(cs_1)$$
$$s_4 = cs_3 = c(c(cs_1))$$
$$s_5 = cs_4 = c(c(c(cs_1)))$$
$$s_6 = cs_5 = cc(c(c(cs_1)))$$
$$\ldots$$
$$s_n = c^{n-1}s_1$$

Equation 14.1 was referenced by substituting s_{n-1} for its difference equation, and continuing this procedure up to s_n. In this manner, a functional equation can be derived. Recall that a functional equation gives the value of a term s_n without using the previous value, s_{n-1}. The following mathematical expression is a general functional equation for geometric growth models.

$$s_n = s_1\, c^{n-1} \tag{14.2}$$

Equation 14.2 gives the value s_n as a function of n for a geometric growth model, with $n \geq 1$. Note that this functional equation includes the fixed value s_1, which is the value of the first term of the data sequence.

A functional equation such as Equation 14.2 is an example of an exponential function because the independent variable, n, is the exponent. This type of growth in the data is also known as *exponential growth*.

Functional equations can be used to answer additional questions about a model. For example: what will the population be 12 years from now? What amount of impurities are left in the water after 8 repetitions of the application of solvents?

When the growth factor does not correspond to the desired unit of time, then instead of n, a more appropriate variable can be used. For example the first population data in Case Study 1, Section 14.2.3, the variable n represents number of years. To deal with months instead of years, a small substitution in the functional equation is needed. Variable t will represent time, and the starting point of the data is at $t = 0$ with an initial value of y_0. This gives meaning to the concept of a continuous model. Because one year has 12 months and using the same growth factor c as before, the following is a modified functional equation can be applied when dealing with months.

$$y(t) = y_0\, c^{t/12} \tag{14.3}$$

Summary

This chapter presented some basic concepts of geometric growth in mathematical models. The data in these models increase or decrease in a constant growth factor for equal intervals. Simple mathematical techniques such as difference equations and functional equations are used in the study of models with geometric growth. The functional equation of geometric growth is basically an exponential function, so this type of growth is also known as exponential growth.

Some important applications involving computational models with geometric growth are: pollution control, human drug treatment, population growth, radioactive decay, and heat transfer.

Key Terms		
geometric growth	exponential growth	average growth
growth factor	exponent	continuous data
exponentiation rules	logarithms	exponent base

Exercises

Exercise 14.1   In the population problem Case Study 1 Section 14.2.3, use the average growth factor already calculated and compute the population up to year 14. For this, modify the corresponding C source file then compile, link, and execute. Draw the graphs with GnuPlot.

Exercise 14.2   In the population problem Case Study 1 Section 14.2.3, use the average growth factor already calculated and compute the population up to year 20. For this, modify the corresponding C source file then compile, link, and execute. Draw the graphs with GnuPlot.

Exercise 14.3   In the population problem Case Study 1 Section 14.2.3, estimate the population for year 18 and month 4. Use the average growth factor already calculated and the modified functional equation, Equation 14.3. Develop a C program for this.

Exercise 14.4   In the population problem Case Study 1 Section 14.2.3, estimate the population for month 50. Use the average growth factor already calculated and the modified functional equation, Equation 14.3. Develop a C program for this.

Exercise 14.5   In the population problem Case Study 1 Section 14.2.3, compute the year when the population reaches 750,000. Use the average growth factor and develop a C program for this.

Exercise 14.6   In the population problem Case Study 1 Section 14.2.3, compute the month when the population reaches 875,000. Use the average growth factor already calculated and the modified functional equation, Equation 14.3. Develop a C program for this.

Exercise 14.7   In a modified water treatment problem Case Study 2 Section 14.2.4, an application of solvents removes 57% of impurities in the water. Compute the levels of impurities after several repetitions of the application of solvents. Use this growth factor and develop a C program for this.

Exercise 14.8   In the original water treatment problem Case Study 2 Section 14.2.4, compute the levels of impurities after 8 repetitions of the application of solvents. Use this growth factor and develop a C program for this.

Exercise 14.9   In the modified water treatment problem, Exercise 14.8, compute the levels of impurities after 8 repetitions of the application of solvents. Use the growth factor and develop a C program for this.

Exercise 14.10   In the original water treatment problem Case Study 2 Section 14.2.4, compute the number of repetitions of the application of solvents that are necessary to reach 0.5 parts per gallon of impurities. Use the growth factor and develop a C program for this.

Chapter 15

Computational Models with Polynomials

15.1 Introduction

Linear and quadratic equations are special cases of polynomial functions. These equations are of higher order than quadratic equations and more general mathematical methods are used to solve them.

This chapter presents general concepts and techniques to evaluate and solve polynomial functions with emphasis on numerical and graphical methods. Solutions implemented with the C programming language and the Gnu Scientific Library (GSL) are presented and discussed.

15.2 General Forms of Polynomial Functions

Linear and quadratic equations are special cases of polynomial functions. The degree of a polynomial function is the highest degree among those in its terms. A linear function such as: $y = 3x + 8$, is a polynomial equation of degree 1 and a quadratic equation, such as: $y = 4.8x^2 + 3x + 7$, is a polynomial function of degree 2; thus the term "second degree equation".

A function such as: $y = 2x^4 + 5x^3 - 3x^2 + 7x - 10.5$, is a polynomial function of degree 4 because 4 is the highest exponent of the independent variable x. A polynomial function has the general form:

$$y = p_1 x^n + p_2 x^{n-1} + p_3 x^{n-2} + \ldots p_{k-1} x + p_k$$

This function is a polynomial equation of degree n, and $p_1, p_2, p_3, \ldots p_k$ are the coefficients of the equation and are constant values.

In addition to the algebraic form of a polynomial function, the graphical form is also important; a polynomial function is represented by a graph. As mentioned previously, *Gnuplot* is used for plotting data.

15.3 GNU Scientific Library

The GNU Scientific Library (GSL) is a collection of functions for numerical computing. The functions have been written from scratch in C, and present a modern Applications Programming Interface (API) for C programmers, allowing wrappers to be written for very high level languages. The source code is distributed under the GNU General Public License.

The library of functions covers a wide range of topics in numerical computing. The following list includes some of these topics:

- Complex Numbers Roots of Polynomials
- Special Functions Vectors and Matrices
- Permutations Combinations
- Sorting BLAS Support
- Linear Algebra
- Fast Fourier Transforms
- Eigensystems
- Random Numbers Quadrature
- Random Distributions Quasi-Random Sequences
- Histograms Statistics

- Monte Carlo Integration N-Tuples
- Differential Equations Simulated Annealing
- Numerical Differentiation Interpolation
- Series Acceleration Chebyshev Approximations
- Root-Finding Discrete Hankel Transforms
- Least-Squares Fitting Minimization
- IEEE Floating-Point
- Basis Splines Wavelets

15.4 Evaluation of Polynomial Functions

A polynomial function is evaluated by using various values of the independent variable x and computing the value of the dependent variable y. In a general mathematical sense, a polynomial function defines y as a function of x. With the appropriate expression, several values of polynomial can be computed and graphs can be produced.

The GSL functions evaluate a polynomial, $P(x)$, using Horner's method for stability. The general form of the polynomial is:

$$P(x) = c_0 + c_1 x + c_2 x^2 + \cdots + c_{n-1} x^{n-1}$$

Function *gsl_poly_eval* is the main function for evaluating polynomials with real coefficients for the real variable x. The prototype of this function is:

```
double gsl_poly_eval (const double c[], const int len, const double x);
```

With a computer program, a relatively large number of values of x can be used to evaluate the polynomial function for every value of x. The set of values of x that are used to evaluate a polynomial function are taken from an interval $a \le x \le b$. Where a is the lower bound of the interval and b is the upper bound. The interval is known as the *domain* of the polynomial function. In a similar manner, the interval of the values of the function y is known as the *range* of the polynomial function.

To evaluate a polynomial function $y = f(x)$ with the C programming language and the GSL, two arrays are defined: one with values of x and the other with values of y.

For example, the polynomial function $y = 2x^3 - 3x^2 - 36x + 14$ is evaluated and the values in the two arrays x and y are computed by the following C program. Only 20 values of x are evaluated in this example. The following listing is the output produced by executing the program.

```
Evaluating polynomial with coefficients:
14.000000 -36.000000 -3.000000 2.000000
Data points calculated of the polynomial
x0  = -6.000000000000000000 y0  = -310.000000000000000000
x1  = -5.315789473684210600 y1  = -179.827525878407950000
x2  = -4.631578947368421200 y2  = -82.326578218399206000
x3  = -3.947368421052631900 y3  = -13.653448024493390000
x4  = -3.263157894736842500 y4  = +30.035573698789893000
x5  = -2.578947368421053100 y5  = +52.584195946931032000
x6  = -1.894736842105263700 y6  = +57.836127715410413000
x7  = -1.210526315789474300 y7  = +49.635077999708429000
x8  = -0.526315789473684850 y8  = +31.824755795305457000
x9  = +0.157894736842104640 y9  = +8.248870097681900000
x10 = +0.842105263157894130 y10 = -17.248870097681856000
x11 = +1.526315789473683600 y11 = -40.824755795305421000
x12 = +2.210526315789473000 y12 = -58.635077999708400000
x13 = +2.894736842105262400 y13 = -66.836127715410413000
x14 = +3.578947368421051800 y14 = -61.584195946931054000
x15 = +4.263157894736841600 y15 = -39.035573698789932000
x16 = +4.947368421052631000 y16 = +4.653448024493316700
x17 = +5.631578947368420400 y17 = +73.326578218399092000
x18 = +6.315789473684209700 y18 = +170.827525878407810000
x19 = +6.999999999999999100 y19 = +300.999999999999830000
```

Evaluating polynomial functions with the GSL library involves calling function *gsl_poly_eval* in a C program. A polynomial function is represented by a vector of *coefficients* in ascending order. Function *gsl_poly_eval* takes the coefficient vector and the vector x as arguments. The function computes the values of the polynomial (values of y) for all the given values of x. For example, for the polynomial equation $y = 2x^3 - 3x^2 - 36x + 14$, the coefficient vector is $[14, -36, -3, 2]$.

Listing 15.1 shows the source code of the C program is stored in file polyval1.c. The program evaluates the polynomial and computes the values in vector *y*. Line 18 defines the array with the polynomial coefficients: *a*, *b*, *c*, and *d*. Line 29 defines array *x* with 20 different values from the interval $-6.0 \leq x \leq 7.0$. Line 30 calls a local function *polyval* that computes the value of the function for every value in array *x* by calling the GSL function *gsl_poly_eval*. Function *polyval* uses a loop to compute the value of a polynomial function for every value of the independent variable *x*.

Listing 15.1: A program that computes the values of a polynomial.

```
 1 /* File: polyval1.c
 2  This program evaluates a polynomial given in the form of
 3  a coefficient vector.
 4  The program uses the GSL library
 5  J. M. Garrido, Feb 23 2013
 6 */
 7 #include <stdio.h>
 8 #include <gsl/gsl_poly.h>
 9 #include "basic_lib.h"
10 #define N 4   // number of coefficients
11 #define M 20
12 void polyval(double p[], double xv[], double yv[], int n, int m);
13 /* const int N = 4;    num of coefficients of polynomial */
14 int main ()   {
15     int i;
16     int j;
17     /* coefficients of Polynomial  */
18     double a[N] = {14.0, -36.0, -3.0, 2.0};  /* coefficients */
19     double xi, xf;
20     double x[M];
21     double y[M];
22     printf("Evaluating polynomial with coefficients: \n");
23     for (i=0; i < N; i++)
24        printf("%f ", a[i]);
25     printf(" \n");
26     /* Evaluate the polynomial at the following points in x */
27     xi = -6.0;  /* first value of x */
28     xf = 7.0;   /* final value of x */
29     linspace(x, xi, xf, M);    // vector of values for x
30     polyval(a, x, y, N, M);
31     printf(" \n");
32     printf("Data points calculated of the polynomial x y \n");
33     for(j=0; j < M; j++)
34        printf("%+.18f %+.18f \n", x[j], y[j]);
35     return 0;
36 } /* end main */
37 void polyval (double p[], double xv[], double yv[], int n, int m) {
38     int j;
39     yv[0] = gsl_poly_eval(p, N, xv[0]);
40     for (j = 1; j < M; j++) {
41        yv[j] = gsl_poly_eval(p, N, xv[j]);
42     }
43     return;
44 } /* end polyval */
```

The following shell commands in Linux are used to compile, link with the GSL, and execute the program.

```
$ gcc -Wall polyval1.c -o polyval1.out basic_lib.o -lgsl -lgslcblas -lm
$ ./polyval1.out
```

To evaluate the polynomial equation $y = 3x^5 - 2$, the corresponding coefficient vector is $[-2, 0, 0, 0, 0, 3]$. The program calls function *polyval()* to evaluate the polynomial in the interval $-2.5 \leq x \leq 2.5$. The program is basically the same as the previous example. Executing this program produces the following output:

```
Polynomial with coefficients:
-2.000000 0.000000 0.000000 0.000000 0.000000 3.000000
Data points calculated of the polynomial
x0  = -2.500000000000000000  y0  = -294.968750000000000000
x1  = -2.236842105263158000  y1  = -169.995597295887620000
x2  = -1.973684210526315900  y2  = -91.848242954441673000
x3  = -1.710526315789473900  y3  = -45.930895369591468000
x4  = -1.447368421052631900  y4  = -21.055340742131090000
x5  = -1.184210526315789800  y5  = -8.986599372137392900
x6  = -0.921052631578947790  y6  = -3.988581951388054900
x7  = -0.657894736842105750  y7  = -2.369745855779596500
x8  = -0.394736842105263660  y8  = -2.028751437745421400
x9  = -0.131578947368421570  y9  = -2.000118318673849600
x10 = +0.131578947368420520  y10 = -1.999881681326150600
x11 = +0.394736842105262610  y11 = -1.971248562254578800
x12 = +0.657894736842104640  y12 = -1.630254144220406600
x13 = +0.921052631578946680  y13 = -0.011418048611957127
x14 = +1.184210526315788700  y14 = +4.986599372137360000
x15 = +1.447368421052630700  y15 = +17.055340742131015000
x16 = +1.710526315789472800  y16 = +41.930895369591326000
x17 = +1.973684210526314800  y17 = +87.848242954441417000
x18 = +2.236842105263157100  y18 = +165.995597295887310000
x19 = +2.499999999999999100  y19 = +290.968749999999490000
```

A graph can easily be produced with Gnuplot using the values in arrays of x and y computed. Figure 15.1 shows the graph of the polynomial function $y = 2x^3 - 3x^2 - 36x + 14$ with the data computed previously. Using Gnuplot and executing the script files polyval1.cgp produces the graph in Figure 15.1.

In a similar manner, executing the command file polyval2.cgp with Gnuplot produces a graph of polynomial equation $y = 3x^5 - 2$. Figure 15.2 shows the graph of this polynomial function.

15.5   Solving Polynomial Equations

The solution to quadratic equation, which is a second degree equation, is relatively straight forward. The solution to this equation involves finding two values of x

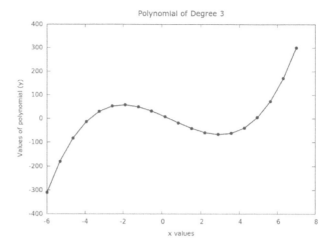

FIGURE 15.1: Graph of the equation $y = 2x^3 - 3x^2 - 36x + 14$.

that give y value zero. These two values of x are known as the *roots* of the function.
For higher order polynomial functions, the degree of the polynomial determines the
number of roots of the function. A polynomial function of degree 7 will have 7 roots.

It is not generally feasible to find the roots of polynomial equations of degree
4 or higher by analytical methods. The GSL implements an algorithm that uses an
iterative method to find the approximate locations of roots of polynomials. The GSL
includes library function *gsl_poly_complex_solve*, that computes the complex roots
of a polynomial function. This function takes the coefficients vector of the polynomial function as one of the arguments, and returns a vector with the roots. This
vector is known as the roots vector. Listing 15.3 shows the C source code of a program that computes the roots of a polynomial function. The program is stored in file
`polyrootsc.c`.

Listing 15.2: A C program that computes the roots of a polynomial function.

```
 1 /* Program: polyrootsc.c
 2   This program solves a polynomial given in the form of
 3   a coefficient vector and uses the GSL library.
 4   The program computes the (complex) roots.
 5   J. M. Garrido, Jan 5 2013
 6 */
 7 #include <stdio.h>
 8 #include <gsl/gsl_poly.h>
 9 #define N 5   // number of coefficients
10 int main () {
11     int i;
12     /* coefficients of Polynomial  */
13     double a[N] = {3.0, 3.0, -14.0, -17.0, 23.0};
14     double z[2*N];                        /* complex roots */
15     printf("Program solves a polynomial");
16     printf(" with coefficients: \n");
```

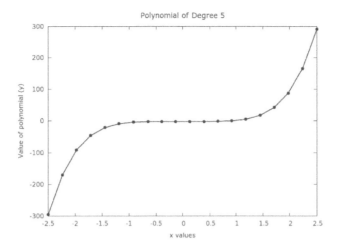

FIGURE 15.2: Graph of the equation $y = 3x^5 - 2$.

```
17      for (i=0; i < N; i++)
18          printf("%f ", a[i]);
19      printf(" \n");
20      gsl_poly_complex_workspace * w
21          = gsl_poly_complex_workspace_alloc (N);
22      gsl_poly_complex_solve (a, N, w, z);
23      gsl_poly_complex_workspace_free (w);
24      printf("\nThe roots of polynomial are: \n");
25      for (i = 0; i < N-1; i++)
26          printf ("z%d %+.18f %+.18f\n", i, z[2*i], z[2*i+1]);
27      return 0;
28 } /* end main */
```

For the polynomial function: $y = 2x^3 - 3x^2 - 36x + 14$, the following output was produced when executing the program. The solution vector, z, has the values of the roots of the polynomial function. The individual values of vector z are: $s_1 = 4.8889$, $s_2 = -3.7688$, and $s_3 = 0.3799$.

```
Program solves a polynomial with coefficients:
14.000000 -36.000000 -3.000000 2.000000
The roots of polynomial are:
z0 = +0.379907627008756910 +0.000000000000000000
z1 = +4.888924027976250000 +0.000000000000000000
z2 = -3.768831654985008100 +0.000000000000000000
```

The solution to the polynomial equation $y = 3x^5 - 2$ is computed in a similar manner executing the C program. Four of the roots computed are complex values, which are known as *complex roots*. Executing the command file produces the following results:

```
Program solves a polynomial with coefficients:
-2.000000 0.000000 0.000000 0.000000 0.000000 3.000000
The roots of polynomial are:
z0 = -0.746000971036307360 +0.542001431391172630
z1 = -0.746000971036307360 -0.542001431391172630
z2 = +0.284947015295443530 +0.876976737942012030
z3 = +0.284947015295443530 -0.876976737942012030
z4 = +0.922107911481727880 +0.000000000000000000
```

The following output was produced after computing the roots of the polynomial function: $y = 23x^4 - 17x^3 - 14x^2 + 3x + 3$.

```
Program solves a polynomial with coefficients:
3.000000 3.000000 -14.000000 -17.000000 23.000000
The roots of polynomial are:
z0 = +0.525288237587778870 +0.000000000000000000
z1 = -0.436656945625196860 +0.194254427938322930
z2 = -0.436656945625196860 -0.194254427938322930
z3 = +1.087156088445222500 +0.000000000000000000
```

Summary

This chapter presented some basic techniques for solving polynomial equations that are very useful in computational modeling. Linear and quadratic equations are special cases of polynomial equations. The concepts discussed apply to polynomial functions of any degree. The main emphases of the chapter are: evaluation of a polynomial function, the graphs of these functions, and solving the functions by computing the roots of the polynomial functions. The C programming language and the GSL are used for several case studies.

Key Terms

polynomial functions	polynomial evaluation	roots
coefficient vector	root vector	evaluation interval
function domain	function range	variable interval

Exercises

Exercise 15.1   Develop a C program to evaluate the polynomial function $y = x^4 + 4x^2 + 7$. Find an appropriate interval of x for which the function evaluation is done and plot the graph.

Exercise 15.2   Develop a C program to evaluate the polynomial function $y = 3x^5 + 6$. Find an appropriate interval of x for which the function evaluation is done and plot the graph.

Exercise 15.3   Develop a C program to evaluate the polynomial function $y = 2x^6 - 1.5x^5 + 5x^4 - 6.5x^3 + 6x^2 - 3x + 4.5$. Find an appropriate interval of x for which the function evaluation is done and plot the relevant data.

Exercise 15.4   Develop a C program to solve the polynomial function $y = x^4 + 4x^2 + 7$.

Exercise 15.5   Develop a C program to solve the polynomial function $y = 3x^5 + 6$.

Exercise 15.6   Develop a C program to solve the polynomial function $y = 2x^6 - 1.5x^5 + 5x^4 - 6.5x^3 + 6x^2 - 3x + 4.5$.

Chapter 16

Models with Interpolation and Curve Fitting

16.1 Introduction

The mathematical model in a computational model can be expressed as a set of polynomial functional equations of any order. If only raw data is available, estimates of the values of the function can be computed for other values of the independent variable, within the bounds of the available set of values of the independent variable. Computing these estimates is carried out using *interpolation* techniques.

If the mathematical model in the form a polynomial function is needed to represent the raw data, then *curve fitting*, also known as *regression*, techniques are used. In the general case, the coefficients of the polynomial can be computed (estimated) for a polynomial function of any degree. This chapter discusses two general numerical techniques that help estimate data values: interpolation and curve fitting. Solutions implemented with the GSL and the C programming language are presented and discussed.

16.2 Interpolation

The given or raw data in a problem usually provides only a limited number of values of data points (x, y). These are normally values of a function y for the corresponding values of variable x. As mentioned previously, using an interpolation technique, one can estimate intermediate data points. These intermediate values of x and y are not part of the original data. The GSL includes a set of functions for performing interpolation of several types. Two well-known interpolation techniques are:

- linear interpolation
- cubic spline interpolation.

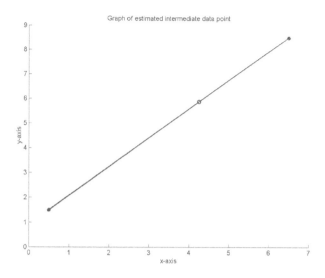

FIGURE 16.1: Linear interpolation of an intermediate data point.

16.2.1   Linear Interpolation

In linear interpolation technique the assumption is that the intermediate data point (value of the function y for a value of x), between two known data points: (x_1, y_1) and (x_2, y_2), can be estimated by a straight line between the known data points. In other words, the intermediate data point (x_i, y_i) falls on the straight line between the known points: (x_1, y_1) and (x_2, y_2).

Assume that there are two data points: $(0.5, 1.5)$ and $(6.5, 8.5)$. An intermediate data point between the given points is to be estimated for $x = 4.25$. Applying a linear interpolation technique, the estimated value computed for y is 5.875. The new intermediate data point is therefore $(4.25, 5.875)$. Figure 16.1 illustrates the technique of estimating an intermediate data point on a straight line between two given data points.

The GSL has several library functions that perform linear interpolation given three vectors: x, y, and *xint*. The first two vectors store the values of the given data points. The third vector, *xint*, stores the new or intermediate values of x for which estimates are to be computed and these are new or intermediate data points. The following C program computes the values of the intermediate data points and stores these in vector *yint*. In the previous example, vector *xint* has only a single value, 4.25 and the value of *yint* computed is: 5.875.

In the following problem, there are eight given data points in vectors: x and y. Vector x has equally spaced values of x starting at 0 and increasing by 1. Linear interpolation is used to estimate intermediate data points for every value of x spaced 0.25, starting at 0 and up to 7. The values given in the problem are stored in the three given vectors, x, y, and *xint*, and the values in vector *yint* are computed using linear interpolation. The values given and the values computed using linear interpolation are shown as follows:

```
Discrete Points for Interpolation
 i          x                        y
 0:    0.0000000000000000e+000    0.0000000000000000e+000
 1:    1.0000000000000000e+000    3.0000000000000000e+000
 2:    2.0000000000000000e+000    6.0000000000000000e+000
 3:    3.0000000000000000e+000    8.0000000000000000e+000
 4:    4.0000000000000000e+000    1.2000000000000000e+001
 5:    5.0000000000000000e+000    1.7000000000000000e+001
 6:    6.0000000000000000e+000    2.3000000000000000e+001
 7:    7.0000000000000000e+000    2.6000000000000000e+001
```

```
           x                   interpolated y
0.0000000000000000e+000    0.0000000000000000e+000
2.5000000000000000e-001    7.5000000000000000e-001
5.0000000000000000e-001    1.5000000000000000e+000
7.5000000000000000e-001    2.2500000000000000e+000
1.0000000000000000e+000    3.0000000000000000e+000
1.2500000000000000e+000    3.7500000000000000e+000
1.5000000000000000e+000    4.5000000000000000e+000
1.7500000000000000e+000    5.2500000000000000e+000
2.0000000000000000e+000    6.0000000000000000e+000
2.2500000000000000e+000    6.5000000000000000e+000
2.5000000000000000e+000    7.0000000000000000e+000
2.7500000000000000e+000    7.5000000000000000e+000
3.0000000000000000e+000    8.0000000000000000e+000
3.2500000000000000e+000    9.0000000000000000e+000
3.5000000000000000e+000    1.0000000000000000e+001
3.7500000000000000e+000    1.1000000000000000e+001
4.0000000000000000e+000    1.2000000000000000e+001
4.2500000000000000e+000    1.3250000000000000e+001
4.5000000000000000e+000    1.4500000000000000e+001
4.7500000000000000e+000    1.5750000000000000e+001
5.0000000000000000e+000    1.7000000000000000e+001
5.2500000000000000e+000    1.8500000000000000e+001
5.5000000000000000e+000    2.0000000000000000e+001
5.7500000000000000e+000    2.1500000000000000e+001
6.0000000000000000e+000    2.3000000000000000e+001
6.2500000000000000e+000    2.3750000000000000e+001
6.5000000000000000e+000    2.4500000000000000e+001
6.7500000000000000e+000    2.5250000000000000e+001
7.0000000000000000e+000    2.6000000000000000e+001
```

The estimated intermediate values were computed in program linterp1f.c, which appears in Listing 16.1. Line 31 in the listing calls function *linterp* to compute the estimated intermediate points using linear interpolation.

Figure 16.2 shows a plot of the data given in the problem. Figure 16.3 illustrates the linear interpolation of several intermediate data points given two arrays of the original data points *x* and *y*.

Listing 16.1: C program that computes linear interpolation of data points.

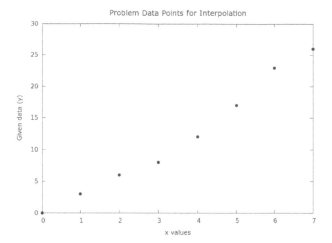

FIGURE 16.2: Graph of given data points.

```
 1 /* Program: linterp1f.c
 2    Interpolation of Discrete Points
 3    Using linear interpolation with GSL
 4 */
 5 #include <stdio.h>
 6 #include <math.h>
 7 #include <gsl/gsl_interp.h>
 8 #include "basic_lib.h"
 9 /* Number of Discrete Points (original data */
10 #define N 8
11 void linterp(double x[], double y[], double nx[], double ny[],
        int m, int n);
12 int main() {
13     int i;
14     int n;
15     double h;
16     double x[N];
17     double xi, xf;
18     double y[N] = {0.0, 3.0, 6.0, 8.0, 12.0, 17.0, 23.0, 26};
19     double xint[30];
20     double yint[30];
21     /* Data for Linear Interpolation */
22     xi = 0.0;
23     xf = 7.0;
24     n = arraycol(x, xi, xf, 1.0);
25     printf("Discrete Points for Interpolation\n");
26     printf("  i                x                        y\n");
27     for(i = 0; i < N; i++)
28         printf("%3d %25.17e %25.17e\n", i, x[i], y[i]);
29     printf("\n");
30     h = 0.25;
31     n = arraycol(xint, xi, xf, h);    /* fill new array */
32     linterp(x, y, xint, yint, N, n); /* linear interpolation */
```

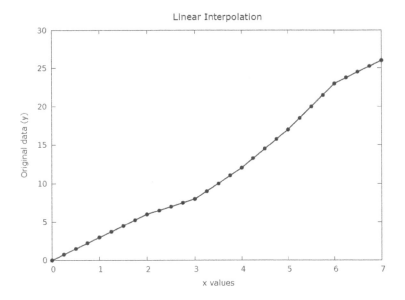

FIGURE 16.3: Graph of linear interpolation of multiple data points.

```
33     printf("                  x                    interpolated y  \n");
34     for(i = 0; i < n; i++)
35          printf("%25.17e %25.17e \n", xint[i], yint[i]);
36     return 0;
37 }
38 void linterp(double x[], double y[], double nx[], double ny[],
                int m, int n) {
39     int i;
40     gsl_interp *workspace;
41     /* Define type of Interpolation: Linear */
42     workspace = gsl_interp_alloc(gsl_interp_linear, m);
43     /* Initialize workspace */
44     gsl_interp_init(workspace, x, y, m);
45     for(i = 0; i < n; i++)
46          ny[i] = gsl_interp_eval(workspace, x, y, nx[i], NULL);
47     gsl_interp_free(workspace);   /* free memory */
48 }
```

16.2.2 Non-Linear Interpolation

Non-linear interpolation can generate more improved estimates for intermediate data points than linear interpolation. The GSL includes functions that implement the *cubic spline* interpolation technique, and smoother curves can be generated using this technique.

The following example applies the *cubic spline* interpolation technique to data provided by the problem, using an array of values of *nx* for intermediate data points.

The C program in file `csinterpf.c` is shown in Listing 16.2. Line 32 calls function *csinterp* to compute the interpolation using the cubic spline technique. This program is very similar to the previous program that performed linear interpolation. The first argument used in calling the library function *gsl_interp_alloc* is the gsl constant *gsl_interp_cspline*, which specifies the type of interpolation to apply.

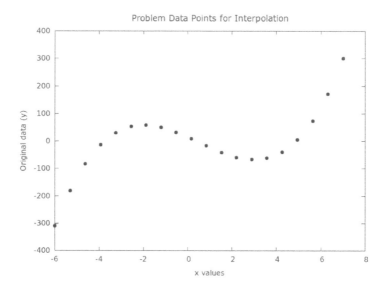

FIGURE 16.4: Graph of given data points.

Listing 16.2: Script to compute cubic spline interpolation of data.

```
1 /* Program: csinterpf.c
2    Interpolation of Discrete Points
3    Using Cubic Spline interpolation with GSL
4    J. M. Garrido, Feb 22, 2013 */
5 #include <stdio.h>
6 #include <math.h>
7 #include <gsl/gsl_interp.h>
8 #include "basic_lib.h"
9 /* Number of Discrete Points (original data) */
10 #define N 20
11 void csinterp(double x[], double y[], double nx[], double ny[],
           int m, int n);
12 int main() {
13     int i;
14     int n;
15     double h;
16     double x[N];
17     double xi, xf;
18     double y[N] = {-310.0, -179.8, -82.3, -13.6, 30.0, 52.6,
           57.8, 49.6, 31.8, 8.2, -17.2, -40.8, -58.6, -66.8, -61.5,
           -39.0, 4.6, 73.3, 170.8, 301.0};
```

```
19      double xint[2*N], yint[2*N];
20      /* Linear Interpolation */
21      xi = -6.0;   /* first value  of x */
22      xf = 7.0;   /* last value */
23      linspace(x, xi, xf, N);
24      printf("Discrete Points for Cubic Spline Interpolation\n");
25      printf("  i            x                      y\n");
26      for(i = 0; i < N; i++)
27          printf("%3d %25.17e %25.17e\n", i, x[i], y[i]);
28      printf("\n");
29      /* Evaluation */
30      h = 0.45;   // x increment
31      n = arraycol(xint, -5.0, xf, h);    /* fill new array */
32      csinterp(x, y, xint, yint, N, n);
33      printf("          x                    interpolated y  \n");
34      for(i = 0; i < n; i++)
35          printf("%25.17e %25.17e \n", xint[i], yint[i]);
36      return 0;
37 }
38 void csinterp(double x[], double y[], double nx[], double ny[],
          int m, int n) {
39      int i;
40      gsl_interp *workspace;
41      /* type of Interpolation: Cubic Spline */
42      workspace = gsl_interp_alloc(gsl_interp_cspline, m);
43      gsl_interp_init(workspace, x, y, m); /* Initialize worksp */
44      for(i = 0; i < n; i++)
45          ny[i] = gsl_interp_eval(workspace, x, y, nx[i], NULL);
46      gsl_interp_free(workspace);   /* free memory */
```

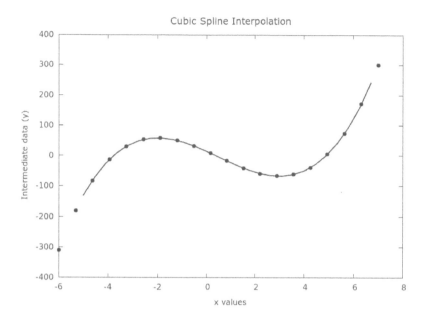

FIGURE 16.5: Cubic spline interpolation of data points.

Figure 16.4 shows the graph of original data points that define a curve. Figure 16.5 shows the graph of the original data points and the computed intermediate data points. Executing the program produces the output shown in the following listing.

```
Discrete Points for Cubic Spline Interpolation
   i             x                         y
   0  -6.00000000000000000e+000  -3.10000000000000000e+002
   1  -5.31578947368421060e+000  -1.79800000000000010e+002
   2  -4.63157894736842120e+000  -8.22999999999999970e+001
   3  -3.94736842105263190e+000  -1.36000000000000000e+001
   4  -3.26315789473684250e+000   3.00000000000000000e+001
   5  -2.57894736842105310e+000   5.26000000000000010e+001
   6  -1.89473684210526370e+000   5.77999999999999970e+001
   7  -1.21052631578947430e+000   4.96000000000000010e+001
   8  -5.26315789473684850e-001   3.18000000000000010e+001
   9   1.57894736842104640e-001   8.19999999999999930e+000
  10   8.42105263157894130e-001  -1.71999999999999990e+001
  11   1.52631578947368360e+000  -4.07999999999999970e+001
  12   2.21052631578947300e+000  -5.86000000000000010e+001
  13   2.89473684210526240e+000  -6.67999999999999970e+001
  14   3.57894736842105180e+000  -6.15000000000000000e+001
  15   4.26315789473684160e+000  -3.90000000000000000e+001
  16   4.94736842105263100e+000   4.59999999999999960e+000
  17   5.63157894736842040e+000   7.32999999999999970e+001
  18   6.31578947368420970e+000   1.70800000000000010e+002
  19   6.99999999999999910e+000   3.01000000000000000e+002
```

```
           x                  interpolated y
-5.00000000000000000e+000  -1.30511433045375100e+002
-4.54999999999999980e+000  -7.27416447500828840e+001
-4.09999999999999960e+000  -2.66706712257389820e+001
-3.64999999999999950e+000   8.22513115126883320e+000
-3.19999999999999930e+000   3.29037511832741460e+001
-2.74999999999999910e+000   4.87185978957122570e+001
-2.29999999999999890e+000   5.66027197527996580e+001
-1.84999999999999900e+000   5.76299663534711540e+001
-1.39999999999999900e+000   5.29894949936771910e+001
-9.49999999999999070e-001   4.37530528217195850e+001
-4.99999999999999060e-001   3.09739878014374380e+001
-4.99999999999990450e-002   1.57392964551063020e+001
 4.00000000000000970e-001  -7.69274148742360550e-001
 8.50000000000000980e-001  -1.74898374122936990e+001
 1.30000000000000090e+000  -3.34378819492079060e+001
 1.75000000000000090e+000  -4.74456021927536400e+001
 2.20000000000000110e+000  -5.83890529200809570e+001
 2.65000000000000120e+000  -6.52157594947413770e+001
 3.10000000000000140e+000  -6.67995125636800400e+001
 3.55000000000000160e+000  -6.20469831049832550e+001
 4.00000000000000180e+000  -4.99229534828566200e+001
 4.45000000000000200e+000  -2.93777940458094270e+001
 4.90000000000000210e+000   8.04590157736302160e-001
 5.35000000000000230e+000   4.18784223404739250e+001
 5.80000000000000250e+000   9.42457931905228180e+001
 6.25000000000000270e+000   1.59837979817333350e+002
 6.70000000000000280e+000   2.41402011410504830e+002
```

16.3 Curve Fitting

Curve fitting techniques attempt to find the best polynomial expression that represents a given sequence of data points. The most widely-used curve fitting technique is the *least squares* technique. Recall that a polynomial function has the general form:

$$y = p_1 x^n + p_2 x^{n-1} + p_3 x^{n-2} + \dots p_{k-1} x + p_k$$

The parameters $p_1, p_2, p_3, \dots p_k$ are the coefficients of the equation, which are constants. If a polynomial function of degree 1 is fitted to the given data, the technique is known as *linear regression*, and a straight line is fitted to the data points. As mentioned previously, if the degree of the polynomial is greater than 1, then a curve, instead of a straight line, is fitted to the given data points.

The GSL provides several functions to compute the coefficients (p) of the polynomial function of degree n. The arguments for the function calls are vectors x and y that define the data points, and the value of the desired degree of the polynomial.

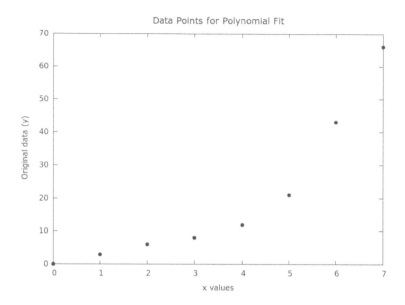

FIGURE 16.6: Graph of given set of data points.

The GSL functions compute a vector that consists of the values of the coefficients of a polynomial function of degree n that best fit the given data points in vectors x and y.

16.3.1   Linear Polynomial Function

The following example illustrates the fitting of a polynomial of degree 1 to a given set of data points. Figure 16.6 shows a given set of data points for which polynomial fit is to be carried out.

Listing 16.4 includes the C program that computes the coefficients of a polynomial of degree 1 by calling function *polynomialfit* in line 27. With the coefficients computed, function *polyval()* is called in line 34 to evaluate the polynomial with additional data points. The C source code of the program code is stored in file `polyfitm1.c`. Using GnuPlot, the computed data points are plotted and Figure 16.7 shows the graph of the fitted line.

Listing 16.4: C program that computes linear regression of given data points.

```
1 /* Program: polyfitm1.c
2    This program fits a polynomial of degree 1 to a set of
3    data points. Then it evaluates the fitted polynomial to
4    generate data for plotting. The program uses the GSL library.
5 */
6 #include <stdio.h>
7 #include <gsl/gsl_poly.h>
8 #include "basic_lib.h"
```

```
 9 #include "polyfitgsl.h"
10 #define NC 2   /* size of coefficient vector */
11 void polyval(double p[], double xv[], double yv[], int n, int m);
12 int main()
13 {
14    const int NP = 8;       // Number of given data points
15    const int M = 30;       // number of computed data points
16    /* const int NC = 2 ;      Number of coefficients */
17    double x[] = {0.0, 1.0, 2.0, 3.0, 4.0, 5.0, 6.0, 7.0};
18    double y[] = {0.0, 3.0, 6.0, 8.0, 12.0, 21.0, 43.0, 66.0};
19    double coeff[NC];  // vector of coefficients
20    int i;
21    double xc[M]; /* computed values for x using polynomial */
22    double yc[M];
23    printf("Program finds best fit polynomial of degree %d \n",
              NC-1);
24    printf("Data points (x,y): \n");
25    for (i = 0; i < NP; i++)
26        printf(" %f %f \n", x[i], y[i]);
27    polynomialfit(NP,  NC, x, y, coeff); /* find coefficients */
28    printf("\n\nCoefficients of polynomial found\n");
29    for(i=0; i < NC; i++) {
30        printf("%lf\n", coeff[i]);
31    }
32    /* Evaluate the fitted polynomial */
33    linspace(xc, 0.0, 7.0, M);
34    polyval(coeff, xc, yc, NC, M);
35    printf("\nData points calculated with the polynomial \n");
36    for(i=0; i < M; i++)
37        printf("%d  %+.18f  %+.18f \n", i, xc[i], yc[i]);
38    return 0;
39 }  /* end main */
40 void polyval (double p[], double xv[], double yv[], int n,
            int m) {
41     int j;
42     yv[0] = gsl_poly_eval(p, n, xv[0]);
43     for (j = 1; j < m; j++)
44         yv[j] = gsl_poly_eval(p, n, xv[j]);
45     return;
46 } /* end polyval */
```

When the program executes, the output produced as shown in the following listing.

```
Program finds best fit polynomial of degree 1
Data points (x,y):
 0.000000 0.000000
 1.000000 3.000000
 2.000000 6.000000
 3.000000 8.000000
 4.000000 12.000000
 5.000000 21.000000
 6.000000 43.000000
 7.000000 66.000000
Coefficients of polynomial found
```

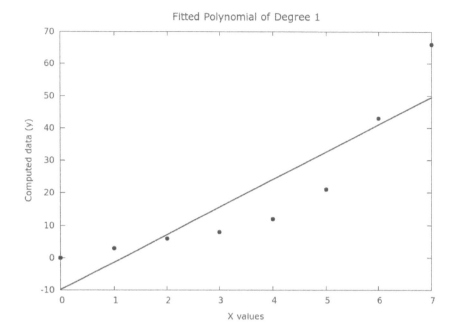

FIGURE 16.7: Polynomial of degree 1 fit from given data points.

```
-9.750000
8.464286
```

16.3.2   Non-Linear Polynomial Function

Executing the program to compute the coefficients of a polynomial of degree 3 by calling function *polynomialfit* using 4 as the value of the second argument (number of coefficients), produces the following output.

```
Program finds best fit polynomial of degree 3
Data points (x,y):
 0.000000 0.000000
 1.000000 3.000000
 2.000000 6.000000
 3.000000 8.000000
 4.000000 12.000000
 5.000000 21.000000
 6.000000 43.000000
 7.000000 66.000000

Coefficients of polynomial found
0.030303
4.908730
```

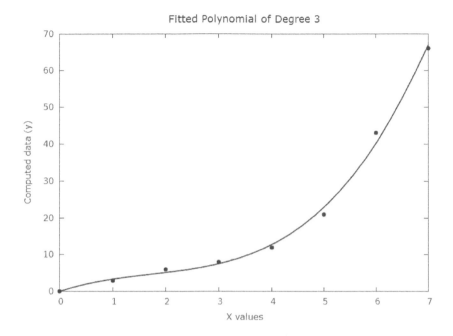

FIGURE 16.8: Curve fitting of polynomial of degree 3.

```
-1.898268
0.366162
```

Figure 16.8 shows the graph of a fitted curve that corresponds to a polynomial of degree 3 with the same data points used with the previous example.

Summary

This chapter presented two important techniques that deal with raw data: interpolation and curve fitting. Interpolation is used to compute estimates of the value of the function for intermediate values of the independent variable, within the bounds of available set of values of the independent variable. Curve fitting or *regression* techniques are used when a polynomial function is needed that would represent the raw data. Solutions implemented with the GSL and the C programming language were presented and discussed.

Key Terms		
intermediate data	estimated data	raw data
linear interpolation	non-linear interpolation	curve fitting
cubic spline interpolation	curve fitting	regression

Exercises

Exercise 16.1   In a typical spring day, the temperature varies according to the data recorded in Table 16.1. Develop a C program and apply linear interpolation to compute estimates of intermediate values of the temperature. Produce a plot of the given and estimated data.

TABLE 16.1: Temperature changes in 12-hour period.

Time	7	8	9	10	11	12	1	2	3	4	5	6
Temp. (F^0)	51	56	60	65	73	78	85	86	84	81	80	70

Exercise 16.2   In a typical spring day, the temperature varies according to the data recorded in Table 16.1. Develop a C program and apply cubic spline interpolation to compute estimates of intermediate values of the temperature. Produce a plot of the given and estimated data.

Exercise 16.3   In a typical spring day, the temperature varies according to the data recorded in Table 16.1. Apply curve fitting to derive the polynomial function of degree 2 that represents the given data. Develop a C program and use the polynomial function to compute estimates of intermediate values of the temperature. Produce a plot of the given and estimated data.

Exercise 16.4   In a typical spring day, the temperature varies according to the data recorded in Table 16.1. Apply curve fitting to derive the polynomial function of degree 3 that represents the given data. Develop a C program and use the polynomial function to compute estimates of intermediate values of the temperature. Produce a plot of the given and estimated data.

Exercise 16.5   In a typical spring day, the temperature varies according to the data recorded in Table 16.1. Apply curve fitting to derive the polynomial function of degree 4 that represents the given data. Develop a C program and use the polynomial function to compute estimates of intermediate values of the temperature. Produce a plot of the given and estimated data.

Exercise 16.6 Develop a C program that uses the data in Table 16.2 and applies linear interpolation to compute estimates of intermediate values of the number of patients. Produce a plot of the given and estimated data.

TABLE 16.2: Number of patients for years 1995–2002.

Year	1995	1996	1997	1998	1999	2000
Patients	5,500	8,500	13,500	20,500	29,500	40,500
Increase	0	3,000	5,000	7,000	9,000	11,000

Year	2001	2002
Patients	53,5000	68,500
Increase	13,000	15,000

Exercise 16.7 Develop a C program and use the data in Table 16.2 and apply non-linear interpolation to compute estimates of intermediate values of the number of patients. Produce a plot of the given and estimated data.

Exercise 16.8 Develop a C program and use the data in Table 16.2 and apply curve fitting to derive a polynomial function of degree 2. Use this function to compute estimates of intermediate values of the number of patients. Produce a plot of the given and estimated data.

Exercise 16.9 Develop a C program and use the data in Table 16.2 and apply curve fitting to derive a polynomial function of degree 3. Use this function to compute estimates of intermediate values of the number of patients. Produce a plot of the given and estimated data.

Chapter 17

Using Vectors with the GSL

17.1   Introduction

This chapter presents an overview of vectors, implementation techniques with programming in C and the GSL, and a summary of computing with vectors. In general, an array is a term used in programming and defined as a data structure that is a collection of values and these values are organized in several ways. The following arrays: X, Y, and Z have their data arranged in different manners. Array X is a one-dimensional array with n elements and it is considered a *row vector* because its elements $x_1, x_2, \ldots, x_n$ are arranged in a single row.

$$X = [x_1 \ x_2 \ x_3 \ \cdots \ x_n] \qquad Z = \begin{bmatrix} z_1 \\ z_2 \\ z_3 \\ \vdots \\ z_m \end{bmatrix}$$

Array Z is also a one-dimensional array; it has m elements organized as a *column vector* because its elements: $z_1, z_2, \ldots, z_m$ are arranged in a single column.

The following array, Y, is a two-dimensional array organized as an $m \times n$ matrix; its elements are arranged in m rows and n columns. The first row of Y consists of elements: $y_{11}, y_{12}, \ldots, y_{1n}$. Its second row consists of elements: $y_{21}, y_{22}, \ldots, y_{2n}$. The last row of Y consists of elements: $y_{m1}, y_{m2}, \ldots, y_{mn}$.

$$Y = \begin{bmatrix} y_{11} & y_{12} & \cdots & y_{1n} \\ y_{21} & y_{22} & \cdots & y_{2n} \\ \vdots & \vdots & \ddots & \vdots \\ y_{m1} & y_{m2} & \cdots & y_{mn} \end{bmatrix}$$

17.2   Vectors and Operations

A vector is a mathematical entity that has magnitude and direction. In physics, it is used to represent items such as the velocity, acceleration, or momentum of a physical object. A vector v can be represented by an n-tuple of real numbers:

$$v = (v_1, v_2, \ldots, v_n)$$

Several operations with vectors are performed with a vector and a scalar or with two vectors.

17.2.1   Addition of a Scalar and a Vector

To add a scalar to a vector involves adding the scalar value to every element of the vector. In the following example, the scalar α is added to the elements of vector Z, element by element.

$$Z = \begin{bmatrix} z_1 \\ z_2 \\ z_3 \\ \vdots \\ z_m \end{bmatrix} \qquad Z + \alpha = \begin{bmatrix} z_1 + \alpha \\ z_2 + \alpha \\ z_3 + \alpha \\ \vdots \\ z_m + \alpha \end{bmatrix}$$

17.2.2   Vector Addition

Vector addition of two vectors that are n-tuple involves adding the corresponding elements of each vector. The following example illustrates the addition of two vectors, Y and Z.

$$Y = \begin{bmatrix} y_1 \\ y_2 \\ y_3 \\ \vdots \\ z_m \end{bmatrix} \qquad Z = \begin{bmatrix} z_1 \\ z_2 \\ z_3 \\ \vdots \\ z_m \end{bmatrix} \qquad Y + Z = \begin{bmatrix} y_1 + z_1 \\ y_2 + z_2 \\ y_3 + z_3 \\ \vdots \\ y_m + z_m \end{bmatrix}$$

17.2.3   Multiplication of a Vector and a Scalar

Scalar multiplication is performed by multiplying the scalar with every element of the specified vector. In the following example, scalar α is multiplied by every element z_i of vector Z.

$$Z = \begin{bmatrix} z_1 \\ z_2 \\ z_3 \\ \vdots \\ z_m \end{bmatrix} \qquad Z \times \alpha = \begin{bmatrix} z_1 \times \alpha \\ z_2 \times \alpha \\ z_3 \times \alpha \\ \vdots \\ z_m \times \alpha \end{bmatrix}$$

17.2.4 Dot Product of Two Vectors

Given vectors $v = (v_1, v_2, \ldots, v_n)$ and $w = (w_1, w_2, \ldots, w_n)$, the dot product $v \cdot w$ is a scalar defined by:

$$v \cdot w = \sum_{i=1}^{n} v_i w_i = v_1 w_1 + v_2 w_2 + \ldots + v_n w_n$$

Therefore, the dot product of two vectors in an n-dimensional real space is the sum of the product of the vectors' components.

When the elements of the vectors are complex, then the dot product of two vectors is defined by the following relation. Note that $\bar{v}_i$ is the complex conjugate of v_i.

$$v \cdot w = \sum_{i=1}^{n} \bar{v}_i w_i = \bar{v}_1 w_1 + \bar{v}_2 w_2 + \ldots + \bar{v}_n w_n$$

17.2.5 Length (Norm) of a Vector

Given a vector $v = (v_1, v_2, \ldots, v_n)$ of dimension n, the Euclidean norm of the vector denoted by $\|v\|_2$, is the length of v and is defined by the square root of the dot product of the vector:

$$\|v\|_2 = \sqrt{v \cdot v} = \sqrt{v_1^2 + v_2^2 + \cdots + v_n^2}$$

In the case that vector v is a 2-dimensional vector, the Euclidean norm of the vector is the value of the hypotenuse of a right angled triangle. When vector v is a 1-dimensional vector, then $\|v\|_2 = |v_1|$, the absolute value of the only component v_1.

17.3   Vector Properties and Characteristics

A vector $v = (v_1, v_2, \ldots, v_n)$ in $\mathscr{R}^n$ (an n-dimensional real space) can be specified as a column or row vector. When v is an n column vector, its transpose v^T is an n row vector.

17.3.1   Orthogonal Vectors

Vectors v and w are said to be orthogonal if their dot product is zero. The angle θ between vectors v and w is defined by:

$$cos(\theta) = \frac{v \cdot w}{\|v\|_2 \; \|w\|_2}$$

where θ is the angle from v to w, non-zero vectors are orthogonal if and only if they are perpendicular to each other, ie when $cos(\theta) = 0$ and θ is equal to $\pi/2$ or 90 degrees. Orthogonal vectors v and w are called *orthonormal* if they are of length one, ie $v \cdot v = 1$, and $w \cdot w = 1$.

17.3.2   Linear Dependence

A set k of vectors $\{x_1, x_2, \ldots, x_k\}$ is linearly dependent if at least one of the vectors can be expressed as a linear combination of the others. Assuming there exists a set of scalars $\{\alpha_1, \alpha_2, \ldots, \alpha_k\}$, vector x_k is defined as follows:

$$x_k = \alpha_1 x_1 + \alpha_2 x_2 + \ldots + \alpha_{k-1} x_{k-1}$$

If a vector w depends linearly on vectors $\{x_1, x_2, \ldots, x_k\}$, this is expressed as follows:

$$w = \alpha_1 x_1 + \alpha_2 x_2 + \ldots + \alpha_k x_k$$

17.4   Implementation of Vectors in C and the GSL

Vectors are created and manipulated in C by calling the various library function in the GSL. Before using a vector, it needs to be created. Function *gsl_vector_alloc* creates a vector of the specified length n, and returns a pointer to the vector. When a vector is no longer needed in the program, it is destroyed by calling the GSL function *gsl_vector_free*.

A vector is manipulated by accessing its elements and changing and/or retrieving the value of the elements. The GSL library function *gsl_vector_set* assigns the value specified to the *i*-th element of a vector. The GSL Function *gsl_vector_set_all* sets the value specified to all elements of a vector.

The value of individual elements of a vector can be retrieved by calling the GSL function *gsl_vector_get*. This function returns the *i*-th element of a vector.

The functions *gsl_vector_get* and *gsl_vector_set* perform portable range checking and report an error if the index value of an element is outside the allowed range.

Listing 17.1 shows the source code of a C program that uses the GSL functions to create and manipulate a vector. Line 6 includes the header file that contains the definitions for calling the GSL functions for vectors. Line 11 calls function *gsl_vector_alloc* to create vector *pv* with *N* elements. In line 12, all elements of vector *pv* are set to value 2.5. In line 17, element *i* of vector *pv* is set to value 5.25. In line 21 vector *pv* is destroyed by calling function *gsl_vector_free*.

Listing 17.1: C program that creates and manipulates a vector.

```
1 /* Program: vector_man.c
2    This program allocates, initializes, sets values to the first
     n elements
3    and retrieves data from a vector using the GSL functions.
4 */
5 #include <stdio.h>
6 #include <gsl/gsl_vector.h>
7 int main (void) {
8    const int N = 8;
9    int i;
10   int n;
11   gsl_vector * pv = gsl_vector_alloc (N); // create vector
12   gsl_vector_set_all(pv, 0.25);  // assign 0.25 to all elements
13   n = N - 3;
14   printf("Program creates and manipulates a vector\n\n");
15   /* set the first n elements of vector to 5.25 */
16   for (i = 0; i < n; i++)
17       gsl_vector_set (pv, i, 5.25); // assign value to element i
18   printf("Data from vector\n\n");
19   for (i = 0; i < N; i++)
20       printf ("element %d = %g\n", i, gsl_vector_get (pv, i));
21   gsl_vector_free (pv);  // release allocated memory
22   return 0;
23 }
```

The following listing shows the Linux shell commands to compile, link, and execute the program, then shows the output produced by the program.

```
gcc -Wall vector_man.c -o vectorm.out -lgsl -lgslcblas -lm
./vectorm.out
Program creates and manipulates a vector
Data from vector
element 0 = 5.25
element 1 = 5.25
element 2 = 5.25
```

```
element 3 = 5.25
element 4 = 5.25
element 5 = 0.25
element 6 = 0.25
element 7 = 0.25
```

17.5   Simple Vector Operations

Operations on vectors are performed with a vector and a scalar or with two vectors. These are performed by calling GSL functions in C.

17.5.1   Arithmetic Operations

To add a scalar to a vector involves adding the scalar value to every element of the vector. With C and GSL, calling function *gsl_vector_add_constant*. This function adds the specified constant value to the elements of the vector specified.

In the following portion of code, vector *pv* is created and then its elements are assigned values. On the fifth line, function *gsl_vector_add_constant* is called to add the constant *x* to all the elements of vector *pv*.

```
const double x = 10.45;
gsl_vector * pv = gsl_vector_alloc (N); // create vector
gsl_vector_set_all(pv, 0.5);   // assign 2.5 to all elements
gsl_vector_set (pv, 3, 5.25);
gsl_vector_add_constant (pv, x);
```

To add two vectors involves adding the corresponding elements of each vector, and a new vector is created. This addition operation on vectors is only possible if the row vectors (or column vectors) are of the same size. In the following example, vectors *pv* and *qv* are both vectors of size *N*. Calling function *gsl_vector_add* adds the elements of vector *qv* to the elements of vector *pv*.

```
gsl_vector * pv = gsl_vector_alloc (N);    // create vector pv
gsl_vector * qv = gsl_vector_alloc (N);    // create vector qv
// ...
gsl_vector_add (pv, qv);        // vector qv added to vector pv
```

In a similar manner, subtracting a scalar from a vector can be specified; subtracting two vectors of the same size. Calling function *gsl_vector_sub*, as in the following line of code, subtracts the elements of vector *qv* from the elements of vector *pv*.

```
gsl_vector_sub (pv, qv);    // vector qv subtracted from vector pv
```

Multiplying a vector by a scalar results in multiplying each element of the vector by the value of the scalar. Function *gsl_vector_scale* multiplies the elements of the specified vector by the constant factor specified. In the following call to this function, the elements of vector *pv* are multiplied by the constant *x*.

```
gsl_vector_scale (pv, x);
```

Element by element multiplication, also known as dot multiplication, multiplies the corresponding elements of two vectors. This operation is applied to two vectors of equal size and is performed by calling the GSL function *gsl_vector_mul*. This function multiplies the elements of the specified vector by the elements of the second specified vector. In the following function call, vectors *pv* and *qv* are dot-multiplied and the results are stored in vector *pv*.

```
gsl_vector_mul (pv, qv);
```

Element by element division, also known as dot division, divides the corresponding elements of two vectors. This operation is applied to two vectors of equal size and is performed by calling the GSL function *gsl_vector_div*. This function divides the elements of the specified vector by the elements of the second specified vector. In the following function call, the elements of vector *pv* are divided by the elements of vector *qv* and the results are stored in vector *pv*.

```
gsl_vector_div (pv, qv);
```

17.5.2 Additional Vector Functions

Various additional functions can be applied to vectors. Copying one vector to another vector is performed by calling the GSL function *gsl_vector_memcpy*. This function copies the elements of the specified source vector into the specified destination vector. The two vectors must have the same length.

In the following function call, the elements of vector *qv* are copied to the corresponding elements of vector *pv* and the results are stored in vector *pv*.

```
gsl_vector_memcpy (pv, qv);
```

The comparison of two vectors for equality is performed by calling the GSL function *gsl_vector_equal*. This function returns 1 if the specified vectors are equal (by comparison of element values) and 0 otherwise.

In the following function call, the elements of vector *qv* are compared to the corresponding elements of vector *pv* and the result is an integer value assigned to variable *vflag*.

```
vflag = gsl_vector_equal (pv, qv);
```

Because there are only two possible outcomes in the call to this function, the return value may be considered a boolean value (true is represented by 1 and false is represented by 0), which can be used in a conditional statement.

The attribute *size* of a vector stores the number of elements in the vector. The following line of C code gets the size of vector *pv* and assigns this value to *n*.

```
n = pv->size;
```

Function *gsl_vector_max* gets the maximum value stored in the specified vector. The following function call gets the maximum value in vector *pv* and assigns this value to *x*, which should be a variable of type *double*.

```
x = gsl_vector_max(pv);
```

In addition to the maximum value in a vector, the index of the element with that value may be desired. Calling function *gsl_vector_max_index* returns the index value of the element with the maximum value in a specified vector. In the following function call, the index value of the element with the maximum value in vector *pv* is returned and assigned to integer variable *idx*.

```
idx = gsl_vector_max_index (pv);
```

In a similar manner, function *gsl_vector_min* gets the minimum value (of type *double*) stored in a vector. Function *gsl_vector_min_index* returns the index value (of type *int*) of the minimum value in the specified vector. Listing 17.2 shows a C source listing of a program that includes the implementation of the operations on vectors that have been discussed. The program is stored in file `vector_ops.c`.

Listing 17.2: C program that shows various operations on vectors.

```
 1 /* Program: vector_ops.c
 2    This program shows various simple operations on vectors
 3    using the GSL functions.
 4    J. M. Garrido. Dec 15, 2013
 5 */
 6 #include <stdio.h>
 7 #include <gsl/gsl_vector.h>
 8 #include "basic_lib.h"
 9 int main (void) {
10    const int N = 8;
11    int i;
12    double x = 10.45;
13    double *yy;
14    double *zz;
15    printf("Program to demonstrate operations on vectors\n\n");
16    gsl_vector * pv = gsl_vector_alloc (N); // create vector pv
```

```
17    gsl_vector * qv = gsl_vector_alloc (N); // create vector qv
18    // gsl_vector_set_all(pv, 8.25); // assign 8.25 to elements
19    gsl_vector_set_all(qv, 10.5);    // assign 10.5 to elements
20    zz = pv->data;
21    linspace (zz, 1.25, 122.45, N);
22    for (i = 0; i < N; i++)
23        printf ("zz[%d] = %g\n", i, zz[i]);
24    printf("\nDifferences in zz \n");
25    yy = diff (zz, N);
26    for (i = 0; i < N-1; i++)
27        printf ("yy[%d] = %g\n", i, yy[i]);
28    for (i = 0; i < N-1; i++)
29        gsl_vector_set(qv, i, yy[i]); // copy elements yy to qv
30    printf("Data from vectors pv and qv\n");
31    for (i = 0; i < N; i++)
32        printf ("element %d =     %6.3f    %6.3f\n", i,
33            gsl_vector_get (pv, i), gsl_vector_get(qv, i));
34    gsl_vector_add_constant (pv, x); // add x to elements of pv
35    printf("\nAddition of scalar %g to vector pv\n", x);
36    for (i = 0; i < N; i++)
37        printf ("element %d = %g\n", i, gsl_vector_get (pv, i));
38    if (gsl_vector_equal(pv, qv) == 1)
39        printf("Vectors pv and qv are equal\n");
40    else
41        printf("Vectors pv and qv are not equal\n");
42    gsl_vector_add (pv, qv);   // vector qv added to vector pv
43    printf("\nAddition of vector qv to vector pv\n");
44    for (i = 0; i < N; i++)
45        printf ("element %d = %g\n", i, gsl_vector_get (pv, i));
46    printf("\nSubtraction of vector qv from vector pv\n");
47    gsl_vector_sub (pv, qv);   // vector qv subtracted from pv
48    for (i = 0; i < N; i++)
49        printf ("element %d = %g\n", i, gsl_vector_get (pv, i));
50    gsl_vector_scale (pv, x);  // multiply vector pv by scalar x
51    printf("\nMultiplication of vector pv by scalar %g\n",x);
52    for (i = 0; i < N; i++)
53        printf ("element %d = %g\n", i, gsl_vector_get (pv, i));
54    gsl_vector_mul (pv, qv);
55    printf("\nMultiplication of vectors pv and qv\n");
56    for (i = 0; i < N; i++)
57        printf ("element %d = %g\n", i, gsl_vector_get (pv, i));
58    gsl_vector_div (pv, qv);
59    printf("Division of vectors pv and qv\n");
60    for (i = 0; i < N; i++)
61        printf ("element %d = %g\n", i, gsl_vector_get (pv, i));
62    gsl_vector_memcpy (pv, qv);
63    printf("\nCopy of vector qv to pv\n");
64    for (i = 0; i < N; i++)
65        printf ("element %d = %g\n", i, gsl_vector_get (pv, i));
66    printf("\nSize of vector pv: %d\n", pv->size);
67    x = 105.25;
68    printf("Assign %g to element 4 of pv\n", x);
69    gsl_vector_set(pv, 4, x);
70    printf ("Max value in pv: %g\n", gsl_vector_max (pv));
71    printf ("Index of max: %d\n", gsl_vector_max_index (pv));
72    gsl_vector_free (pv);  // release allocated memory
73    gsl_vector_free (qv);  // release allocated memory
```

```
74    return 0;
75 }
```

17.5.3   Vector Views

The operation to handle part of a vector is typically useful when a vector needs
to be created that is a subvector of another vector. A GSL vector view is a temporary object, stored on the stack, which can be used to operate on a subset of vector
elements. Function *gsl_vector_subvector*, when called, returns a vector view defined
as a subvector of another vector. The arguments are: the vector, the start of the new
vector is offset by the number of elements from the start of the original vector, and
the number of elements of the new vector. The following C statement calls function
gsl_vector_subvector and defines *asv* as a subvector of vector *av*.

```
asv = gsl_vector_subvector(av, offset, nsize);
```

A subvector gas type *gsl_vector_view* and can be manipulated as a vector using
various GSL vector functions. The address of component *vector* (with the & operator) has top used as argument. The following statement calls function *gsl_vector_get*
to retrieve element *i* of subvector *asv*.

```
x = gsl_vector_get (&asv.vector, i);
```

A vector view of an array is created by calling function *gsl_vector_view_array*.
The arguments are: the base or pointer to the start of the array and the number of
elements for the vector. Other similar vector view functions are provided by the GSL.
In the following C statements, an array *a* is declared and initialized, then a vector
view is created using the array with 8 elements.

```
double a[] = {4.45, 1.65, 6.25, 2.48, 1.22, 10.59, 11.22, 8.55};
avv = gsl_vector_view_array(a, 8);
```

Listing 17.3 shows the C source code of a program two vector views and performs
several vector operations. The C program is stored in file `vector_v2.c`.

Listing 17.3: Creating and using vector views.

```
1 /* Program: vector_v2.c
2    This program uses vector views to set up vector
3    of an array. Then creates a subvector.
4    Uses GSL functions.
5    J. M. Garrido, January 4, 2013
6 */
7 #include <stdio.h>
8 #include <gsl/gsl_vector.h>
9 int main (void) {
```

```
10    const int N = 8;
11    int i;
12    double a[] = {4.45, 1.65, 6.25, 2.48, 1.22, 10.59, 11.22,
          8.55};
13    int offset;  /* starting element of subvector */
14    int nsize;   /* size of subvector */
15    gsl_vector_view avv = gsl_vector_view_array(a, N); // view
16    printf("Program creates and manipulates a subvector\n\n");
17    printf("Data from vector\n");
18    for (i = 0; i < N; i++)
19        printf ("element %d = %g\n", i, gsl_vector_get
                  (&avv.vector, i));
20    offset = 2;
21    nsize = 4;
22    gsl_vector_view bvv = gsl_vector_subvector(&avv.vector,
            offset, nsize);
23    printf("\nSubvector: \n");
24    for (i = 0; i < nsize; i++)
25        printf ("element %d = %g\n", i, gsl_vector_get
                  (&bvv.vector, i));
26    return 0;
27 }
```

17.5.4 Complex Vectors

Complex vectors store collections of complex values. Using GSL, a complex variable is of type *gsl_complex* and several GSL functions can be called to manipulate complex numbers. A complex value can be defined with its rectangular or polar representation. The following C statements declare complex variable $z1$, then assigns its component values using the rectangular representation by calling function *gsl_complex_rect*.

```
gsl_complex z1;
z1 = gsl_complex_rect(4.75, 9.6);
```

There are several functions to perform complex arithmetical operations and other functions to perform complex trigonometric operations. The following C statements perform a complex multiplication of complex variables $z1$ and $z2$, the complex product is assigned to complex variable $z3$. The real and imaginary components, which are of type *double* by default, are retrieved with the macros: *GSL_REAL* and *GSL_IMAG*.

```
z3 = gsl_complex_mul(z1, z2);
real3 = GSL_REAL(z3);
imag3 = GSL_IMAG(z3);
```

A complex vector is created by calling function *gsl_vector_alloc* with an integer argument that indicates the number of complex elements of the vector. To set a complex value to an element of a complex vector, function *gsl_vector_complex_set* is called. Function *gsl_vector_complex_get* is called to retrieve the complex value of an element of the specified vector. There are several functions that perform arithmetical operations of complex vectors. Listing 17.4 shows the C source code of a program defines and manipulates complex variables, and performs several complex vector operations. The C program is stored in file `vector_man2.c`.

Listing 17.4: C program that uses complex variables and complex vectors.

```
 1 /* Program: vector_man2.c
 2    This program allocates, initializes, complex vectors. It sets
 3    complex values to the first n elements and retrieves data from
 4    a complex vector using the GSL functions.
 5    J. M. Garrido, January 12, 2013
 6 */
 7 #include <stdio.h>
 8 #include <gsl/gsl_complex.h>
 9 #include <gsl/gsl_complex_math.h>
10 #include <gsl/gsl_vector.h>
11 int main (void) {
12    const int N = 8;
13    gsl_complex z1; // complex values
14    gsl_complex z2;
15    gsl_complex z3;
16    gsl_complex z4;
17    gsl_complex zz;
18    double xarg;
19    double xabs;
20    double z3r;
21    double z3i;
22    double z4r;
23    double z4i;
24    int i;
25    int n;
26    printf("Program creates and manipulates a complex vector\n");
27    z1.dat[0] = 2.5;  // real part
28    z1.dat[1] = 1.25; // imaginary part
29    z2 = gsl_complex_rect (4.5, 8.5); // in rectangular form
30    xarg = gsl_complex_arg(z1);
31    xabs = gsl_complex_abs(z2);
32    z3 = gsl_complex_add(z1, z2);
33    z4 = gsl_complex_mul(z1, z2);
34    z3r = GSL_REAL (z3);
35    z3i = GSL_IMAG(z3);
36    z4r = GSL_REAL(z4);
37    z4i = GSL_IMAG(z4);
38    printf("Argument of z1: %lf   absolute val of z2: %lf \n",
              xarg, xabs);
39    printf("z3= %lf+i%lf   z4= %lf+i%lf \n", z3r, z3i, z4r, z4i);
40    /* Using complex vectors */
41    gsl_vector_complex * pv = gsl_vector_complex_alloc (N);
42    gsl_vector_complex_set_all(pv, z1); // assign z1 to elements
43    n = N - 3;
44    /* set the first n elements of vector to z2 */
45    for (i = 0; i < n; i++)
46        gsl_vector_complex_set (pv, i, z2); // value to element i
47    printf("Data in complex vector\n");
48    for (i = 0; i < N; i++) {
```

```
49          zz = gsl_vector_complex_get (pv, i);
50          printf ("element %d  %g + i%g \n", i, GSL_REAL (zz),
            GSL_IMAG(zz));
51      }
52      gsl_vector_complex_free (pv);   // release allocated memory
53      return 0;
54  }
```

When this program executes, the following output is produced:

```
Program creates and manipulates a complex vector
Argument of z1: 0.463648    absolute val of z2: 9.617692
z3= 7.000000+i9.750000   z4= 0.625000+i26.875000
Data in complex vector
element 0  4.5 + i8.5
element 1  4.5 + i8.5
element 2  4.5 + i8.5
element 3  4.5 + i8.5
element 4  4.5 + i8.5
element 5  2.5 + i1.25
element 6  2.5 + i1.25
element 7  2.5 + i1.25
```

Summary

Arrays are data structures that store collections of data. To refer to an individual element, an integer value, known as the index, is used to indicate the relative position of the element in the array. C and the GSL manipulate arrays as vectors and matrices. Many operations and functions are defined for creating and manipulating vectors and matrices.

	Key Terms	
arrays	elements	index
vectors	array elements	matrices
column vector	row vector	two-dimensional array
vector operations	vector functions	complex vectors

Exercises

Exercise 17.1 Develop a C program that computes the values in a vector V that are the sines of the elements of vector T. The program must assign to T a vector with 75 elements running from 0 to 2π. Plot the elements of V as a function of the elements of T; use GnuPlot.

Exercise 17.2 Develop a C program that finds the index of the first negative number in a vector. If there are no negative numbers, it should set the result to -1.

Exercise 17.3 The Fibonacci series is defined by: $F_n = F_{n-1} + F_{n-2}$. Develop a C program that computes a vector with the first n elements of a Fibonacci series. A second vector should also be computed with the ratios of consecutive Fibonacci numbers. The program must plot this second vector using GnuPlot.

Exercise 17.4 Develop a C program that reads the values of a vector P and computes the element values of vector Q with the cubes of the positive values in vector P. For every element in P that is negative, the corresponding element in Q should be set to zero.

Exercise 17.5 Develop a C program that reads the values of a vector P and assigns the element values of vector Q with every other element in vector P.

Exercise 17.6 Develop a C program that reads the values of a matrix M of m rows and n columns. The program must create a column vector for every column in matrix M, and a row vector for every row in matrix M.

Exercise 17.7 A trapezoid is a four-sided region with two opposite sides parallel. The area of a trapezoid is the average length of the parallel sides, times the distance between them. The area of the trapezoid with width $\Delta x = x_2 - x_1$, is computed as the width, Δx, times the average height, $(y_2 + y_1)/2$.

$$A_t = \Delta x \frac{y_1 + y_2}{2}$$

For the interval of $[a,b]$ on variable x, it is divided into $n-1$ equal segments of length Δx. Any value of y_k is defined as $y_k = f(x_k)$. The trapezoid sum to compute the area under the curve for the interval of $[a,b]$, is defined by the summation of the areas of the individual trapezoids and is expressed as follows.

$$A = \sum_{k=2}^{k=n} [\Delta x \frac{1}{2}(f(x_{k-1}) + f(x_k))]$$

The larger the number of trapezoids, the better the approximation of the area. The area from a to b, with segments: $a = x_1 < x_2 < \ldots < x_n = b$ is given by:

$$A = \frac{b-a}{2n}[f(x_1) + 2f(x_2) + ... + 2f(x_{n-1}) + f(x_n)]$$

Develop a C function that computes an estimate of the area under a curve given by a sequence of values y with a corresponding sequence of values x. The function must be called from *main*. The program must use the GSL.

Chapter 18

Matrices and Sets of Linear Equations

18.1 Introduction

As mentioned in previous chapters, in scientific computing, data is conveniently organized as collections of values known as **vectors** and **matrices** and are used to implement data lists and sequences of values. This chapter presents basic concepts of matrices, programming in C and the GSL, and a summary of computing with matrices. Systems of linear equations are solved using the GSL.

18.2 Matrices

In general, a matrix is a two-dimensional array of data values and is organized in rows and columns. The following array, Y, is a *two-dimensional array* organized as an $m \times n$ matrix; its elements are arranged in m rows and n columns.

$$
Y = \begin{bmatrix} y_{11} & y_{12} & \cdots & y_{1n} \\ y_{21} & y_{22} & \cdots & y_{2n} \\ \vdots & \vdots & \ddots & \vdots \\ y_{m1} & y_{m2} & \cdots & y_{mn} \end{bmatrix}
$$

The first row of Y consists of elements: $y_{11}, y_{12}, \ldots, y_{1n}$. The second row consists of elements: $y_{21}, y_{22}, \ldots, y_{2n}$. The last row of Y consists of elements: $y_{m1}, y_{m2}, \ldots, y_{mn}$. In a similar manner, the elements of each column can be identified.

18.2.1 Basic Concepts

A matrix is defined by specifying the rows and columns of the array. An m by n matrix has m rows and n columns. A *square* matrix has the same number of rows and columns, n rows and n columns, which is denoted as $n \times n$. The following example shows a 2×3 matrix, which has two rows and three columns:

$$\begin{bmatrix} 0.5000 & 2.3500 & 8.2500 \\ 1.8000 & 7.2300 & 4.4000 \end{bmatrix}$$

A matrix of dimension $m \times 1$ is known as a *column vector* and a matrix of dimension $1 \times n$ is known as a *row vector*. A vector is considered a special case of a matrix with one row or one column. A row vector of size n is typically a matrix with one row and n columns. A column vector of size m is a matrix with m rows and one column.

The elements of a matrix are denoted with the matrix name in lower-case and two indices, one corresponding to the row of the element and the other index corresponding to the column of the element. For matrix Y, the element at row i and column j is denoted by y_{ij} or by $y_{i,j}$.

The *main diagonal* of a matrix consists of those elements on the diagonal line from the top left and down to the bottom right of the matrix. These elements have the same value of the two indices. The diagonal elements of matrix Y are denoted by y_{ii} or by y_{jj}. For a square matrix, this applies for all values of i or all values of j. For a square matrix X (an $n \times n$ matrix), the elements of the main diagonal are:

$$x_{1,1}, \; x_{2,2}, \; x_{3,3}, \; x_{4,4}, \; \ldots, \; x_{n,n}$$

An *identity matrix* of size n, denoted by I_n is a square matrix that has all the diagonal elements with value 1, and all other elements (off-diagonal) with value 0. The following is an identity matrix of size 3 (of order n):

$$I = \begin{bmatrix} 1 & 0 & 0 \\ 0 & 1 & 0 \\ 0 & 0 & 1 \end{bmatrix}$$

18.2.2   Arithmetic Operations

The arithmetic matrix operations are similar to the ones discussed previously for vectors. The multiplication of a matrix Y by a *scalar* λ calculates the multiplication of every element of matrix Y by the scalar λ. The result defines a new matrix.

$$Y = \begin{bmatrix} y_{11} & y_{12} & \cdots & y_{1n} \\ y_{21} & y_{22} & \cdots & y_{2n} \\ \vdots & \vdots & \ddots & \vdots \\ y_{m1} & y_{m2} & \cdots & y_{mn} \end{bmatrix} \qquad \lambda Y = \begin{bmatrix} \lambda y_{11} & \lambda y_{12} & \cdots & \lambda y_{1n} \\ \lambda y_{21} & \lambda y_{22} & \cdots & \lambda y_{2n} \\ \vdots & \vdots & \ddots & \vdots \\ \lambda y_{m1} & \lambda y_{m2} & \cdots & \lambda y_{mn} \end{bmatrix}$$

In a similar manner, the addition of a scalar λ to matrix Y, calculates the addition of every element of the matrix with the scalar λ. The subtraction of a scalar λ from a matrix Y, denoted by $Y - \lambda$, computes the subtraction of the scalar λ from every element of matrix Y

The *matrix addition* is denoted by $Y + Z$ of two $m \times n$ matrices Y and Z, calculates the addition of every element of matrix Y with the corresponding element of matrix Z. The result defines a new matrix. This operation requires that the two matrices have the same number of rows and columns.

$$Y + Z = \begin{bmatrix} y_{11} + z_{11} & y_{12} + z_{12} & \cdots & y_{1n} + z_{1n} \\ y_{21} + z_{21} & y_{22} + z_{22} & \cdots & y_{2n} + z_{2n} \\ \vdots & \vdots & \ddots & \vdots \\ y_{m1} + z_{m1} & y_{m2} + z_{m2} & \cdots & y_{mn} + z_{mn} \end{bmatrix}$$

Similarly, the subtraction of two matrices Y and Z, denoted by $Y - Z$, subtracts every element of matrix Z from the corresponding element of matrix Y. The result defines a new matrix.

The *element-wise multiplication* (also known as the Hadamard product or Schur product) of two matrices, multiplies every element of a matrix X by the corresponding element of the second matrix Y and is denoted by $X \circ Y$. The result defines a new matrix, Z. This operation requires that the two matrices have the same number of rows and columns. The following is the general form of the element-wise multiplication of matrix X multiplied by matrix Y.

$$Z = X \circ Y = \begin{bmatrix} x_{11} \times y_{11} & x_{12} \times y_{12} & \cdots & x_{1n} \times y_{1n} \\ x_{21} \times y_{21} & x_{22} \times y_{22} & \cdots & x_{2n} \times y_{2n} \\ \vdots & \vdots & \ddots & \vdots \\ x_{m1} \times y_{m1} & x_{m2} \times y_{m2} & \cdots & x_{mn} \times y_{mn} \end{bmatrix}$$

The *determinant* of a matrix A, denoted by $\det A$ or by $|A|$, is a special number that can be computed from the matrix. The determinant is useful to describe various properties of the matrix that are applied in systems of linear equations and in calculus.

The determinant provides important information when the matrix consists of the coefficients of a system of linear equations. If the determinant is nonzero the system of linear equations has a unique solution. When the determinant is zero, there are either no solutions or many solutions.

One of several techniques to compute the determinant of a matrix is applying Laplace's formula, which expresses the determinant of a matrix in terms of its minors. The minor $M_{i,j}$ is defined to be the determinant of the submatrix that results from matrix A by removing the ith row and the jth column. Note that this technique is not very efficient. The expression $C_{i,j} = (1)^{i+j} M_{i,j}$ is known as a *cofactor*. The determinant of A is computed by

$$\det A = \sum_{j=1}^{n} C_{i,j} \times a_{i,j} \times M_{i,j}$$

The *matrix multiplication* of an $m \times n$ matrix X and an $n \times p$ matrix Y, denoted by XY, defines another matrix Z of dimension m by p and the operation is denoted by $Z = XY$. In matrix multiplication, the number of rows in the first matrix has to equal the number of columns in the second matrix. An element of the new matrix Z is determined by:

$$z_{ij} = \sum_{k=1}^{n} x_{ik} \, y_{kj}$$

that is, element z_{ij} of matrix Z is computed as follows:

$$z_{ij} = x_{i1} \, y_{1j} + x_{i2} \, y_{2j} + \ldots + x_{in} \, y_{nj}$$

The matrix multiplication is not normally commutative, that is, $XY \neq YX$. The following example defines a 2 by 3 matrix, X, and a 3 by 3 matrix, Y. The matrix multiplication of matrix X and Y defines a new matrix Z.

$$X = \begin{bmatrix} 1 & 2 & 1 \\ 0 & 2 & 1 \end{bmatrix} \quad Y = \begin{bmatrix} 1 & 2 & 0 \\ 0 & 3 & 1 \\ -2 & 1 & 1 \end{bmatrix}$$

$$Z = XY = \begin{bmatrix} -1 & 9 & 3 \\ -2 & 7 & 3 \end{bmatrix}$$

The *transpose* of an $m \times n$ matrix X is an $n \times m$ matrix X^T formed by interchanging the rows and columns of matrix X. For example, for the given matrix X, the transpose (X^T) is:

$$X = \begin{bmatrix} 1 & 2 & 1 \\ 0 & 5 & 3 \end{bmatrix} \quad X^T = \begin{bmatrix} 1 & 0 \\ 2 & 5 \\ 1 & 3 \end{bmatrix}$$

The *conjugate* of an $m \times n$ complex matrix Z is a matrix $\overline{Z}$ with all its elements conjugate of the corresponding elements of matrix Z. The *conjugate transpose* of an $m \times n$ matrix Z, is an $n \times m$ matrix that results by taking the transpose of matrix Z and then the complex conjugate. The resulting matrix is denoted by Z^H or by Z^* and is also known as the *Hermitian transpose*, or the *adjoint matrix* of matrix Z. For example:

$$Z = \begin{bmatrix} 1.5+2.3i & 2.1-1.4i & 1+0.7i \\ 0+3.2i & 5.2-1.5i & 3.7+3.5i \end{bmatrix} \quad Z^H = \begin{bmatrix} 1.5-2.3i & 0-3.2i \\ 2.1+1.4i & 5.2+1.5i \\ 1-0.7i & 3.7-3.5i \end{bmatrix}$$

A square matrix Y is *symmetric* if $Y = Y^T$. The following example shows a 3×3 symmetric matrix Y.

$$Y = \begin{bmatrix} 1 & 2 & 7 \\ 2 & 3 & 4 \\ 7 & 4 & 1 \end{bmatrix}$$

A square matrix X is *triangular* if all the elements above or below its diagonal have value zero. A square matrix Y is *upper triangular* if all the elements below its diagonal have value zero. A square matrix Y is *lower triangular* if all the elements above its diagonal have value zero. The following examples shows a matrix Y that is upper triangular.

$$Y = \begin{bmatrix} 1 & 2 & 7 \\ 0 & 3 & 4 \\ 0 & 0 & 1 \end{bmatrix}$$

The *rank* of a matrix Y is the number of independent columns in matrix Y, or the number of linearly independent rows of matrix Y. The *inverse* of an $n \times n$ matrix X is another matrix X^{-1} (if it exists) of dimension $n \times n$ such that their matrix multiplication results in an identity matrix or order n. This relation is expressed as:

$$X^{-1}X = XX^{-1} = I_n$$

A square matrix X is *orthogonal* if for each column x_i of X, $x_i^T x_j = 0$ for any other column x_j of matrix X. If the rows and columns are orthogonal unit vectors (the norm of each column x_i of X has value 1), then X is *orthonormal*. A matrix X is orthogonal if its transpose X^T is equal to its inverse X^{-1}. This implies that for orthogonal matrix X,

$$X^T X = I$$

Given matrices A, X, and B, a general matrix equation is expressed by $AX = B$. This equation can be solved for X by pre-multiplying both sides of the matrix equation by the inverse of matrix A. The following the expression shows this:

$$A^{-1}AX = A^{-1}B$$

This results in

$$X = A^{-1}B$$

18.2.3   Matrix Manipulation in C and the GSL

Matrices are created and manipulated in C by calling the various library function in the GSL. Before using a matrix, it needs to be created. Function *gsl_matrix_alloc* creates a matrix of the specified length *m* rows and *n* columns, and returns a pointer to the matrix. Matrices are stored in row-major order, the elements of each row form a contiguous block in memory. This is the standard C-language ordering of two-dimensional arrays.

When a matrix is no longer needed in the program, it can be destroyed by calling the GSL function *gsl_matrix_free*. The first of the following C statements shows matrix *mmat* created with *M* rows and *N* columns. The second statement destroys matrix *mmat*.

```
gsl_matrix * mmat = gsl_matrix_alloc (M, N);
gsl_matrix_free (mmat);
```

A matrix is manipulated by accessing its elements and changing and/or retrieving the value of the elements. The GSL library function *gsl_matrix_set* assigns the value specified to the element indexed with row *i* and column *j* of a matrix. The GSL Function *gsl_matrix_set_all* sets the value specified to all elements of a matrix. Function *gsl_matrix_set_zero* sets all elements of the specified matrix to zero.

Function *gsl_matrix_set_identity* creates an identity matrix and sets it to the specified matrix. Therefore, the elements of the main diagonal will have a value of 1, and all other elements will have a value of zero.

The value of individual elements of a matrix can be retrieved by calling the GSL function *gsl_matrix_get*. This function returns the value of the (i, j)-th element of the specified matrix. The first of the following C statements calls function *gsl_matrix_set* to assign the value of variable *val* to the element i, j of matrix *mmat*. The second statement retrieves the value of element i, j of matrix *mmat* and assigns this value to variable *y*.

```
gsl_matrix_set (mmat, i, j, val);
y = gsl_matrix_get (mmat, i, j);
```

The functions *gsl_matrix_get* and *gsl_matrix_set* and similar functions perform portable range checking and report an error if the index value of an element is outside the allowed range.

Listing 18.1 shows the source code of a C program that uses GSL functions to create and manipulate a matrix. Line 7 includes the header file that contains the declarations for calling the GSL functions for matrices. Line 14 calls function *gsl_matrix_alloc* to create matrix *mmat* with $M \times N$ elements. In line 23, element *i* of matrix *mmat* is set to the specified value. In line 35 matrix *mmat* is destroyed by calling function *gsl_matrix_free*.

Listing 18.1: C program that creates and manipulates a matrix.

```
 1 /* Program: matrix_bops.c
 2    This program allocates, initializes (stores) and retrieves
 3    data from a matrix using the functions gsl_matrix_alloc,
 4    gsl_matrix_set and gsl_matrix_get.
 5 */
 6 #include <stdio.h>
 7 #include <gsl/gsl_matrix.h>
 8 int main (void)
 9 {
10    const int M = 10;   // number of rows
11    const int N = 3;    // number of columns
12    int i, j;
13    double val;
14    gsl_matrix * mmat = gsl_matrix_alloc (M, N);
15    printf ("Simple operations on a matrix \n\n");
16    // generate values of matrix elements
17    printf ("Generate values of elements for matrix\n");
18    for (i = 0; i < M; i++) {
19    printf ("\nRow %d \n", i);
20       for (j = 0; j < N; j++) {
21          val = 12.25 + 100*i + j;
22          printf ("%g ", val);
23          gsl_matrix_set (mmat, i, j, val);
24       }
25    }
26    // retrieve element data from the matrix
27    printf ("\n\nRetrieve value of elements of matrix\n");
28    for (i = 0; i < M; i++) {
29       for (j = 0; j < N; j++) {
30          val = gsl_matrix_get (mmat, i, j);
31          printf ("mmat(%d,%d) =  %5.2f ", i, j, val);
32       }
33       printf ("\n");
34    }
35    gsl_matrix_free (mmat);
36    return 0;
37 }
```

The following listing includes the Linux shell commands in a terminal window to compile, link, and execute the program; it also includes the output results produced by the program execution.

```
$ gcc -Wall -o matrixbops.out matrix_bops.c   -lgsl -lm
$ ./matrixbops.out
Simple operations on a matrix
Generate values of elements for matrix
Row 0
12.25 13.25 14.25
Row 1
112.25 113.25 114.25
Row 2
212.25 213.25 214.25
Row 3
```

```
312.25 313.25 314.25
Row 4
412.25 413.25 414.25
Row 5
512.25 513.25 514.25
Row 6
612.25 613.25 614.25
Row 7
712.25 713.25 714.25
Row 8
812.25 813.25 814.25
Row 9
912.25 913.25 914.25
Retrieve value of elements of matrix
mmat(0,0) =   12.25 mmat(0,1) =   13.25 mmat(0,2) =   14.25
mmat(1,0) =  112.25 mmat(1,1) =  113.25 mmat(1,2) =  114.25
mmat(2,0) =  212.25 mmat(2,1) =  213.25 mmat(2,2) =  214.25
mmat(3,0) =  312.25 mmat(3,1) =  313.25 mmat(3,2) =  314.25
mmat(4,0) =  412.25 mmat(4,1) =  413.25 mmat(4,2) =  414.25
mmat(5,0) =  512.25 mmat(5,1) =  513.25 mmat(5,2) =  514.25
mmat(6,0) =  612.25 mmat(6,1) =  613.25 mmat(6,2) =  614.25
mmat(7,0) =  712.25 mmat(7,1) =  713.25 mmat(7,2) =  714.25
mmat(8,0) =  812.25 mmat(8,1) =  813.25 mmat(8,2) =  814.25
mmat(9,0) =  912.25 mmat(9,1) =  913.25 mmat(9,2) =  914.25
```

18.3   Simple Matrix Operations

The basic operations on matrices are performed with a matrix and a scalar or with two matrices. These are carried out by calling several GSL functions in C.

18.3.1   Arithmetic Operations

To add a scalar to a matrix, involves adding the scalar value to every element of the matrix. With C and GSL, this is performed by calling function *gsl_matrix_add_constant*. This function adds the specified constant value to the elements of the matrix specified.

In the following portion of code, matrix *pm* is created then all its elements are assigned the value 3.5. On the fourth line, function *gsl_matrix_add_constant* is called to add the constant *x*, which has value 10.45 to all the elements of matrix *pm*. In a similar manner, subtracting a scalar from a matrix can be specified.

```
const double x = 10.45;
gsl_matrix * pm = gsl_matrix_alloc (M, N); // create matrix pm
gsl_matrix_set_all (pm, 3.5);       // assign 3.5 to all elements
gsl_matrix_add_constant (pm, x);    // add x to all elements
```

Matrix addition adds two matrices and involves adding the corresponding elements of each matrix. This addition operation on matrices is only possible if the two matrices are of the same size. In the following example, matrix *pm* and *qm* are both matrices of size $M \times N$. Calling function *gsl_matrix_add* adds the elements of matrix *qm* to the elements of matrix *pm*.

```
gsl_vector * pm = gsl_matrix_alloc (M, N);   // create matrix pm
gsl_vector * qm = gsl_matrix_alloc (M, N);   // create matrix qm
// ...
gsl_matrix_add (pm, qm);       // add matrix qm to matrix pm
```

Matrix subtraction is carried out by subtracting the corresponding elements of each of two matrices. The matrices must be of the same size and the matrix subtraction is performed by calling function *gsl_matrix_sub*. In the following line of code, the elements of matrix *qm* are subtracted from the elements of matrix *pm*.

```
gsl_matrix_sub (pm, qm);    // matrix qm subtracted from matrix pm
```

Matrix scalar multiplication involves multiplying each element of the matrix by the value of the scalar. Calling function *gsl_matrix_scale* multiplies the elements of the specified matrix by the constant factor specified. In the following call to this function, the elements of matrix *pm* are multiplied by the constant *x*.

```
gsl_matrix_scale (pm, x);
```

Element by element matrix multiplication involves multiplying the corresponding elements of two matrices. This operation is applied to two matrices of equal size ($m \times n$) and is performed by calling the GSL function *gsl_matrix_mul_elements*. The function call multiplies the elements of the first matrix specified by the elements of the second specified matrix. In the following function call, matrices *pm* and *qm* are element multiplied and the results are stored in vector *pm*.

```
gsl_matrix_mul_elements (pm, qm);
```

Element by element matrix division involves dividing the corresponding elements of two matrices. This operation is applied to two matrices of equal size ($m \times n$) and is performed by calling the GSL function *gsl_matrix_div_elements*. This function divides the elements of the first specified matrix by the elements of the second specified matrix. In the following function call, the elements of matrix *pm* are divided by the elements of matrix *qm* and the results are stored in matrix *pm*.

```
gsl_matrix_div_elements (pm, qm);
```

Listing 18.2 shows the C source code of a program that copies the elements from C two-dimensional arrays into GSL matrices *ga* and *gb*. Then it computes scalar addition and subtraction, scalar multiplication, and element addition, subtraction, multiplication, and division of matrices *ga* and *gb*. The program is stored in file matrix_arith.c.

Lines 22–29 of the program copy the values of the elements from the C two-dimensional arrays *a* and *b* to the GSL matrices *ga* and *gb*. Lines 31–48 perform the various arithmetic operations on matrices *ga* and *gb*. The prototype of function *display_mat* is specified in line 10 and the function is defined in the program on lines 53–65. Lines 54–55 get the number of rows and the number of columns of a GSL matrix.

Listing 18.2: C program that several arithmetic matrix operations.

```
 1 /* Program: matrix_arith.c
 2 This program computes matrix arithmetic operations using the
 3 GSL library. It copies data from C arrays into gsl matrices.
 4 J. M. Garrido. December 21, 2012
 5 */
 6 #include <stdio.h>
 7 #include <gsl/gsl_blas.h>
 8 #define M 2   // number of rows matrix a
 9 #define N 3   // number of columns matrix a
10 void display_mat(gsl_matrix * gslmat);
11 int main (void) {
12   int i, j;
13   double val;
14   double a[M][N] = { {0.11, 0.12, 0.13},
15                      {0.21, 0.22, 0.23} };
16   double b[M][N] = { {1.11, 10.12, 8.75},
17                      {9.21, 3.22, 6.55}};
18   double x = 7.5;
19   gsl_matrix * ga = gsl_matrix_alloc (M, N);
20   gsl_matrix * gb = gsl_matrix_alloc (M, N);
21   printf ("GSL Arithmetic Matrix Operations\n");
22   for (i = 0; i < M; i++) {
23      for (j = 0; j < N; j++) {
24         val = a[i][j];
25         gsl_matrix_set (ga, i, j, val);
26         val = b[i][j];
27         gsl_matrix_set (gb, i, j, val);
28      }
29   }
30   /* Compute scalar addition */
31   gsl_matrix_add_constant (ga, x);  // add x to elements of ga
32   printf("\nMatrix ga after scalar addition: \n");
33   display_mat(ga);  // display matrix
34   /* matrix addition and subtraction of ga and gb  */
35   gsl_matrix_add (ga, gb);     // add matrix gb to matrix ga
36   printf("\nMatrix ga after matrix addition: \n");
37   display_mat(ga);
38   gsl_matrix_sub (ga, gb);     // subtract matrix qm to matrix pm
39   gsl_matrix_add_constant (ga, -x); // add -x to elements of ga
40   printf("\nMatrix ga after matrix add and scalar sub: \n");
```

```
41   display_mat(ga);
42   /* Compute scalar multiplication of matrix ga */
43   gsl_matrix_scale (ga, x);
44   printf("\nMatrix ga after scalar mul: \n");
45   display_mat(ga);
46   /* matrix element multiplication and div of ga and gb */
47   gsl_matrix_mul_elements (ga, gb);
48   gsl_matrix_div_elements (ga, gb);
49   printf("\nMatrix ga after matrix element mul and div: \n");
50   display_mat(ga);
51   return 0;
52 } // end main
53 void display_mat(gsl_matrix * gslmat) {
54   int m = gslmat->size1; // get number of rows
55   int n = gslmat->size2; // get number of cols
56   int i, j;
57   double val;
58   for (i = 0; i < m; i++) {
59     for (j = 0; j < n; j++) {
60         val = gsl_matrix_get (gslmat, i, j);
61         printf("%g ", val);
62     }
63     printf("\n");
64   }
65 } // end display_mat
```

The following listing shows the output results produced by the program execution.

```
GSL Arithmetic Matrix Operations
Matrix ga after scalar addition:
7.61 7.62 7.63
7.71 7.72 7.73
Matrix ga after matrix addition:
8.72 17.74 16.38
16.92 10.94 14.28
Matrix ga after matrix add and scalar sub:
0.11 0.12 0.13
0.21 0.22 0.23
Matrix ga after scalar mul:
0.825 0.9 0.975
1.575 1.65 1.725
Matrix ga after matrix element mul and div:
0.825 0.9 0.975
1.575 1.65 1.725
```

18.3.2 Additional Matrix Functions

Various additional functions can be applied to matrices. To create an identity matrix and set it to the specified matrix, function *gsl_matrix_set_identity* is called. The function takes the number of rows and the number of columns of the specified matrix. In the following example, the first statement creates matrix *mymat*, the second statement sets the matrix to an identity matrix.

```
gsl_matrix * mymat = gsl_matrix_alloc (M, N);
gsl_matrix_set_identity (mymat);
```

Copying one matrix to another matrix is performed by calling the GSL function *gsl_matrix_memcpy*. This function copies the elements of the specified source matrix into the specified destination matrix. The two matrices must have the same length $m \times n$. In the following example, the elements of matrix *qm* are copied to the corresponding elements of matrix *pm* and the results are stored in matrix *pm*.

```
gsl_matrix_memcpy (pm, qm);
```

Copying rows and columns from matrices to vectors and copying rows and columns from vectors to matrices is performed by functions: *gsl_matrix_get_row*, *gsl_matrix_get_col*, *gsl_matrix_set_row*, and *gsl_matrix_set_col*. In the following example, the first statement copies the elements of column *j* of matrix *mymat*, to vector *pv*. The second statement copies vector *qv* to row *i* of matrix *mymat*.

```
gsl_matrix_get_col (pv, mymat, j);
gsl_matrix_set_row (mymat, i, qv);
```

To exchange the rows and columns of a matrix, the following functions are defined: *gsl_matrix_swap_rows* and *gsl_matrix_swap_columns*. The following example exchanges rows *ia* and *ib* of matrix *mymat*.

```
gsl_matrix_swap_rows (mymat, ia, ib);
```

Function *gsl_matrix_swap_rowcol* exchanges row *i* and column *j* of the specified square matrix. Function *gsl_matrix_transpose_memcpy* gets the transpose of the second specified matrix and sets it to the first matrix specified.

The comparison of two matrices for equality is performed by calling the GSL function *gsl_matrix_equal*. This function returns 1 if the specified vectors are equal (by comparison of element values) and 0 otherwise.

In the following function call, the elements of matrix *qm* are compared to the corresponding elements of *pm* and the result is an integer value assigned to variable *vflag*.

```
vflag = gsl_matrix_equal (pm, qm);
```

Because there are only two possible outcomes in the call to this function, the return value may be considered a boolean value (true is represented by 1 and false is represented by 0), which can be used in a conditional statement.

The attributes *size1* and *size2* of a matrix store the number of rows and the number of columns of the corresponding matrix. The following lines of C code get the size of matrix *pm* and assigns these values to variables *m* and *n*.

```
m = pm->size1;
n = pm->size2;
```

Function *gsl_matrix_max* gets the maximum value stored in the specified matrix. The following function call gets the maximum value in matrix *pm* and assigns this value to *x*, which should be a variable of type *double*.

```
x = gsl_matrix_max(pm);
```

In addition to the maximum value in a matrix, the index of the element with that value may be desired. Calling function *gsl_matrix_max_index* gets the index values (row and column) of the element with the maximum value in the specified matrix. In the following function call, the index values of the element with the maximum value in matrix *pm* are stored in the reference arguments *im* and *jm*.

```
gsl_matrix_max_index (pm, &im, &jm);
```

In a similar manner, function *gsl_matrix_min* gets the minimum value (of type *double*) stored in the specified matrix. Function *gsl_matrix_min_index* gets the index values (row and column) of the minimum value in the specified matrix.

Listing 18.3 shows a C source listing of a program that performs several of the matrix manipulation operations discussed. The program is stored in file matrix_ops2.c. In lines 20–22 the program creates two matrices and one vector. In line 31, the program creates an identity matrix to the size of matrix *gb*. Line 35 copies matrix *ga* to matrix *gb*. Line 40 copies row *i* to vector *pv*. Line 46 changes the values in vector *pv* by adding a constant to its elements. Line 47 copies vector *pv* to column *j* of matrix *ga*. Line 53 exchanges rows 2 and 1 of matrix *ga*. Line 57 copies the transpose of matrix *ga* to matrix *gb*. Line 63 gets the index values of the row and column of the element with minimum value in matrix *gb*.

Listing 18.3: C program that includes several matrix operations.

```
1 /* Program: matrix_ops2.c
2 This program perform various matrix operations using the
3 GSL library.
4 J. M. Garrido. January 3, 2013
5 */
6 #include <stdio.h>
7 #include <gsl/gsl_blas.h>
8 #define M 3  // number of rows matrix a
9 #define N 3  // number of columns matrix a
10 void display_mat(gsl_matrix * gslmat);
11 int main (void) {
12    int i, j;
13    size_t ii, jj; // GSL int type
14    int mflag;     // for comparison of matrices
15    double val;
16    double a[M][N] = { {3.11, 5.12, 2.13},
17                       {1.21, 8.22, 5.23},
```

```
18                      {6.77, 2.88, 7.55}}};
19     double x = 11.75;
20     gsl_matrix * ga = gsl_matrix_alloc (M, N);
21     gsl_matrix * gb = gsl_matrix_alloc (M, N);
22     gsl_vector * pv = gsl_vector_alloc (N); // create vector pv
23     printf ("GSL Matrix Operations\n");
24     for (i = 0; i < M; i++) {
25        for (j = 0; j < N; j++) {
26           val = a[i][j];
27           gsl_matrix_set (ga, i, j, val);
28        }
29     }
30     /* Set identity matrix to gb */
31     gsl_matrix_set_identity (gb);
32     printf("\nMatrix gb identity matrix: \n");
33     display_mat(gb);   // display matrix
34     /* copy matrix ga to gb  */
35     gsl_matrix_memcpy (gb, ga);
36     printf("\nMatrix gb after copied from ga: \n");
37     display_mat(gb);
38     /* copy row i of matrix ga to vector pv */
39     i = 1; // row 1
40     gsl_matrix_get_row (pv, ga, i);
41     printf("\nVector pv from matrix ga row: %d \n", i);
42     for (i = 0; i < N; i++)
43        printf ("element %d = %g\n", i, gsl_vector_get (pv, i));
44     /* copy vector pv to column j of matrix ga */
45     j = 2;
46     gsl_vector_add_constant (pv, x); // add x to vector pv
47     gsl_matrix_set_col (ga, j, pv);
48     printf("\nMatrix ga after setting col: %d\n", j);
49     display_mat(ga);
50     /* Exchange matrix row i of ga and gb */
51     i = 2;
52     j = 1;
53     gsl_matrix_swap_rows (ga, i, j);
54     printf("\nMatrix ga after exchanging row: %d and %d \n",
                i, j);
55     display_mat(ga);
56     /* perform transpose of matrix ga into gb */
57     gsl_matrix_transpose_memcpy (gb, ga);
58     printf("\nMatrix gb after setting transpose of ga \n");
59     display_mat(gb);
60     /* get maximum value in matrix gb */
61     x = gsl_matrix_max (gb);
62     printf("Maximum value in matrix gb: %g \n", x);
63     gsl_matrix_min_index (gb, &ii, &jj);
64     printf("Minimum value in matrix gb, row: %d col %d \n",
                ii, jj);
65     /* Compare matrices ga and gb */
66     mflag = gsl_matrix_equal (ga, gb);
67     if (mflag == 1)
68        printf("Matrices ga and gb are equal\n");
69     else
70        printf("Matrices ga and gb are not equal \n");
71     /* get the size of matrix ga */
72     ii = ga->size1; // rows
```

```
73   jj = ga->size2; // cols
74   printf("Size of matrix ga: %d rows, %d cols \n", ii, jj);
75   return 0;
76 } // end main
77 void display_mat(gsl_matrix * gslmat) {
78   int m = gslmat->size1; // get number of rows
79   int n = gslmat->size2; // get number of cols
80   int i, j;
81   double val;
82   for (i = 0; i < m; i++) {
83     for (j = 0; j < n; j++) {
84         val = gsl_matrix_get (gslmat, i, j);
85         printf("%g  ", val);
86       }
87     printf("\n");
88   }
89 } // end display_mat
```

The following listing shows the output results produced by the program execution.

```
GSL Matrix Operations
Matrix gb identity matrix:
1  0  0
0  1  0
0  0  1
Matrix gb after copied from ga:
3.11   5.12   2.13
1.21   8.22   5.23
6.77   2.88   7.55
Vector pv from matrix ga row: 1
element 0 = 1.21
element 1 = 8.22
element 2 = 5.23
Matrix ga after setting col: 2
3.11   5.12   12.96
1.21   8.22   19.97
6.77   2.88   16.98
Matrix ga after exchanging row: 2 and 1
3.11   5.12   12.96
6.77   2.88   16.98
1.21   8.22   19.97
Matrix gb after setting transpose of ga
3.11   6.77   1.21
5.12   2.88   8.22
12.96  16.98  19.97
Maximum value in matrix gb: 19.97
Minimum value in matrix gb, row: 0 col 2
Matrices ga and gb are not equal
Size of matrix ga: 3 rows, 3 cols
```

18.3.3   GSL Matrix Views

To use a submatrix of a matrix, the GSL provides matrix views. These are of type *gsl_matrix_view* and the elements of the views can be accessed with the address of component *matrix* of the view (using the & operator). In the following example, matrix *a* is created with 7 rows and 5 columns. Then a matrix view *suba* of matrix *a* is created to get a submatrix of *a*. The upper left element of the submatrix is on row 2 and column 2 of matrix *a*. The submatrix, *suba*, has 4 rows and 3 columns. The third statement illustrates how to use a matrix view to add 12.4 to the elements of the submatrix *suba*.

```
gsl_matrix * a = gsl_matrix_alloc (7, 5);
gsl_matrix_view suba = gsl_matrix_submatrix (a, 2, 2, 4, 3);
gsl_matrix_add_constant (&suba.matrix, 12.4);
```

Listing 18.4 shows the C source code of a program that creates a matrix view of a matrix to get a submatrix and perform operations on this submatrix. Line 19 creates a matrix view, *subga*, of matrix *ga*. This defines a submatrix that starts on row 1 and column 1 of matrix *ga*. The submatrix has one less ($M - 1$) row and one less ($N - 1$) column than matrix *ga*. Line 31 calls function *display_mat* to display submatrix *subga*. Line 33 adds the value of variable *x* to the elements of submatrix *subga*. Similarly, line 37 multiplies the value of variable *x* to the elements of submatrix *subga*. The program is stored in file matrix_views.c

Listing 18.4: C program matrix views.

```
 1 /* Program: matrix_views.c
 2    This program computes matrix arithmetic operations using the
 3    GSL matrix views to get a submatrix.
 4    J. M. Garrido. December 21, 2012
 5 */
 6 #include <stdio.h>
 7 #include <gsl/gsl_blas.h>
 8 #define M 3   // number of rows matrix a
 9 #define N 3   // number of columns matrix a
10 void display_mat(gsl_matrix * gslmat);
11 int main (void) {
12    int i, j;
13    double val;
14    double a[M][N] = { {3.11, 5.12, 2.13},
15                       {1.21, 8.22, 5.23},
16                       {7.25, 8.65, 2.66} };
17    double x = 11.75;
18    gsl_matrix * ga = gsl_matrix_alloc (M, N);
19    gsl_matrix_view subga =
                    gsl_matrix_submatrix( ga, 1, 1, M-1, N-1);
20    printf ("GSL Sub-Matrix Operations\n");
21    for (i = 0; i < M; i++) {
22       for (j = 0; j < N; j++) {
23          val = a[i][j];
24          gsl_matrix_set (ga, i, j, val);
25       }
```

```
26   }
27   printf("Matrix ga \n");
28   display_mat(ga);
29   /* display submatrix subga */
30   printf("Submatrix subga \n");
31   display_mat(&subga.matrix);
32   /* Compute scalar addition */
33   gsl_matrix_add_constant (&subga.matrix, x);   // add x to ga
34   printf("\nSubmatrix subga after scalar addition: \n");
35   display_mat(&subga.matrix);   // display matrix
36   /* Compute scalar multiplication of submatrix subga */
37   gsl_matrix_scale (&subga.matrix, x);
38   printf("\nSubmatrix subga after scalar multiplication: \n");
39   display_mat(&subga.matrix);
40   return 0;
41 } // end main
42 void display_mat(gsl_matrix * gslmat) {
43   int m = gslmat->size1; // get number of rows
44   int n = gslmat->size2; // get number of cols
45   int i, j;
46   double val;
47   for (i = 0; i < m; i++) {
48     for (j = 0; j < n; j++) {
49         val = gsl_matrix_get (gslmat, i, j);
50         printf("%6.2f ", val);
51     }
52     printf("\n");
53   }
54 } // end display_mat
```

The following listing is the output produced by the program execution.

```
GSL Sub-Matrix Operations
Matrix ga
   3.11    5.12    2.13
   1.21    8.22    5.23
   7.25    8.65    2.66
Submatrix subga
   8.22    5.23
   8.65    2.66
Submatrix subga after scalar addition:
  19.97   16.98
  20.40   14.41
Submatrix subga after scalar multiplication:
234.65 199.52
239.70 169.32
```

When a C array has already been declared and initialized, it is convenient to create a GSL matrix view of the array, instead of copying the elements of the array to a GSL matrix. Function *gsl_matrix_view_array* gets a matrix view of the specified array. The size of the array ($n \times m$) has to be specified also. This functions takes a linear array, so if the array is two-dimensional a pointer conversion is necessary. In the following example, the first statement declares and initializes array *myarray*. The second statement creates a matrix view of the array, with 2 rows and 4 columns.

```
double myarray [] = {3.5, 4.56, 1.25, 12.55, 6.45, 10.05, 7.85, 5.25};
gsl_matrix_view mav = gsl_matrix_view_array (myarray, 2, 4);
```

Listing 18.5 shows a C source program that applies array views. This program includes a small change compared to the previous C program; the only changes in the program are lines 18–19. The results of the program execution are the same. The C source code of the program is stored in file matrix_views2.c.

Listing 18.5: C program matrix views.

```
 1 /* Program: matrix_views2.c
 2   This program creates a matrix view of an array and computes
 3   matrix arithmetic operations using the GSL matrix views.
 4   J. M. Garrido. December 21, 2012
 5 */
 6 #include <stdio.h>
 7 #include <gsl/gsl_blas.h>
 8 #define M 3  // number of rows matrix a
 9 #define N 3  // number of columns matrix a
10 void display_mat(gsl_matrix * gslmat);
11 int main (void) {
12   double a[M][N] = { {3.11, 5.12, 2.13},
13                      {1.21, 8.22, 5.23},
14                      {7.25, 8.65, 2.66} };
15   double *aa;
16   double x = 11.75;
17   aa = (double *) a;
18   gsl_matrix_view gav = gsl_matrix_view_array (aa, M, N);
19   gsl_matrix_view subga =
             gsl_matrix_submatrix( &gav.matrix, 1, 1, M-1, N-1);
20   printf ("GSL Matrix Views\n");
21
22   printf("Matrix array view gav \n");
23   display_mat(&gav.matrix);
24   /* display submatrix subga */
25   printf("Submatrix subga \n");
26   display_mat(&subga.matrix);
27   /* Compute scalar addition */
28   gsl_matrix_add_constant (&subga.matrix, x);  // add x to ga
29   printf("\nSubmatrix subga after scalar addition: \n");
30   display_mat(&subga.matrix);  // display matrix
31   /* Compute scalar multiplication of submatrix subga */
32   gsl_matrix_scale (&subga.matrix, x);
33   printf("\nSubmatrix subga after scalar multiplication: \n");
34   display_mat(&subga.matrix);
35   return 0;
36 } // end main
```

Other GSL functions that define matrix views and vector views are:

- function *gsl_matrix_view_vector* gets the matrix view of the specified vector;

- function *gsl_matrix_row* gets the vector view of a given row number of the specified matrix;

- function *gsl_matrix_subrow* gets the vector view of a given row number of the specified matrix, starting at element *offset* and with *n* elements;

- function *gsl_matrix_column* gets the vector view of a given column of the specified matrix;

- function *gsl_matrix_subcolumn* gets the vector view of a given column of the specified matrix, starting at element *offset* and with *n* elements;

- function *gsl_matrix_diagonal* gets the vector of the given diagonal of the specified matrix.

The following lines of C code statements first create vector *pv* and matrix *pm*. The next statement creates a matrix view *mv* with *m* rows and *n* columns of vector *pv*. The next source statement creates a vector view *bvv* of row 3 of matrix *pm*. The next statement creates a vector view *bvvs* of a subrow of matrix *pm* from row 2, starting at element 3, size of 6. The next statement creates a vector view *cvv* of column 2 of matrix *pm*. The next statement creates a vector view *cvvs* of a subcolumn of matrix *pm* from row 2, starting at element 1, size of 8. The last statement creates vector view *dv* of the diagonal of matrix *pm*.

```
gsl_vector * pv = gsl_vector_alloc (N);
gsl_matrix * pm = gsl_matrix_alloc (M, N);
gsl_matrix_view mv = gsl_matrix_view_vector (pv, m, n);
gsl_vector_view bvv = gsl_matrix_row (pm, 3);      // row 3
gsl_vector_view bvvs = gsl_matrix_subrow (pm, 2, 3, 6);
gsl_vector_view cvv = gsl_matrix_column (pm, 2); // column 2
gsl_vector_view cvvs = gsl_matrix_subcolumn (pm, 2, 1, 8);
gsl_vector_view dv = gsl_matrix_diagonal (pm);
```

18.4 Solving Systems of Linear Equations

Several methods exist on how to solve systems of linear equations applying vectors and matrices. Some of these methods are: substitution, cancellation, or matrix manipulation. A system of *n* linear equations can be expressed by the general equations:

$$
\begin{aligned}
a_{11}x_1 &+ a_{12}x_2 + \ldots + a_{1n}x_n = b_1 \\
a_{21}x_1 &+ a_{22}x_2 + \ldots + a_{2n}x_n = b_2 \\
&\vdots \qquad \vdots \qquad\qquad \vdots \qquad \vdots \\
a_{n1}x_1 &+ a_{n2}x_2 + \ldots + a_{nn}x_n = b_n
\end{aligned}
$$

This system of linear equations is more conveniently expressed in matrix form in the following manner:

$$
\begin{bmatrix}
a_{11} & a_{12} & \cdots & a_{1n} \\
a_{21} & a_{22} & \cdots & a_{2n} \\
\vdots & \vdots & \ddots & \vdots \\
a_{m1} & a_{2m} & \cdots & a_{mn}
\end{bmatrix}
\begin{bmatrix}
x_1 \\
x_2 \\
\vdots \\
x_n
\end{bmatrix}
=
\begin{bmatrix}
b_1 \\
b_2 \\
\vdots \\
b_m
\end{bmatrix}
\tag{18.1}
$$

This can also be expressed in a more compact matrix form as: $AX = B$. Matrix A is the coefficients matrix (of the variables x_i for $i = 1 \ldots n$). X is the vector of unknowns x_i, and B is the vector of solution. Consider a simple linear problem that consists of a system of three linear equations ($n = 3$):

$$
\begin{array}{rrrr}
5x_1 & +2x_2 & +x_3 & = 25 \\
2x_1 & +x_2 & +3x_3 & = 12 \\
-x_1 & +x_2 & +2x_3 & = 5
\end{array}
$$

In matrix form, this system of three linear equations can be written in the following manner:

$$
\begin{bmatrix}
5 & 2 & 1 \\
2 & 1 & 3 \\
-1 & 1 & 2
\end{bmatrix}
\begin{bmatrix}
x_1 \\
x_2 \\
x_3
\end{bmatrix}
=
\begin{bmatrix}
25 \\
12 \\
5
\end{bmatrix}
\tag{18.2}
$$

Matrix A is a square ($n \times n$) of coefficients, X is a vector of size n, and B is the solution vector also of size n.

$$
A = \begin{bmatrix}
5 & 2 & 1 \\
2 & 1 & 3 \\
-1 & 1 & 2
\end{bmatrix}
\quad
X = \begin{bmatrix}
x_1 \\
x_2 \\
x_3
\end{bmatrix}
\quad
B = \begin{bmatrix}
25 \\
12 \\
5
\end{bmatrix}
\tag{18.3}
$$

Decomposing the coefficient matrix, A is the main technique used in GSL to compute the determinant, matrix inversion, and the solution a set of linear equations. Common numerical methods used are:

- Gaussian Elimination

- LU Decomposition

- SV Decomposition

- QR Decomposition

With LU Decomposition, a general square matrix A is factorized into upper and lower triangular matrices, applying the matrix relation: $PA = LU$, where P is a permutation matrix, L is unit lower triangular matrix and U is upper triangular matrix.

Function *gsl_linalg_LU_decomp* factorizes the specified square matrix A into the LU decomposition. The function returns matrix A with the diagonal and upper triangular part of the original matrix A. In the following example, the first C statement creates a GSL square ($M \times M$) matrix a. The second statement creates a permutation of size M. The third statement performs the LU decomposition of matrix a.

```
gsl_matrix * a = gsl_matrix_alloc (M, M);
gsl_permutation * p = gsl_permutation_alloc (M);
gsl_linalg_LU_decomp (a, p, &s);
```

To compute the determinant of a matrix A, first the LU decomposition must be performed, then function *gsl_linalg_LU_det* is called. To compute the inverse of a matrix A, function *gsl_linalg_LU_invert* is called. In the following example, the first C statement performs the LU decomposition of matrix a. The second statement computes the determinant d of matrix a. The third statement computes the inverse matrix *invm* of matrix a.

```
gsl_linalg_LU_decomp (a, p, &s);
d = gsl_linalg_LU_det (a, s);
gsl_linalg_LU_invert (a, p, invm);
```

For computing matrix multiplication of two matrices A and B, the GSL provides a set of general matrix multiplication functions for several types of data. For type *double*, the GSL function is *gsl_blas_dgemm* and it calculates the value of matrix C using the following relation in which α and β are scalar coefficients:

$$C = \alpha\, AB + \beta\, C$$

The following statements in the C language, assign values to the scalars *alpha* and *beta* then call the function to compute matrix c by multiplying matrix a by matrix b.

```
alpha = 1.0;
beta = 0.0;
gsl_blas_dgemm (CblasNoTrans, CblasNoTrans, alpha, a, b, beta, c);
```

Listing 18.6 shows the C source program that computes the determinant of a matrix, the matrix inversion of a matrix, and the matrix multiplication of two matrices. In line 41, matrix *ga* is copied to matrix *mymat* to retain the original matrix (*ga*). Line 44 performs the LU decomposition of matrix *mymat*. In line 45, the determinant of matrix *mymat* is computed and stored in variable *determ*. In line 47, the inverse matrix of *mymat* is computed and stored in matrix *invmat*. Lines 61–62 calls function is *gsl_blas_dgemm* to compute the matrix multiplication of matrices *ga* and *gb* and the result matrix is *gc*.

Listing 18.6: Computing matrix determinant, inversion, and multiplication.

```
 1 /* Program: matrix_mul4.c
 2    This program computes matrix determinant, matrix inversion,
 3    and matrix multiplication using the GSL library. It copies data
 4    from C arrays into GSL matrices, instead of using matrix views.
 5    J. Garrido. December 22, 2012
 6 */
 7 #include <stdio.h>
 8 #include <gsl/gsl_linalg.h>
 9 #include <gsl/gsl_blas.h>
10 #define M 3   // number of rows matrix A
11 #define N 3   // number of columns matrix A
12 #define K 2
13 void display_mat(gsl_matrix * gslmat);
14 int main (void) {
15    int i, j;
16    double alpha;
17    double beta;
18    double val;
19    double a[M][N] = { {2.15, 1.10, 5.15},
20                       {1.25, 2.35, 7.55},
21                       {2.44, 3.55, 8.25}};
22    double b[N][K] = { {10.21, 9.22},
23                       {12.15, 7.22},
24                       {1.35, 3.25}};
25    double determ;   // determinant
26    gsl_matrix * ga = gsl_matrix_alloc (M, N);
27    gsl_matrix * mymat = gsl_matrix_alloc (M, N);
28    gsl_matrix * invmat = gsl_matrix_alloc(M, N);
29    gsl_matrix * gb = gsl_matrix_alloc (N, K);
30    gsl_matrix * gc = gsl_matrix_alloc (M, K);
31    printf ("Program computes matrix determinant, inversion\n");
32    printf ("and multiplication ( C = A B)\n\nMatrix A: \n\n");
33    for (i = 0; i < M; i++) {
34       for (j = 0; j < N; j++) {
35          val = a[i][j];
36          gsl_matrix_set (ga, i, j, val);
37       }
38    }
39    display_mat(ga);
40    int s;
41    gsl_matrix_memcpy(mymat, ga);
42    // create a permutation
43    gsl_permutation * p = gsl_permutation_alloc (M);
44    gsl_linalg_LU_decomp (mymat, p, &s);
45    determ = gsl_linalg_LU_det(mymat, s); // determinant
46    printf("\nDeterminant of matrix A: %g \n", determ);
47    gsl_linalg_LU_invert(mymat, p, invmat);
48    printf("Inverse matrix of A:\n");
49    display_mat(invmat);
50    printf ("\nMatrix B: \n\n");
51    for (i = 0; i < N; i++) {
52       for (j = 0; j < K; j++) {
53          val = b[i][j];
54          gsl_matrix_set (gb, i, j, val);
55       }
56    }
57    display_mat(gb);
```

```
58    /* Compute C = alpha A B + beta C */
59    alpha = 1.0;
60    beta = 0.0;
61    gsl_blas_dgemm (CblasNoTrans, CblasNoTrans,
62                              alpha, ga, gb, beta, gc);
63    printf("\nMatrix C = AB: \n\n");
64    display_mat(gc);
65    return 0;
66 } // end main
```

The following listing is the output produced by the program execution.

```
Program computes GSL Matrix determinant, inversion, and
multiplication ( C = A B)
Matrix A:
   2.15    1.10    5.15
   1.25    2.35    7.55
   2.44    3.55    8.25
Determinant of matrix A: -13.6988
Inverse matrix of A:
   0.54   -0.67    0.28
  -0.59   -0.38    0.72
   0.09    0.36   -0.27
Matrix B:
  10.21    9.22
  12.15    7.22
   1.35    3.25
Matrix C = AB:
  42.27   44.50
  51.51   53.03
  79.18   74.94
```

To solve a system of linear equations expressed in matrix form $AX = B$, GSL provides function *gsl_linalg_LU_solve*. This function uses the LU decomposition of matrix A by first calling function *gsl_linalg_LU_decomp*. The following example shows three C statements that solve a system of linear equations $ax = b$. The first statement creates a permutation p of size M. The second statement performs an LU decomposition of matrix a, and changes the matrix. The third statement solves the system of linear equations by using LU decomposition; the results are stored in vector x.

```
gsl_permutation * p = gsl_permutation_alloc (M);
gsl_linalg_LU_decomp (a, p, &s);
gsl_linalg_LU_solve (a, p, b, x);   // solve linear equations
```

Listing 18.7 shows the C source code of a program that solves a system of linear equations, $AX = B$, using LU decomposition of matrix A. The program reads as input data all values it requires to execute.

Listing 18.7: C program that solves a system of linear equations.

```
1 /* Program: lineqs1.c
2    The following program solves the linear system A x = b.
3    The system is solved using LU decomposition of the matrix A.
4    Modified December 10, 2012 by J. Garrido
5 */
6 #include <stdio.h>
7 #include <gsl/gsl_linalg.h>
8 int main (void) {
9      int m;   // number of rows in A
10     int n;   // number of columns in A
11     double val;
12     int i;
13     int j;
14     int s;
15     scanf("%d ", &m); // read number of rows
16     scanf("%d ", &n); // read number of columns
17     printf("\nProgran to solve system of linear equations with
                LU decomposition\n");
18     printf("Matrix A, number of rows: %d \n", m);
19     printf("Matrix A, number of cols: %d \n", n);
20     gsl_matrix * mmat = gsl_matrix_alloc (m, n);
21     gsl_vector * b = gsl_vector_alloc(n);
22     gsl_vector *x = gsl_vector_alloc (n);
23     printf("\nMatrix A: ");
24     // input values into matrix A
25     for (i = 0; i < m; i++) {
26         printf("\nRow %d \n", i);
27         for (j = 0; j < n; j++) {
28             scanf("%lf ", &val);
29             printf("%g ", val);
30             gsl_matrix_set (mmat, i, j, val);
31         }
32     }
33     printf("\n\nvector b: \n");
34     // Input values into vector b
35     for (i = 0; i < n; i++) {
36         scanf("%lf ", &val);
37         printf("%6.2f ", val);
38         gsl_vector_set(b, i, val);
39     }
40     gsl_permutation * p = gsl_permutation_alloc (n);
41     gsl_linalg_LU_decomp (mmat, p, &s);
42     gsl_linalg_LU_solve (mmat, p, b, x);
43     printf ("\n\nSolution  Vector x: \n");
44     gsl_vector_fprintf (stdout, x, "%g");
45     gsl_matrix_free (mmat);
46     gsl_permutation_free (p);
47     gsl_vector_free (x);
48     return 0;
```

The following output listing is produced by executing the program.

```
Progran to solve system of linear equations with LU decomposition
```

```
Matrix A, number of rows: 4
Matrix A, number of cols: 4

Matrix A:
Row 0
3.25 1.6 8.55 4.25
Row 1
1.45 3.25 2.75 1.55
Row 2
2.25 5.35 1.3 7.5
Row 3
3.5 7.15 3.85 3.6
vector b:
   1.20    2.55    3.65    4.75
Solution Vector x:
-0.858919
0.915252
0.273682
0.0440242
```

Summary

Computations that involve single numbers are known as scalars. Matrices are data structures that store collections of data in two dimensions: rows and columns. To refer to an individual element, two index values are used: one to indicate the row and the other to indicate the column of the element in the array. With the C programming language and the GSL, several functions are available to create and manipulate matrices. Several case studies are presented that show for each problem, the C source program and the output produced by the program execution.

Key Terms

arrays	elements	index
matrices	array elements	GSL matrices
column vector	row vector	double-dimension array
matrix operations	matrix functions	determinant
inverse matrix	system of linear equations	LU decomposition

Exercises

Exercise 18.1   Develop a C program that reads the values of a matrix M of m rows and n columns. The program must use the GSL and create a new matrix that has the same number of rows and columns, from the appropriate elements in matrix M. Hint: if $m < n$ then the second matrix would be an $m \times m$ square matrix.

Exercise 18.2   Develop a computational model that inputs and processes the rainfall data for the last five years. For every year, four quarters of rainfall are provided, measured in inches. Hint: use a matrix to store these values. The attributes are: the precipitation (in inches), the year, and the quarter. The program must compute the average, minimum, and maximum rainfall per year and per quarter (for the last five years). Implement with C using a GSL matrix.

Exercise 18.3   A computational model has a mathematical representation as a set of three linear equations. Find the solution to the following set of linear equations using LU elimination, and implement with C and the GSL.

$$
\begin{aligned}
3x_1 &+ 5x_2 &+ 2x_3 &= 8 \\
2x_1 &+ 3x_2 &- x_3 &= 1 \\
x_1 &- 2x_2 &- 3x_3 &= -1
\end{aligned}
$$

Chapter 19

Introduction to Dynamical Systems

19.1 Introduction

Computational models of dynamical systems are used to study the behavior of systems over time. The foundations for modeling dynamical systems are based on the mathematical concepts of derivatives, integrals, and differential equations. Models of dynamical systems use difference and differential equations to describe the behavior of the systems they represent. This chapter discusses models of dynamical systems and the computer (numerical) solution to the corresponding types of equations using C and the GSL.

19.2 Continuous and Discrete Models

As mentioned in Chapter 1, from the perspective of how a computational model represents the changes of the system state with time, computational models can be divided into two general categories:

1. Continuous models

2. Discrete-event models

A continuous model is one in which the changes of state in the model occur continuously with time. Often the state variables in the model are represented as continuous functions of time.

For example, a model that represents the temperature in a boiler as part of a power plant can be considered a continuous model because the state variable that represents the temperature of the boiler is implemented as a continuous function of time. These types of models are usually modeled as a set of differential equations.

19.3   Derivative of a Function

A mathematical function defines a relationship between two (or more) variables. A simple relation is expressed as: $y = f(x)$. In this expression, variable y is a function of variable x that is the *independent variable*; for a given value of x there is a corresponding value of y.

The derivative of a function is used to study some relevant properties of the function, for example the *rate of change*. The derivative of a function $y = f(x)$ at a particular point is the slope of the tangent line at that point and can be computed with the value of $\Delta y / \Delta x$ when Δx is *infinitely small*. In mathematics, this is the limit when Δx approaches zero or $\Delta x \to 0$. A description of a graphical interpretation of the slope of a tangent to a function at some specified point appears in Chapter 4. This was explained as the rate of change of y with respect to x. The exact slope, m, of the tangent is expressed as:

$$m = \lim_{\Delta x \to 0} \frac{\Delta y}{\Delta x}$$

The *derivative* of a variable y with respect to variable x is expressed as:

$$\frac{dy}{dx} = \lim_{\Delta x \to 0} \frac{\Delta y}{\Delta x}$$

It is assumed that y is a function of x, expressed as $y = f(x)$, and that y is a *continuous* function in an interval of interest. The function $f(x)$ is continuous in an interval if its limit exists for every value of x in the interval.

The derivative of y with respect to x is the instantaneous rate of change of y with respect to x and is denoted as:

$$\frac{dy}{dx} \quad \text{or} \quad y'$$

The second derivative of y with respect to x is denoted as:

$$y'' \quad \text{or} \quad \frac{d^2 y}{dx^2}$$

The third and higher order derivatives are similarly denoted. A derivative of order n is denoted as:

$$y^{(n)} \quad \text{or} \quad \frac{d^n y}{dx^n}$$

When a variable q is a function of two independent variables, $q = f(x,y)$, the derivative of q has to be specified with respect to x or with respect to y. This concept is known as a *partial derivative*. The partial derivative of q with respect to x is denoted by $\delta q/\delta x$ and the partial derivative of q with respect to y is denoted by $\delta q/\delta y$.

19.3.1 Computing the Derivative of a Function

The derivative of a curve given by $y = f(x)$ at some specified point $x = c$ can be approximated by the use of *finite differences*. Figure 19.1 shows the curve given by $y = f(x)$ at point $x = c$. A finite difference or change of the values of variable y at $x = c$ is denoted by $\Delta y = f(c+h) - f(c)$ and a finite difference or change of the values of variable x is denoted by $\Delta x = h$. An approximation of the derivative of $f(x)$, denoted by $f'(x)$, at $x = c$ can be computed by:

$$f'(x)\big|_{x=c} \approx \frac{f(c+h) - f(c)}{h}$$

This expression is known as the *forward difference* of $f(x)$ at $x = c$. A similar expression allows computing an approximation to the derivative of $f(x)$ at $x = c$ using the *backward difference*.

$$f'(x)\big|_{x=c} \approx \frac{f(c) - f(c-h)}{h}$$

Another similar expression allows computing an approximation to the derivative of $f(x)$ at $x = c$ using the *central difference*.

$$f'(x)\big|_{x=c} \approx \frac{f(c+h) - f(c-h)}{2h}$$

The value of h can be chosen smaller and smaller to improve the approximation of the derivative of $f(x)$ at $x = c$. Because of roundoff and truncation errors, care must be taken in applying the previous expressions of finite differences. There are finite difference expressions for higher order derivatives. For example an estimate of the second derivative of $f(x)$ at $x = c$ can be computed with the central finite difference expression:

$$f''(x)\big|_{x=c} \approx \frac{f(c+h) - 2f(c) + f(c-h)}{h^2}$$

19.3.2 Computing the First Derivative Using C

Listing 19.1 shows the source program `deriv1.c` that computes the approximations of the numerical values of $f(x) = x^2$ using forward and central differences for several values of h.

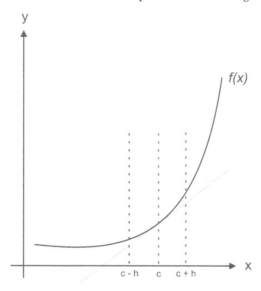

FIGURE 19.1: The slope of $x = c$.

Listing 19.1: C program that computes the approximation of a derivative

```c
/* The following program estimates the derivative of the function
   f(x) = x^2 at x=2.
   The derivative is computed using forward and central differences.
   J M Garrido, April, 2013.
   File: deriv1.c */
#include <stdio.h>
#include <math.h>
const int n = 20;
double myf (double x) {
    return pow (x, 2);
}
//
int main (void) {
    double resultf, resultc;
    double abserrf, abserrc; // relative errors
    double h, x;
    int i;
    printf ("Derivative of f(x) = x^2; exact value = %.10f\n");
    h = 0.5;
    x = 2.0;
    for (i = 0; i < n; i++) {
        resultf = (myf(x+h) - myf(x)) / h;          // forward diff
        resultc = (myf(x+h) - myf(x-h))/(2.0*h); // central diff
        abserrf = (resultf - 2.0 * x)/resultf;
        abserrc = (resultc - 2.0 * x)/resultc;
        printf ("x = 2.0  h = %.10f \n", h);
        printf ("f'(x) = %.10f %.10f +/- %.10e +/- %.10e \n", resultf,
            resultc, abserrf, abserrc);
        h = h / 5.0;
```

```
    }
    return 0;
}
```

The following listing is a partial execution output of the program. The best value of the derivative is 4.0000000361 using $h = 0.0000000512$ with forward differences. The best value using central differences appears using $h = 0.1$

```
Computing derivative of f(x) = x^2; exact value = 4.00000
x = 2.0   h = 0.50000
f'(x)=4.500000000 4.000000000 +/- 1.1111111111e-01 +/- 0.000000000e+00
x = 2.0   h = 0.1000000000
f'(x)=4.100000000 4.000000000 +/- 2.4390243902e-02 +/- 2.220446049e-16
x = 2.0   h = 0.0200000000
f'(x)=4.020000000 4.000000000 +/- 4.9751243781e-03 +/- 8.881784197e-16
x = 2.0   h = 0.0040000000
f'(x)=4.004000000 4.00000000 +/- 9.9900099898e-04 +/- -1.298960938e-14
x = 2.0   h = 0.0008000000
f'(x)=4.000800000 4.00000000 +/- 1.9996000778e-04 +/- -1.795230630e-13
x = 2.0   h = 0.0001600000
f'(x)=4.000160000 4.000000000 +/- 3.999840159e-05 +/- 6.530331830e-13
x = 2.0   h = 0.0000320000
f'(x)=4.000032000 4.000000000 +/- 7.999940437e-06 +/- 2.734701354e-12
x = 2.0   h = 0.0000064000
f'(x)=4.000006400 4.00000000 +/- 1.6000310462e-06 +/- 1.1408207712e-11
x = 2.0   h = 0.0000012800
f'(x)=4.000001280 4.0000000005 +/- 3.202239121e-07 +/- 1.241653446e-10
x = 2.0   h = 0.0000002560
f'(x)=4.000000254 3.999999999 +/- 6.365840848e-08 +/- -9.267520085e-11
x = 2.0   h = 0.0000000512
f'(x)=4.000000036 3.999999993 +/- 9.014622966e-09 +/- -1.827398791e-09
x = 2.0   h = 0.0000000102
f'(x)=3.999999984 3.99999998 +/- -3.995803037e-09 +/- -3.995803037e-09
x = 2.0   h = 0.0000000020
f'(x)=3.999999984 3.99999998 +/- -3.995803037e-09 +/- -3.995803037e-09
x = 2.0   h = 0.0000000004
f'(x)=3.999999116 3.999999116 +/- -2.20836286e-07 +/- -2.20836286e-07
```

19.3.3 Numerical Differentiation

The GSL functions that can be used for numerical differentiation are (taken from the GSL Reference manual):

```
int gsl_deriv_central (const gsl_function *ufunc,
    double x, double h, double *result, double *abserr)
```

This function computes the derivative of the function `ufunc` at the point `x` using an adaptive central difference algorithm with a step-size of `h`. The value of the derivative is returned by the function and the estimate of its absolute error is returned in `abserr`.

```
int gsl_deriv_forward (const gsl_function *ufunc,
       double x, double h, double *result, double *abserr)
```

This function computes the derivative of the function ufunc at the point x using an adaptive forward difference algorithm with a step-size of h. The function is evaluated only at points greater than x. The derivative is returned by
the function and an estimate of its absolute error is returned in abserr. This
function should be used if ufunc has a discontinuity at x, or is undefined for
values less than x.

```
int gsl_deriv_backward (const gsl_function *ufunc,
       double x, double h, double *result, double *abserr)
```

This function computes the derivative of the function ufunc at the point x
using an adaptive backward difference algorithm with a step-size of h. The
function is evaluated only at points less than x, and not at x itself. The derivative is returned by the function and an estimate of its absolute error is returned
in abserr. This function should be used if ufunc has a discontinuity at x,
or is undefined for values greater than ufunc.

Listing 19.2 shows the source program deriv1b.c that uses the GSL and computes the approximations of the numerical values of $f(x) = x^2$ with forward and
central differences.

Listing 19.2: C program with GSL that approximates a derivative

```
 1 /* This program computes approximate derivative of the function
 2     f(x) = x^2 at x=2 using gsl_deriv_central and
               gsl_deriv_forward.
 3     J M garrido, April 2013,  File: deriv1b.c    */
 4 #include <stdio.h>
 5 #include <gsl/gsl_math.h>
 6 #include <gsl/gsl_deriv.h>
 7 double myf (double x, void * params)  {
 8        return pow (x, 2);
 9 }
10 int main (void) {
11       gsl_function F;
12       double result, abserr;
13       double x = 2.0;
14       F.function = &myf;
15       F.params = 0;
16       printf ("Computing the derivative of f(x) = x^2 ");
17       printf ("at x = 2.0\n");
18       printf ("Exact value: f'(x) = %.10f\n\n", 2 * x);
19       gsl_deriv_central (&F, x, 1e-8, &result, &abserr);
20       printf ("Central diff: f'(x) = %.10f +/- %.10f\n", result,
               abserr);
21       gsl_deriv_forward (&F, x, 1e-8, &result, &abserr);
22       printf ("Forward diff: f'(x) = %.10f +/- %.10f\n", result,
               abserr);
```

```
23      return 0;
24 }
```

The following listing shows the Linux shell commands to compile, link, and execute the program, followed by the execution output of the program.

```
$ gcc -Wall deriv1b.c -o deriv1b.out -lgsl -lgslcblas -lm
$ ./deriv1b.out
Computing the derivative of f(x) = x^2 at x = 2.0
Exact value: f'(x) = 4.0000000000
With central diff: f'(x) = 3.9999999757 +/- 0.0000007105
With forward diff: f'(x) = 3.9999959493 +/- 0.0000188910
```

19.4   Numerical Integration

Several methods exist for the approximation of the integration of functions. The simplest methods are the Trapezoid method and Simpson's method. A more advanced method is that of Gauss Quadrature. The integral of function $f(x)$ is formulated as:

$$I = \int_{x_a}^{x_b} f(x)\,w(x)\,dx$$

The function $w(x)$ is a weight function of $f(x)$, x_a is the lower bound, and x_b is the upper bound of the integration interval. In the practical cases presented in this chapter, $w(x) \equiv 1$.

19.4.1   Area under a Curve

A general method for approximating the area under a curve in the interval $x = x_a$ and $x = x_b$ is the trapezoid method. It consists of dividing the interval $[x_a, x_b]$ into several *trapezoids*, computing the areas of the trapezoids, and adding these areas.

A trapezoid is a geometric figure with four sides and only two parallel opposite sides. The area of a trapezoid with width $\Delta x = x_{i+1} - x_i$, is computed as:

$$q = \Delta x \frac{y_i + y_{i+1}}{2}$$

There are $n - 1$ equal subintervals, Δx, in the interval $[x_a, x_b]$ on variable x and $y_i = f(x_i)$. The sum of the areas of the trapezoids is:

$$A = \sum_{k=1}^{k=n-1} [\Delta x \frac{1}{2}(y_k + y_{k+1})]$$

The approximation of the area can be improved with smaller value of the width (Δx) of the trapezoids. Figure 19.2 shows a segment of a curve divided into $n-1$ trapezoids. The area from x_a to x_b with: $x_a = x_1 < x_2 < \ldots < x_n = x_b$ is:

$$A = \frac{x_b - x_a}{2n}[y_1 + 2y_2 + \ldots + 2y_{n-1} + y_n]$$

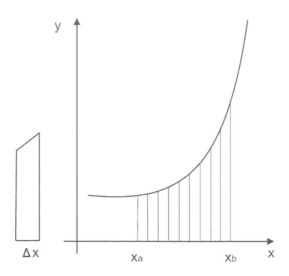

FIGURE 19.2: The area under a curve.

19.4.2 Trapezoid Method

Numerical integration consists of approximating the integral of a function in a given interval. The area under a curve on an interval is the basic concept used for computing an approximation of the integral of a function. This applies for functions continuous and nonnegative on the given interval. For functions that have negative values in a subinterval, the area of this subinterval is given a negative sign. The program in Listing 19.3, `integ.c`, implements the Trapezoid method to compute an approximate integral of function $f(x) = e^x$ in the interval $(0.0, 1.0)$ of x.

Listing 19.3: C program that computes an integral of $f(x) = e^x$

```
1  /* This program computes the numerical integral of f(x)=e^x
2       on the interval (0, 1). It uses the trapezoid method starting
3       with small number of trapezoids and increasing.
4       J M Garrido, may 7, 2013.
5       File: intg.c
6  */
7  #include <stdio.h>
8  #include <math.h>
9  double exactint = 1.718282;
```

```
10 double myf (double x) {
11      return (exp(x));
12 }
13 int main (void) {
14      double xa, xb;
15      double deltax;
16      double sum;
17      double x, y;
18      int i;
19      double error;
20      int n = 4;
21      xa = 0.0;
22      xb = 1.0;
23      printf ("Computing the integral of f(x) = e^x \n");
24      while (n <= 8182) {
25        deltax = (xb - xa)/n;
26        sum = 0.0;
27        x = xa;
28        for (i = 1; i < n; i++) {
29          x = x + deltax;
30          y = myf(x);
31          sum = sum + y;
32        }
33        sum = deltax * ((myf(xa)+myf(xb))/2.0 + sum);
34        error = (sum - exactint);
35        printf ("n = %d Sum = %.15f error = %.15e \n", n, sum,
                 error);
36        n = n * 2;
37      } // end while
38      return 0;
39 }
```

The following listing shows the compilation, linkage, and execution of the program. Note that with $n = 256$ (number of trapezoids), the computed approximation is reasonably close to the exact value of the integral, 1.8282.

```
$ gcc -Wall integ.c -lm
$ ./a.out
Computing the integral of f(x) = e^x
n = 4 Sum = 1.727221904557517 error = 8.939904557516476e-003
n = 8 Sum = 1.720518592164302 error = 2.236592164301721e-003
n = 16 Sum = 1.718841128579995 error = 5.591285799944057e-004
n = 32 Sum = 1.718421660316327 error = 1.396603163270616e-004
n = 64 Sum = 1.718316786850094 error = 3.478685009405957e-005
n = 128 Sum = 1.718290568083478 error = 8.568083477822341e-006
n = 256 Sum = 1.718284013366820 error = 2.013366819708651e-006
n = 512 Sum = 1.718282374686094 error = 3.746860941511443e-007
n = 1024 Sum = 1.718281965015813 error = -3.498418710279339e-008
n = 2048 Sum = 1.718281862598238 error = -1.374017626343260e-007
n = 4096 Sum = 1.718281836993845 error = -1.630061547963635e-007
```

19.4.3 Numerical Integration with C and the GSL

The GSL provides several functions that implement methods for adaptive and non-adaptive integration of general functions, and some functions with specialized methods for specific cases. In the function call, the user specifies the absolute and relative error bounds required.

In partitioning the integration interval, the adaptive functions tend to adjust according to the behavior of the function $f(x)$.

The most common GSL functions that can be used for numerical integration are (taken from the GSL Reference Manual):

```
int gsl_integration_qag (const gsl_function *ufunc,
    double a, double b, double epsabs, double epsrel,
    size_t limit, int key,
    gsl_integration_workspace *workspace, double *result,
    double *abserr)
```

This function implements an adaptive integration method that requires a workspace structure. It returns the approximation of the integral in parameter `result`, and an estimate of the absolute error in `abserr`. The function to integrate is supplied in the program and referenced by `ufunc`. Parameters a and b are the lower and upper bounds of the integration interval. Parameter `limit` is the maximum number of subintervals to use and may not exceed the allocated size of the workspace. The value of parameter `key` depends on the integration rule (Gauss-Kronrod rules) applied and varies from 1 to 6.

```
int gsl_integration_qng (const gsl_function *ufunc,
    double a, double b, double epsabs, double epsrel,
    double *result, double *abserr, size_t *neval)
```

This function implements a non-adaptive integration method. The parameters are similar to the previous function.

```
int gsl_integration_qags (const gsl_function *ufunc,
    double a, double b, double epsabs, double epsrel,
    size_t limit, gsl_integration_workspace *workspace,
    double *result, double * abserr)
```

This function implements an adaptive integration method with singularities.

Listing 19.4 shows a C program that uses the GSL to compute an approximation of the integral of the function $f(x) = e^x$, which is supplied as function `myf` in the program.

Listing 19.4: C program that computes integral of $f(x) = e^x$ and uses the GSL

```
1 /* Compute the numerical integral of f(x) = e^x in the
2    interval (0.0, 1.0), using the GSL adaptive integration
3    J M Garrido, May 12, 20123
4    File: gslints1.c
5 */
6 #include <stdio.h>
```

```
 7 #include <math.h>
 8 #include <gsl/gsl_integration.h>
 9 double myf (double x, void * params) {
10     return (exp(x));
11 }
12 int main (void) {
13     gsl_integration_workspace * w
14         = gsl_integration_workspace_alloc (1000);
15     double intres, error;
16     double exactint = 1.718282;
17     double alpha = 1.0;
18     double xa = 0.0;  // integration interval
19     double xb = 1.0;
20     int key = 3; // integration rule: Gauss31
21     gsl_function F;
22     F.function = &myf;
23     F.params = &alpha;
24     gsl_integration_qag (&F, xa, xb, 0.0, 1e-7, 1000, key,
25         w, &intres, &error);
26     printf ("Computed integral of f(x) = e^x: % .15f\n", intres);
27     printf ("exact integral   = % .15e\n", exactint);
28     printf ("estimated error = % .15e\n", error);
29     printf ("actual error    = % .15e\n", intres - exactint);
30     gsl_integration_workspace_free (w);
31     return 0;
32 }
```

The following listing shows compilation and linkage of the previous C program with the GSL using the script file gslblas. This is followed by the execution results.

```
$ ./gslblas gslints1.c
$ ./a.out
Computed integral of f(x) = e^x:  1.718281828459045
exact integral   =  1.718282000000000e+00
estimated error =  1.907676048750245e-14
actual error    = -1.715409549962743e-07
```

19.5   Differential Equations

An ordinary differential equation (ODE) is one that includes derivatives of a function with a single independent variable. For example the following equation is a differential equation of order one.

$$\frac{dy}{dx} + 20y - 6x = 23.5$$

The order of a differential equation is determined by the highest derivative. The following expression is an example of a differential equation of order two.

$$y'' + 20y'x - 3y'y = 12$$

The general form of a differential equation of order n is expressed as:

$$\frac{d^n z(t)}{dt^n} + a_1 \frac{d^{n-1} z(t)}{dt^{n-1}} + \cdots + a_{n-1} \frac{dz(t)}{dt} + a_n z(t) = Q \tag{19.1}$$

The solution to a differential equation is an expression for the function $y = f(x)$. It is convenient when solving a differential equation of a higher order with numerical methods, to reduce the order of the equation to order one.

A differential equation of order n can be reduced to n first-order differential equations. The following variable substitutions are necessary to reduce Equation (19.1) to a system of n first-order equations:

$$x_1 = z(t), \quad x_2 = \frac{dz(t)}{dt}, \quad x_3 = \frac{d^2 z(t)}{dt^2}, \quad \ldots, \quad x_n = \frac{d^{n-1} z(t)}{dt^{n-1}} \tag{19.2}$$

Substituting the Equations (19.2) in Equation (19.1), the following first-order equations result:

$$\frac{dx_1}{dt} = x_2$$

$$\frac{dx_2}{dt} = x_3$$

$$\frac{dx_3}{dt} = x_4$$

$$\cdots$$

$$\frac{dx_n}{dt} = Q - a_1 x_n - a_2 x_{n-1} - \cdots - a_{n-1} x_2 - a_n x_1 \tag{19.3}$$

19.6   Dynamical Systems and Models

Models of dynamical systems describe the behavior of systems varying with time. Differential equations are used to model this behavior as it changes continuously with time. A model of a dynamical system has two major components:

- The state vector that indicates the current state at a particular time instance

- A set of (linear) differential equations that describe the continuous change of state.

Figure 19.3 illustrates the high-level view of a dynamical system and includes the variables used in the modeling: $U(t)$ is the vector of input variables, $Y(t)$ is the vector of output variables, and $X(t)$ is the vector of state variables. All these variables are functions of time.

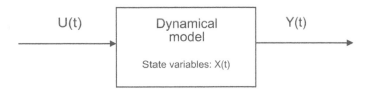

FIGURE 19.3: High-level view of a dynamical system.

The dynamic behavior of a continuous linear system is described by the following set of linear differential equations.

$$\begin{aligned}
\dot{x}_1 &= a_{11}x_1 + a_{12}x_2 + \ldots + a_{1n}x_n + b_{11}u_1 + b_{12}u_2 + \cdots + b_{1m}u_m \\
\dot{x}_2 &= a_{21}x_1 + a_{22}x_2 + \ldots + a_{2n}x_n + b_{21}u_1 + b_{22}u_2 + \cdots + b_{2m}u_m \\
&\ldots \\
\dot{x}_n &= a_{n1}x_1 + a_{n2}x_2 + \ldots + a_{nn}x_n + b_{n1}u_1 + b_{n2}u_2 + \cdots + b_{nm}u_m
\end{aligned} \tag{19.4}$$

19.6.1 State Equations

Equations (19.4) are known as *state equations*, and are expressed in matrix form as:

$$\dot{X} = AX + BU \tag{19.5}$$

This state equation uses the following matrix definitions:

$$\dot{X} = \begin{bmatrix} \dot{x}_1 \\ \dot{x}_2 \\ \vdots \\ \dot{x}_n \end{bmatrix} \quad A = \begin{bmatrix} a_{11} & a_{12} & \cdots & a_{1n} \\ a_{21} & a_{22} & \cdots & a_{2n} \\ \vdots & \vdots & \ddots & \vdots \\ a_{m1} & a_{2m} & \cdots & a_{mn} \end{bmatrix} \quad X = \begin{bmatrix} x_1 \\ x_2 \\ \vdots \\ x_n \end{bmatrix}$$

$$B = \begin{bmatrix} b_{11} & b_{12} & \cdots & b_{1m} \\ b_{21} & b_{22} & \cdots & b_{2m} \\ \vdots & \vdots & \ddots & \vdots \\ b_{n1} & b_{n2} & \cdots & b_{nm} \end{bmatrix} \quad U = \begin{bmatrix} u_1 \\ u_2 \\ \vdots \\ u_m \end{bmatrix} \tag{19.6}$$

In the state equations, $\dot{x}$ that denotes dx/dt, A is an $m \times n$ matrix, X is a column vector of size n, $\dot{X}$ is a column vector of size n, B is an $n \times m$ matrix, and U is a column vector of size m.

19.6.2   Output Equations

The output equations of a model of a dynamical system are expressed as follows:

$$
\begin{aligned}
y_1 &= c_{11}x_1 + c_{12}x_2 + \ldots + c_{1n}x_n + d_{11}u_1 + d_{12}u_2 + \cdots + d_{1m}u_m \\
y_2 &= c_{21}x_1 + c_{22}x_2 + \ldots + c_{2n}x_n + c_{21}u_1 + d_{22}u_2 + \cdots + d_{2m}u_m \\
&\ldots \\
y_n &= a_{k1}x_1 + a_{k2}x_2 + \ldots + a_{kn}x_n + d_{k1}u_1 + d_{k2}u_2 + \cdots + d_{km}u_m \quad (19.7)
\end{aligned}
$$

Equation (19.7) can also be written in matrix form as: $Y = CX + DU$, in which C is an $k \times m$ matrix, Y is a column vector of size k, D is an $k \times m$ matrix, and U is a column vector of size m.

In a more compact form, the model of a dynamical system can be expressed with two matrix equations:

$$
\begin{aligned}
\dot{X} &= AX + BU \\
Y &= CX + DU
\end{aligned} \quad (19.8)
$$

19.7   Formulating Simple Examples

This section describes the formulation of problems using state variables with differential equations and applying basic laws of physics.

19.7.1   Free-Falling Object

An object is released from a certain height, h_0, and falls freely until it reaches ground level. The problem is to study the changes in the vertical location of the object and its velocity as time progresses.

Assume that the mass of the object is m, and the only force applied on the object is that due to gravity, g. Let x denote the vertical displacement of the object, that is, its height as a function of time, and v its vertical velocity. Applying Newton's law of force that relates mass, acceleration and force, gives the following expression:

$$
-mg = m\frac{d^2x}{dt^2}
$$

This differential equation of order 2 can be reduced to two first-order differential equations. Because the velocity is the instantaneous rate of change of the displacement and the acceleration is the instantaneous rate of change of the velocity, the two first-order differential equations are:

$$v = \frac{dx}{dt}$$

$$-mg = m\frac{dv}{dt}$$

The two state variables are the velocity, v, and the displacement, x. There is only one input variable, u_1, and its value is the constant g. The two state equations in the form of general state equations in (19.4), are the following:

$$\frac{dx}{dt} = 0x + v + 0$$

$$\frac{dv}{dt} = 0x + 0v + -g$$

Following the general matrix and vector form in Equation (19.6), the state vector X, matrix A, matrix B, and vector U are:

$$X = \begin{bmatrix} x \\ v \end{bmatrix} \quad A = \begin{bmatrix} 0 & 1 \\ 0 & 0 \end{bmatrix} \quad B = \begin{bmatrix} 0 \\ -1 \end{bmatrix} \quad U = [g]$$

The output equations are:

$$y_1 = x + 0v$$

$$y_2 = 0x + v$$

Matrix C and matrix D are:

$$C = \begin{bmatrix} 1 & 0 \\ 0 & 1 \end{bmatrix} \quad D = \begin{bmatrix} 0 \\ 0 \end{bmatrix}$$

19.7.2   Object on Horizontal Surface

A force, F, is applied to an object on a horizontal surface. The resistance of the surface on the object due to friction is proportional to the velocity of the object, and its value is $-kv$. The constant k is the coefficient of friction of the surface. The horizontal displacement of the object is denoted by x. The problem is to find the instantaneous change in the displacement of the object.

The dynamic behavior of the object is defined by Newton's law, and is expressed as follows:

$$m\frac{d^2x}{dt^2} = F - kv$$

Because the velocity is the instantaneous change of the displacement, and the acceleration is the instantaneous change in the velocity of the object, the previous equation is expressed as follows:

$$v = \frac{dx}{dt}$$

$$m\frac{dv}{dt} = F - kv$$

For this problem, the two state variables are displacement x, and the velocity v, of the object. The input variable, u_1, is the force F. The two state equations are expressed as follows:

$$\frac{dx}{dt} = 0x + v + 0$$

$$\frac{dv}{dt} = 0x - \frac{kv}{m} + F/m$$

In this problem, the state vector X, matrix A, matrix B, and vector U are:

$$X = \begin{bmatrix} x \\ v \end{bmatrix} \quad A = \begin{bmatrix} 0 & 1 \\ 0 & -k/m \end{bmatrix} \quad B = \begin{bmatrix} 0 \\ 1/m \end{bmatrix} \quad U = [F]$$

The output equations are:

$$y_1 = x + 0v$$

$$y_2 = 0x + v$$

Matrix C and matrix D are:

$$C = \begin{bmatrix} 1 & 0 \\ 0 & 1 \end{bmatrix} \quad D = \begin{bmatrix} 0 \\ 0 \end{bmatrix}$$

19.7.3 Object Moving on an Inclined Surface

A force F is applied to an object on an inclined surface. The elevation angle is θ. The frictional force that resists movement is proportional to the velocity of the object. The problem is to derive the change of the displacement and the velocity of the object on the inclined surface.

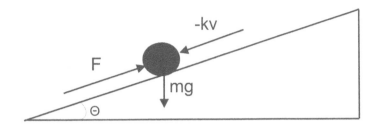

FIGURE 19.4: Object on inclined surface.

The projection of the force of gravity of the object on the inclined surface is $mg \sin \theta$. As explained in previous problems, applying the law of Newton, the following equations are derived:

$$v = \frac{dx}{dt}$$

$$m \frac{dv}{dt} = F - kv - mg \sin \theta$$

For this problem, the two state variables are displacement x, and the velocity v, of the object. The input variables are: the force F and the acceleration due to the gravity. The two state equations are expressed as follows:

$$\frac{dx}{dt} = 0x + v + 0$$

$$\frac{dv}{dt} = 0x - \frac{kv}{m} + F/m - g \sin \theta$$

In this problem, the state vector X, matrix A, matrix B, and vector U are:

$$X = \begin{bmatrix} x \\ v \end{bmatrix} \quad A = \begin{bmatrix} 0 & 1 \\ 0 & -k/m \end{bmatrix} \quad B = \begin{bmatrix} 0 & 0 \\ 1/m & -\sin \theta \end{bmatrix} \quad U = \begin{bmatrix} F \\ g \end{bmatrix}$$

The output equations are:

$$y_1 = x + 0v$$

$$y_2 = 0x + v$$

Matrix C, vector Y, and matrix D are:

$$C = \begin{bmatrix} 1 & 0 \\ 0 & 1 \end{bmatrix} \qquad Y = \begin{bmatrix} x \\ v \end{bmatrix} \qquad D = \begin{bmatrix} 0 & 0 \\ 0 & 0 \end{bmatrix}$$

19.8   Solution of Differential Equations

There are many ordinary differential equations (ODEs) that cannot be solved analytically. Numerical methods are techniques that compute estimates using software implementations. Euler's method is the simplest technique for solving differential equations numerically. With this method, the time step is constant from one iteration to the next. However, this may not be feasible for many functions or may result in an inaccurate solution.

Methods that adjust the time step as the computation proceeds are known as *adaptive* methods. The Dormand–Prince pair of Runge–Kutta is one of the best adaptive methods.

Part of the description of using GSL for ordinary differential equations is taken from the GNU Reference Manual. With the GSL, a system of differential equations is defined using the *gsl_odeiv2_system* data type. The following functions are available with the GSL:

`gsl_odeiv2_step_rk2`
   This function uses the Runge–Kutta (2, 3) method.

`gsl_odeiv2_step_rk4`
   This function uses the 4th order Runge–Kutta. Error estimation is carried out by the step doubling method.

`gsl_odeiv2_step_rkf45`
   This function uses the Runge–Kutta–Fehlberg (4, 5) method, which is a good general-purpose integrator.

`gsl_odeiv2_step_rkck`
   This function uses the Runge–Kutta Cash–Karp (4, 5) method.

`gsl_odeiv2_step_rk8pd`
   This function uses the Runge–Kutta Prince–Dormand (8, 9) method.

`gsl_odeiv2_step_rk1imp`
   This function uses a Gaussian first order Runge–Kutta (Euler or backward Euler method). Error estimation is carried out by the step doubling method. It requires the Jacobian and access to the driver object with `gsl_odeiv2_step_set_driver`.

`gsl_odeiv2_step_rk2imp`

>   This function uses a Gaussian second order Runge–Kutta (mid-point rule). Error estimation is carried out by the step doubling method. It requires the Jacobian and access to the driver object with `gsl_odeiv2_step_set_driver`.

`gsl_odeiv2_step_rk4imp`

>   This function uses a Gaussian 4th order Runge–Kutta. Error estimation is carried out by the step doubling method. It requires the Jacobian and access to the driver object with `gsl_odeiv2_step_set_driver`.

`gsl_odeiv2_step_bsimp`

>   This function uses the Bulirsch–Stoer method of Bader and Deuflhard. It is generally suitable for stiff problems and requires the Jacobian.

`gsl_odeiv2_step_msadams]`

>   This function uses the Adams method in Nordsieck form (a variable-coefficient linear multistep). It uses explicit Adams–Bashforth (predictor) and implicit Adams–Moulton (corrector) methods in $P(EC)^m$ functional iteration mode. Method order varies dynamically between 1 and 12. It requires the access to the driver object with `gsl_odeiv2_step_set_driver`.

`gsl_odeiv2_step_msbdf`

>   This function uses a variable-coefficient linear multistep backward differentiation formula (BDF) method in Nordsieck form. It uses the explicit BDF formula as predictor and the implicit BDF formula as corrector.
>
>   A modified Newton iteration method is used to solve the system of non-linear equations. It is generally suitable for stiff problems and requires the Jacobian and the access to the driver object via `gsl_odeiv2_step_set_driver`.

In a *stiff problem*, some methods for solving the equation are numerically unstable, and the step size is forced to be unacceptably small in a region where the solution curve is very smooth.

The general structure of a C program that solves numerically a set of first-order ordinary differential equations with the GSL has the following sequence of parts in its code:

1. Set up the ODE system of type *gsl_odeiv2_system*. The components of the ODE system are: the programmer-defined function with the right-hand side of the first-order differential equations to solve, the dimension or number of differential equations to solve, and a pointer to an array of parameters used in the equations.

2. A call to one of the GSL driver functions that starts the ODE solver, for example, function *gsl_odeiv2_driver_alloc_y_new*. The arguments to the function call are: the name of the ODE system structure; the name of the GSL stepper method, for example *gsl_odeiv2_step_rkf45*; the starting step size for ode solver; the absolute error requested; the relative error requested.

3. A loop that performs the evolution of the system by advancing the system from the initial time to the final time, at each step, from time t and position y using the stepping function step. The new time and position are stored in t and y on output. An evolution function is called in the loop, for example *gsl_odeiv2_evolve_apply*.

4. A programmer-defined function that specifies the set of differential equations to solve. This function defines an array and expressions of the right-hand side of the equations.

19.8.1 Model with a Single Differential Equation

This case study is a very simple model represented by only a single first-order differential equation. The problem consists of solving numerically the following differential equation:

$$\frac{dx}{dt} = \alpha x [1 + \sin(\omega t)] \qquad (19.9)$$

This equation is solved with $\alpha = 0.015$ and $\omega = 2\pi/365$, an initial value $x(0) = 2$, and an interval of t from 0.0 to 365.0.

Listing 19.5 shows the C program that solves numerically the single differential equation and it is stored in file tode1.c. The structure for the ode system is declared in line 33 and the values of the components are set up in lines 38–40. The GSL driver function is called in lines 46–47. The loop that performs the evolution of the system is coded in lines 58–66. The programmer-defined function *dfunc* that specifies the differential equation (19.9) appears in lines 74–82.

Listing 19.5: C program that solves a single differential equation

```
 1 /* Program: tode1.c
 2    This program solves a single differential equation
 3    dxdt = a x [1+ sin(w t)]
 4    This problem is the C/GSL version from original Matlab version
 5    J M Garrido, Feb 2013. CS dept, KSU
 6    Uses GSL_odeiv2 driver ODE functions
 7    This program solves the equation
 8        x'(t) - alpha x(t) (1+sin(omega t)) = 0
 9        ==>  dy[0]/dt = f[0] = alpha * y[0] * (1+ sin (omega*t))
10 */
11 #include <stdio.h>
12 #include <math.h>
13 #include <gsl/gsl_errno.h>
14 #include <gsl/gsl_matrix.h>
15 #include <gsl/gsl_odeiv2.h>
16
17 /* function prototype */
18 /* dfunct - defines the first order differencial equation */
19 int dfunc (double t, const double y[], double f[],
           void *params_ptr);
```

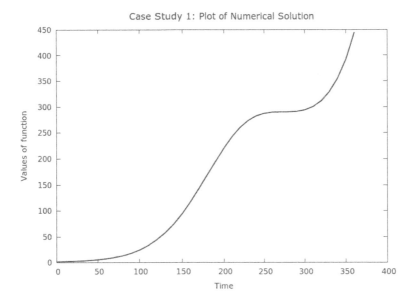

FIGURE 19.5: Case Study 1: Model with single differential equation.

```
20 int main () {
21    const int dimension = 1; /* number of differential equations */
22    int status;                /* status of driver function */
23    const double eps_abs = 1.e-8;   /* absolute error requested  */
24    const double eps_rel = 1.e-10; /* relative error requested */
25    double alpha = 0.015;           /* parameter for the diff eq */
26    double omega = 2.0*M_PI/365.0; /* parameter for the diff eq */
27    double myparams[2];             /* array for parameters      */
28
29    double y[dimension];   /* current solution vector */
30    double t, t_next;      /* current and next independent variable */
31    double tmin, tmax, delta_t; /* range of t and output step size */
32    double h = 1.0e-6;     /* starting step size for ode solver */
33    gsl_odeiv2_system ode_system;
34    myparams[0] = alpha;        /* problem parameters */
35    myparams[1] = omega;
36    printf("\n Solving a system with a single diff  equation\n\n");
37    /* load values into the ode_system structure */
38    ode_system.function = dfunc;    /* the right-hand-side */
39    ode_system.dimension = dimension; /* number of diffeq's */
40    ode_system.params = myparams; /* parameters pass to dfunc */
41
42    tmin = 0.0; /* starting t value */
43    tmax = 365.0; /* final t value */
44    delta_t = 1.0;
45    y[0] = 2.0; /* initial value of x */
46    gsl_odeiv2_driver * drv =
47       gsl_odeiv2_driver_alloc_y_new(&ode_system,
            gsl_odeiv2_step_rkf45, h, eps_abs, eps_rel);
```

```
48    printf("Input data: \n");
49    printf(" alpha = %g; omega = %g\n", alpha, omega);
50    printf(" Starting step size (h): %0.5e\n", h);
51    printf(" Time parameters: %f %f %f \n", tmin, tmax,
          delta_t);
52    printf(" Absolute and relative error requested:
              %0.6e %0.6e \n", eps_abs, eps_rel);
53    printf(" Number of equations (dimension): %d \n\n", dimension);
54    printf("    Time          dx/dt          \n");
55    t = tmin;              /* initialize t */
56    printf ("%.5e  %.5e  \n", t, y[0]); /* initial values */
57    /* step from tmin to tmax */
58    for (t_next = tmin + delta_t; t_next <= tmax; t_next += delta_t)
59    {
60        status = gsl_odeiv2_driver_apply (drv, &t, t_next, y);
61        if (status != GSL_SUCCESS) {
62            printf("Error: status = %d \n", status);
63            break;
64        }
65        printf ("%.5e %.5e \n", t, y[0]); /* print at t=t_next */
66    } // end for
67    gsl_odeiv2_driver_free (drv);
68    return 0;
69 }
70 /*
71    Define array of right-hand-side functions y[i].
72    params is a void pointer that is used in many GSL routines
73 */
74 int dfunc (double t, const double y[], double f[],
          void *params_ptr) {
75    double *lparams = (double *) params_ptr;
76    /* get parameter(s) from params_ptr */
77    double alpha = lparams[0];
78    double omega = lparams[1];
79    /* evaluate the right-hand-side functions at t */
80    f[0] = alpha * y[0] * (1+ sin (omega*t));
81    return GSL_SUCCESS; /* GSL_SUCCESS in gsl/errno.h as 0 */
82 }
```

The resulting output is an array that includes a vector with the values of time and a vector with the values of x. Figure 19.5 shows the graph generated with Gnuplot of the values of x with time and is a visual representation of the numerical solution of the differential equation (19.9).

19.8.2   Model with a System of Differential Equations

Most practical models are represented by a system of first-order and/or higher-order differential equations. For example, consider the following system of linear first-order differential equation:

$$\frac{dx_1}{dt} = -x_1 - x_2$$

$$\frac{dx_2}{dt} = x_1 - 2x_2 \qquad (19.10)$$

These equations can be expressed in the form of state equations and for this, state vector X and matrix A are:

$$X = \begin{bmatrix} x_1 \\ x_2 \end{bmatrix} \qquad A = \begin{bmatrix} -1 & -1 \\ 1 & -2 \end{bmatrix} \qquad (19.11)$$

For this problem, matrix B, and vector U are empty. The output equations are:

$$y_1 = x_1 + 0x_2$$
$$y_2 = 0x_1 + x_2$$

Matrix C, and vector Y are expressed in the following form:

$$C = \begin{bmatrix} 1 & 0 \\ 0 & 1 \end{bmatrix} \qquad Y = \begin{bmatrix} x_1 \\ x_2 \end{bmatrix} \qquad (19.12)$$

This problem can be solved numerically using C and the GSL in a similar manner as with the previous case study. The C code of the programmer-defined function *dfunc* that specifies the two first-order differential equations (19.10) is shown as follows.

```
int dfunc (double t, const double y[], double f[], void *params_ptr)
{
  /* evaluate the right-hand-side functions at t */
  f[0] = - y[0] - y[1];
  f[1] =   y[0] - 2.0 * y[1] ;
  return GSL_SUCCESS;    /* GSL_S7UCCESS in gsl/errno.h as 0 */
}
```

The initial conditions are set as $x_1 = 1.0$ and $x_2 = 1.0$ at $t = 0$; the time span is set for values of t from 0.0 to 5.0. The C program that implements the solution of the model with the two differential equations (19.10) is stored in file tode2.c.

Figure 19.6 shows the graph generated by Gnuplot of the values of variables x and v with time, and represents visually the numerical solution of the differential equations (19.10).

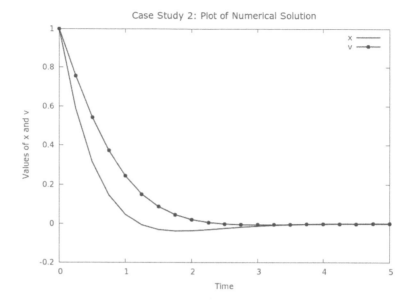

FIGURE 19.6: Numerical solution to a system of two differential equations.

19.8.3 Model with Drag Force

An object is released from a specified height, h_0, and falls freely until it reaches ground level. The problem is to formulate and solve a model to study the changes in the vertical position of the object and its velocity as time progresses.

Assume that the mass of the object is m, and there are two forces applied on the object: one due to gravity with acceleration g and the second is the drag force against the direction of movement. Let x denote the vertical displacement of the object, and v its vertical velocity. The drag force is cv^2, in which c is the drag constant. Applying Newton's law of force that relates mass, acceleration and force, gives the following expression:

$$-mg + cv^2 = m\frac{d^2x}{dt^2}$$

The vertical velocity is the instantaneous rate of change of the vertical position and the acceleration is the instantaneous rate of change of the vertical velocity. The second-order differential equation can be reduced to two first-order differential equations:

$$v = \frac{dx}{dt}$$
$$-mg + cv^2 = m\frac{dv}{dt}$$

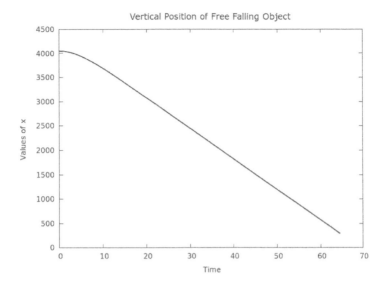

FIGURE 19.7: Vertical position of free-falling object.

The two state variables are the vertical position, x and the vertical velocity, v. The two state equations are the following:

$$\frac{dx}{dt} = v$$

$$\frac{dv}{dt} = (c/m)v^2 - g$$

The output equations are:

$$y_1 = x$$

$$y_2 = v$$

This problem is solved with the parameters $m = 80.0$ and $c = 0.2$; the acceleration constant g is always 9.8 (m/s^2). The C program that implements the solution to the model of the free-falling object is stored in file `odefall.c`. The C code of the programmer-defined function *dfunc* that specifies the two first-order differential equations is shown as follows.

```
int dfunc (double t, const double y[], double f[], void *params_ptr)
{
    /* get parameter(s) from params_ptr */
```

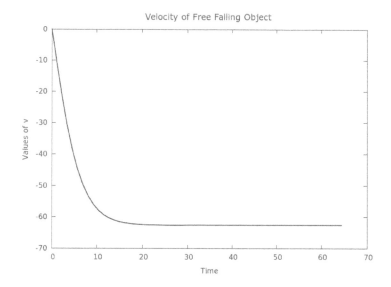

FIGURE 19.8: Vertical velocity of free-falling object.

```
double *lparams = (double *) params_ptr;
double m = lparams[0];   /* mass of object */
double c = lparams[1];   /* drag constant */
/* evaluate the right-hand-side functions at t */
f[0] = y[1];
f[1] = (c/m) * y[1] * y[1] - G;
return GSL_SUCCESS; /* GSL_SUCCESS defined in gsl/errno.h as 0 */
}
```

Figure 19.7 shows the graph of the vertical displacement with time generated by Gnuplot. Observe that after about 10 seconds, the displacement changes linearly. Figure 19.8 shows the graph of vertical velocity of the object with time. Observe that the velocity of the object increases negatively at an exponential rate until about 20 seconds. After this time instant, the velocity remains constant.

19.8.4   Prey and Predator Model

This model helps to study how the population of two animal species changes over time. The prey (rabbits) population is represented by $x_1(t)$. The predator (wolves) population is represented by $x_2(t)$. Without the predator, the prey population will grow as:

$$\frac{dx_1}{dt} = ax_1, \qquad a > 0$$

Without the prey, the population of the predator will decrease as:

$$\frac{dx_2}{dt} = -bx_2, \qquad b > 0$$

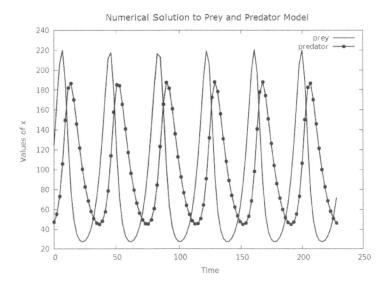

FIGURE 19.9: Population changes of prey and predator.

In these expressions, a and b are constants. When the two species live and interact in the same environment, the prey population will decline when the predator population increases. This interaction produces additional changes in the population of the two species that are given by the following state equations:

$$\frac{dx_1}{dt} = ax_1 - cx_1x_2$$
$$\frac{dx_2}{dt} = -bx_2 + dx_1x_2$$

The output equations are:

$$y_1 = x_1$$
$$y_2 = x_2$$

For a numerical solution, assume $a = 0.25$, $b = 0.12$, $c = 0.0025$, and $d = 0.0013$. At the beginning, there are 125 rabbits and 47 wolves, so the initial conditions for the problem are: $x_1(0) = 125$ and $x_2(0) = 47$.

The C function that implements the evaluation of the two differential equations is shown in the following listing and the complete C program is stored in file `predprey.c`.

```
int dfunc (double t, const double y[], double f[], void *params_ptr)
{
    /* No need to get parameter(s) from params_ptr
       because these are declared as global  */
    /* evaluate the right-hand-side functions at t */
    f[0] =   a * y[0]   -   c * y[0] * y[1];
    f[1] = - b * y[1] + d * y[0] * y[1];
    return GSL_SUCCESS;   /* defined as 0 */
} // end of dfunct
```

Figure 19.9 shows the graph of the population changes of the prey and predator over time. Observe that after about 10 seconds, the displacement changes linearly.

Figure 19.10 shows the phase plot of the two population changes. Because this shows a closed curve, it implies that the prey and predator populations follow periodic cycles.

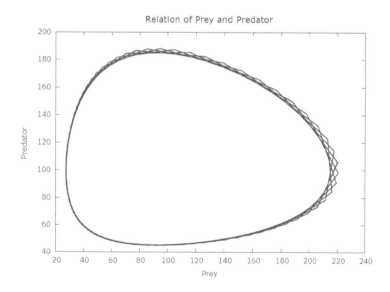

FIGURE 19.10: Phase plot of the population changes.

Summary

Function differentiation and integration are used to better understand the behavior of functions that represent relationships in the properties of a system and are important in formulating computational models. Many systems in science and engineering can be modeled with ordinary differential equations. This chapter presented how to implement mathematical models that are formulated with differential equations, and their numerical solutions. Several case studies were presented and discussed with solutions implemented with the GSL and the C programming language .

Key Terms		
functions	derivative	integrals
rate of change	area under a curve	finite differences
continuous models	dynamical systems	state variables
state equations	output equations	differential equations
initial values	Runge-Kutta methods	

Exercises

Exercise 19.1 Develop a C program that computes the derivate of function $f(x) = (5x - 3)^2 e^x$ at the point $x = 2$. Use finite differences.

Exercise 19.2 Develop a C program that uses the GSL and computes the derivate of function $f(x) = (5x - 3)^2 e^x$ at the point $x = 2$.

Exercise 19.3 An object falls freely, the vertical position of the object is: $y(t) = 16t^2 + 32t + 6$. Compute the velocity and the acceleration at various values of time. To solve this problem, develop a C program that applies finite differences (central, backward, and forward).

Exercise 19.4 An object falls freely, the vertical position of the object is: $y(t) = 16t^2 + 32t + 6$. Compute the velocity and the acceleration at various values of time. To solve this problem, develop a C program that uses the GSL and applies finite differences.

Exercise 19.5 Develop a C program that computes an approximation of the integral of the function $f(x) = x^2$ in the interval $(0.0, 2.0)$ of x. In the program, apply the Trapezoid method. Use a value $n = 6$, then a value $n = 20$.

Exercise 19.6 Develop a C program that uses the GSL and computes an approximation of the integral of the function $f(x) = x^2$ in the interval $(0.0, 2.0)$ of x.

Exercise 19.7 Develop a C program that implements and solves numerically the following mathematical model, and generate a graph with Gnuplot using the values of the variables in the numerical solution. Use a time interval from $t = 0$ to $t = 8.0$.

$$\frac{dy}{dt} = t + e^{t/2\pi} \cos(2\pi t)$$

Exercise 19.8 This problem presents a simplified model of a water heater. The temperature of the water in a tank is increased by the heating elements and the temperature varies with time. The differential equation is:

$$cm\frac{dT}{dt} = q - hA(T - t_s)$$

where T is the temperature of the water as a function of time, m is the mass of the water (in Kg), A is the area of the surface of the heater, q is the constant rate at which the elements produce heat, h is the Newton cooling coefficient, and t_s is the temperature of the surroundings (a constant). Develop a C program that implements and solves numerically the mathematical model, and generate a graph with Gnuplot of the temperature with time. Use the following values: $c = 4175$, $m = 245$, $A = 2.85$, $h = 12.0$, $q = 3600$, $t_s = 15$. The initial condition is $T(0) = 15$ and perform computations over an interval from $t = 0.0$ to $t = 2750.0$.

Exercise 19.9 Edward N. Lorenz was a pioneer of Chaos Theory; the following simplified differential equations are known as the Lorenz Equations. Develop a C program that implements and solves numerically the mathematical model:

$$\frac{dx}{dt} = \sigma(y - x)$$
$$\frac{dy}{dt} = x(\rho - z)$$
$$\frac{dz}{dt} = xy - \beta z$$

Use the following values for the constants: $\sigma = 10$, $\beta = 8/3$, and $\rho = 28$. Use initial conditions: $x(0) = 8$, $y(0) = 8$, and $z(0) = 27$ and a time span from $t = 0$ to $t = 20$. Draw a plot with Gnuplot of the variables in the numerical solution; first plot z with respect to x, then the three variables with respect to time. Investigate and plot a three dimensional plot (x, y, z).

Exercise 19.10 Develop a C program that implements and solves numerically the following mathematical model:

$$\frac{d^2u}{dx^2} + e^x\frac{dv}{dx} + 3u = e^{2x}$$

$$\frac{d^2v}{dx^2} + \cos(x)\frac{du}{dx} + u = \sin(x)$$

Use the initial values: $u(0) = 1$, $u'(0) = 2$, $v(0) = 3$, and $v'(0) = 4$. Solve over the interval from $x = 0$ to $x = 3$. Generate a plot with Gnuplot of the variables in the numerical solution.

Exercise 19.11 Investigate the SIR model for disease spread, which was proposed by W. O. Kermack and A. G. McKendrick. In this model, three groups of population are considered: $S(t)$ is the population group not yet infected and is susceptible, $I(t)$ is the population group that has been infected and is capable of spreading the disease, and $R(t)$ is the population group that has recovered and is thus immune. The mathematical model consists of the following differential equations:

$$\frac{ds}{dt} = -\alpha si$$

$$\frac{di}{dt} = \alpha si - \beta i$$

$$\frac{dr}{dt} = \beta i$$

Use the values of the constants: $\alpha = 1$ and $\beta = 0.3$. The initial conditions are: $s = 0.999$, $i = 0.001$, and $r = 0.0$. Develop a C program that implements and solves numerically the mathematical model and plot with Gnuplot the values of the variables in the numerical solution.

Chapter 20

Linear Optimization Modeling

20.1 Introduction

Mathematical optimization consists of finding the best possible values of variables from a given set that can maximize or minimize a real function. An optimization problem can be formulated in such a form that it can be possible to find an optimal solution; this is known as mathematical modeling and is used in almost all areas of science, engineering, business, industry, and defense. The goal of optimization modeling is of formulating a mathematical model of the system and attempting to optimize some property of the model. The actual optimization is carried out by executing the computer implementation of the model. This chapter discusses the formulation of simple linear optimization models.

The formulating of a problem for linear constrained optimization is also known as *linear optimization* modeling or the mathematical modeling of a linear programming problem (LPP). Linear optimization modeling consists of four general steps, and these are as follows:

1. Identify a linear function, known as the *objective function*, to be maximized or minimized. This function is expressed as a linear function of the decision variables.

2. Identify the *decision variables* and assign to them symbolic names, x, y, etc. These decision variables are those whose values are to be computed.

3. Identify the set of *constraints* and express them as linear equations and inequations in terms of the decision variables. These constraints are derived from the given conditions.

4. Include the restrictions on the non-negative values of decision variables.

The objective function, f, to be maximized or minimized is expressed by:

$$f(x_1, x_2, \ldots, x_n) = c_1 x_1 + c_2 x_2 + \ldots + c_n x_n \tag{20.1}$$

The set of m constraints are expressed in the form:

$$a_{i1} x_1 + a_{i2} x_2 + \ldots + a_{in} x_n \leq b_i \quad i = 1, \ldots m \tag{20.2}$$

Or, of the form:

$$a_{i1}x_1 + a_{i2}x_2 + \ldots + a_{in}x_n \geq b_i \quad i = 1,\ldots m \tag{20.3}$$

The sign restrictions for variables are denoted by: $x_j \geq 0$, or $x_j \leq 0$, or x_j unrestricted in sign, $j = 1,\ldots n$. Many problems are formulated with a mixed of m constraints with $\leq$, $=$, and $\geq$ forms.

20.2   Formulation of Problems

There are many real and practical problems to which the linear optimization modeling may be applied. The following examples, although very simple because they use only two variables, help to illustrate the general method involved in linear optimization modeling.

20.2.1   Case Study 1

An industrial chemical plant produces two products, A and B. The market price for a pound of A is \$12.75, and that of B is \$15.25. Each pound of substance A produced requires 0.25 lbs of material P and 0.125 lbs of material Q. Each pound of substance B produced requires 0.15 lbs of material P and 0.35 lbs of material Q. The amounts of materials available in a week are: 21.85 lbs of material P and 29.5 lbs of material Q. Management estimates that at the most, 18.5 pounds of substance A can be sold in a week. The goal of this problem is to compute the amounts of substance A and B to manufacture in order to optimize sales.

20.2.1.1   Understanding the Problem

For easy understanding and for deriving the mathematical formulation of the problem, the data given is represented in a table as follows. As stated previously, the main resource required in the production of the chemical substances A and B are the amounts of material of type P and Q.

Material	Available	Substance of type A	Substance of type B
P	21.85	0.250	0.15
Q	29.50	0.125	0.35

20.2.1.2 Mathematical Formulation

Let x_1 denote the amount of substance (lbs) of type A to be produced, and x_2 denote the amount of substance (lbs) of type B to be produced. The total sales is $12.75x_1 + 15.25x_2$ (to be maximized). The objective function of the linear optimization model formulation of the given problem is:

Maximize $S = 12.75x_1 + 15.25x_2$
Subject to the constraints:

$$
\begin{aligned}
0.25x_1 + 0.15x_2 &\leq 21.85 \\
0.125x_1 + 0.35x_2 &\leq 29.5 \\
x_1 &\leq 18.5 \\
x_1 &\geq 0 \\
x_2 &\geq 0
\end{aligned}
$$

20.2.2 Case Study 2

A manufacturer of toys produces two types of toys: X and Y. In the production of these toys, the main resource required is machine time and three machines are used: M1, M2 and M3. The machine time required to produce a toy of type X is 4.5 hours of machine M1, 6.45 hours of machine M2, and 10.85 hours of machine M3. The machine time required to produce a toy of type Y is 7.25 hours of machine M1, 3.65 hours of machine M2, and 4.85 hours of machine M3. The maximum available machine time for the machines M1, M2, M3 are 415, 292 and 420 hours respectively. A toy of type X gives a profit of 4.75 dollars, and a toy of type Y gives a profit of 3.55 dollars. Find the number of toys of each type that should be produced to get maximum profit.

20.2.2.1 Understanding the Problem

For easy understanding and for deriving the mathematical formulation of the problem, the data given is represented in a table as follows. As stated previously, the main resource required in the production of toys is machine time of machines M1, M2, and M3.

Machine	Total time available	Req time toy type X	Req time toy type Y
M1	415	4.5	7.25
M2	292	6.45	3.64
M3	420	10.85	4.85

20.2.2.2 Mathematical Formulation

Let x denote the number of toys of type X to be produced, and y denote the number of toys of the type Y to be produced. The total profit is: $= 4.75x + 3.55y$ (to be maximized). The objective function of the linear optimization model formulation of the given problem is:

Maximize: $P = 4.75x + 3.55y$
Subject to the constraints:

$$
\begin{aligned}
4.5x + 7.25y &\leq 415 \\
6.45x + 3.65y &\leq 292 \\
10.85x + 4.85y &\leq 420 \\
x &\geq 0 \\
y &\geq 0
\end{aligned}
$$

20.2.3 Case Study 3

A person needs to follow a diet that has at least 5,045 units of carbohydrates, 450.75 units of fat and 325.15 units of protein. Two types of food are available: P and Q. A unit of food of type P costs 2.55 dollars and a unit of food of type Q costs 3.55 dollars. A unit of food of type P contains 9.75 units of carbohydrates, 18.15 units of fat and 13.95 units of protein. A unit of food type Q contains 22.95 units of carbohydrates, 12.15 units of fat and 18.85 units of protein. A mathematical linear model is needed to find the minimum cost for a diet that consists of a mixture of the two types of food and that meets the minimum diet requirements.

20.2.3.1 Understanding the Problem

For easy understanding and for deriving the mathematical formulation of the problem, the data given is represented in a table. For each type of food, the data include the cost per unit of food, and the contents, carbohydrates, fat, and proteins. The data for the diet problem is represented as follows:

Food type	Cost	Carbohydrates	Fat	Protein
P	2.55	9.75	18.15	13.95
Q	3.55	22.95	12.15	18.85

20.2.3.2 Mathematical Formulation

Let x_1 denote the amount of units of food type P and x_2 the units of food type Q contained in the diet. The total cost of the diet is: $2.55x_1 + 3.55x_2$. As stated previously, the main limitation is the lower bound requirement of carbohydrates, fat, and proteins. The combination of units of type P and of type Q should have the minimum specified units of carbohydrates, fat, and proteins. The objective function of linear optimization model formulation of the given diet problem is:

Minimize: $C = 2.55x_1 + 3.55x_2$
Subject to the following constraints:

$$
\begin{array}{rcl}
9.75x_1 + 22.95x_2 & \geq & 5045 \\
18.15x_1 + 12.15x_2 & \geq & 450.75 \\
13.95x_1 + 18.85x_2 & \geq & 325.15 \\
x_1 & \geq & 0 \\
x_2 & \geq & 0
\end{array}
$$

20.2.4 Case Study 4

The owners of a farm acquired a loan of $16,850.00 to produce three types of crops: corn, barley, and wheat, on 140 acres of land. An acre of land can produce an average of 135 bushels of corn, 45 of barley, or 100 bushels of wheat. The net profit per bushel of barley is $3.05, for corn is $1.70, and for wheat is $2.25. After the harvest, these crops must be stored in relatively large containers. At the present, the farm can store 3895 bushels. The total expenses to plant an acre of land is: $95.00 for corn, $205.00 for barley, and $115.00 for wheat. What amount of land should be the farm plan to dedicate to each crop in order to optimize profit?

20.2.4.1 Understanding the Problem

As in previous case studies, the data given is represented in a table for easy understanding and for deriving the mathematical formulation of the problem. There are three types of resources that will impose constraints on the problem formulation: the total storage capacity of the farm for the crops, the total available funds, and the total amount of land. The data for this problem is represented as follows:

Resource	Total	Corn	Barley	Wheat
Storage (bushels)	3895	135	45	100
Funds ($)	16,850.00	95.00	205.00	115.00
Land (acres)	140	x_1	x_2	x_3

20.2.4.2 Mathematical Formulation

Let x_1 denote the amount of land in acres allotted to corn, x_2 the amount of land allotted to barley, and x_3 the amount of land dedicated to wheat. This problem is to optimize profit. The total net profit is denoted by P and for each crop it consists of: net profit per bushel times the number of bushels per acre, times the amount of acres to plant. The constraints are derived from the resource limitations of the problem. The arithmetic expression for P is given by:

$$P = 135 \times 1.70 \times x_1 + 45 \times 3.05 \times x_2 + 100 \times 2.25 \times x_3$$

The objective function of the linear optimization formulation of the given problem is:

Maximize: $P = 229.5x_1 + 137.25x_2 + 225.00x_3$
subject to the following constraints:

$$
\begin{array}{rrrcr}
135x_1 & +45x_2 & +100x_3 & \leq & 3895 \\
95x_1 & +205x_2 & +115x_3 & \leq & 16850.00 \\
x_1 & +x_2 & +x_3 & \leq & 140 \\
& & x_1 & \geq & 0 \\
& & x_2 & \geq & 0 \\
& & x_3 & \geq & 0 \\
\end{array}
$$

Summary

Linear optimization modeling consists of formulating a mathematical linear model that includes: the linear function to be optimized (maximized or minimized), a set of decision variables, and a set of constraints. The computational models of this type are also known as Linear Programming Models.

Key Terms		
linear optimization	decision variables	constraints
linear programming	linear problems	problem formulation
sign restriction	objective function	technological coefficient

Exercises

Exercise 20.1   A factory produces three types of bed, A, B, and C. The company that owns the factory sells beds of type A for $250.00 each, beds of type B for $320.00 each, and beds of type C for $625.00 each. Management estimates that all beds produced of types A and C will be sold. The number of beds of type B that can be sold is at the most 45. The production of different types of bed requires different amounts of resources such as: basic labor hours, specialized hours, and materials. The following table provides this data.

Type A	Type B	Type C	Resource
10 ft.	60ft.	80 ft.	Material
15 h	20 h	40 h	Basic labor
5 h	15 h	20 h	Specialized labor

The amounts of resources available are: 450 ft. of material, 210 hours of basic labor, and 95 hours of specialized labor. The problem is to optimize profit. Write the formulation of the mathematical optimization problem.

Exercise 20.2 An automobile factory needs to have a different number of employees working for every day of the week. Each employee has to work 5 consecutive days a week and have two days of rest. The following are the requirements of the factory: Monday needs 330 employees, Tuesday needs 270 employees, Wednesday needs 300 employees, Thursday needs 390 employees, Friday needs 285 employees, Saturday needs 315 employees, and Sunday needs 225 employees. To improve profitability, the factory is required to optimize the number of employees. (Hint: the key decision is how many employees begin work on each day. Employees who begin work on Monday are those who do not begin on Tuesday or Wednesday). Write the formulation of the mathematical optimization problem.

Exercise 20.3 Consider the problem in Exercise 20.2 with the following changes: the problem is to optimize the labor costs of the auto factory based of amount of labor time in hours worked by employees. Full-time employees work 8 hours per day and cost the factory $12.50 per hour. Part-time employees work 4 hours a day and cost $8.50 per hour. The total amount of part-time labor should be at the most 30% of total labor time. Write the formulation of the mathematical optimization problem.

Exercise 20.4 A factory manufactures 5 different products: P_1, P_2, P_3, P_4, and P_5. The factory needs to maximize profit. Each product requires machine time on three different devices: A, B, and C, each of which is available 135 hours per week. The following table provides the data (machine time in minutes):

Product	Device A	Device B	Device C
P_1	20	15	8
P_2	13	17	12
P_3	15	7	11
P_4	16	4	7
P_5	18	14	6

The unit sale price for products P_1, P_2, and P_3 is $7.50, $6.00, and $8.25 respectively. The first 26 units of P_4 and P_5 have a sale prices $5.85 each, all excess units have a sale price of $4.50 each. The operational costs of devices A and B are $5.25 per hour, and $5.85 for device C. The cost of materials for products P_1 and P_4 are $2.75. For products P_2, P_3, and P_5 the materials cost is $2.25. Hint: assume that x_i,

$i = 1 \ldots 5$ are the number of units produced of the various products and use two new variables y_4 and y_5 for the units produced of P_4 and P_5 respectively in excess of 26. Write the formulation of the mathematical optimization problem.

Exercise 20.5 A company produces a certain number of products and distributes these to various customer distribution centers that request specified numbers of units of the product. There are two production facilities P_1 and P_2. There are three storage facilities: S_1, S_2, and S_3. The company also has four distribution centers: D_1, D_2, D_3, and D_4. This problem can be represented as a network in which there is a cost per unit to ship products from a production facility to a storage facility, and from a storage facility to a distribution center. The problem must optimize the cost of distribution of the products. The following table provides the costs per unit shipped from a source node to a destination node.

Source node	S_1	S_2	S_3	D_1	D_2	D_3	D_4
P_1	4	6	0	0	0	0	0
P_2	10	4	7	0	0	0	0
S_1	0	0	0	17	20	0	0
S_2	0	0	0	20	15	25	0
S_3	0	0	0	0	30	18	11

Chapter 21

Solving Linear Optimization Problems

21.1 Introduction

The *Simplex algorithm*, due to George B. Dantzig, is used to solve linear optimization problems. It is a tabular solution algorithm and is a powerful computational procedure that provides fast solutions to relatively large-scale applications. There are many software implementations of this algorithm, or variations of it. The basic algorithm is applied to a linear programming problem that is in standard form, in which all constraints are equations and all variables non-negative. This chapter presents the general principles and the basic concepts of numerical solution to linear optimization problems. Several case studies are presented using the software solvers: LP_solve and GLPK.

21.2 General Form

The following is a general form of a linear optimization problem that is basically organized in three parts.

1. The objective function, f, to be maximized or minimized, mathematically expressed as:

$$f(x_1, x_2, \ldots, x_n) = c_1 x_1 + c_2 x_2 + \ldots + c_n x_n \tag{21.1}$$

2. The set of m constraints, which is of the form:

$$a_{i1} x_1 + a_{i2} x_2 + \ldots + a_{in} x_n \leq b_i \quad i = 1, \ldots m \tag{21.2}$$

The other form is:

$$a_{i1} x_1 + a_{i2} x_2 + \ldots + a_{in} x_n \geq b_i \quad i = 1, \ldots m \tag{21.3}$$

3. The sign restriction for variables: $x_j \geq 0$, or $x_j \leq 0$, or x_j unrestricted in sign, $j = 1, \ldots n$.

Many problems are formulated with a mix of m constraints with $\leq$, $=$, and $\geq$ forms. Note that the objective function, which is expressed mathematically in (21.1), and the constraints, which are expressed mathematically in (21.2) and (21.3), are linear mathematical (algebraic) expressions.

An important assumption included in the general formulation of a linear optimization problem is that the variables, $x_i, i = 1, \ldots n$, take numeric values that are real or fractional. In the case that one or more variables only take integer values, then other techniques and algorithms are used. These methods belong to the class of *Integer Programming* or *Mixed Integer Programming*.

21.3   Foundations of the Algorithm

21.3.1   Basic Concepts

For a given linear optimization problem, *a point* is the set of values corresponding to one for each decision variable. The *feasible region* for the problem, is the set of all points that satisfy the constraints and all sign restrictions. If there are points that are not in the feasible region, they are said to be in an *infeasible region*.

The *optimal solution* to a linear maximization problem is a point in the feasible region with the largest value of the objective function. In a similar manner, the *optimal solution* to a linear minimization problem is a point in the feasible region with the smallest value of the objective function.

There are four cases of solutions to consider in a linear optimization problem.

1. A unique optimal solution

2. An infinite number of optimal solutions

3. No feasible solutions

4. An unbounded solution

In a linear maximization problem, a constraint is *binding* at an optimal solution if it holds with equality when the values of the variables are substituted in the constraint.

21.3.2   Problem Formulation in Standard Form

Because the Simplex algorithm requires the problem to be formulated in *standard form*, the general form of the problem must be converted to standard form.

- For each constraint of $\leq$ form, a *slack variable* is defined. For constraint i, slack variable s_i is included. Initially constraint i has the general form:

$$a_{i1}x_1 + a_{i2}x_2 + \ldots + a_{in}x_n \leq b_i \qquad (21.4)$$

To convert constraint i of the general form of the expression in (21.4) to an equality, slack variable s_i is added to the constraint, and $s_i \geq 0$. The constraint will now have the form:

$$a_{i1}x_1 + a_{i2}x_2 + \ldots + a_{in}x_n + s_i = b_i \qquad (21.5)$$

- For each constraint of $\geq$ form, an *excess variable* is defined. For constraint i, excess variable e_i is included. Initially constraint i has the general form:

$$a_{i1}x_1 + a_{i2}x_2 + \ldots + a_{in}x_n \geq b_i \qquad (21.6)$$

To convert constraint i of the general form of the expression in (21.6) to an equality, excess variable e_i is subtracted from the constraint, and $e_i \geq 0$. The constraint will now have the form:

$$a_{i1}x_1 + a_{i2}x_2 + \ldots + a_{in}x_n - e_i = b_i \qquad (21.7)$$

Consider the following formulation of a numerical example:

Maximize: $5x_1 + 3x_2$
Subject to:

$$
\begin{array}{rrcl}
2x_1 & + & x_2 & \leq 40 \\
x_1 & + & 2x_2 & \leq 50 \\
& & x_1 & \geq 0 \\
& & x_2 & \geq 0
\end{array}
$$

After rewriting the objective function and adding two slack variables s_1 and s_2 to the problem, the transformed problem formulation in standard form is:

Maximize: $z - 5x_1 - 3x_2 = 0$.
Subject to the following constraints:

$$
\begin{array}{rcll}
2x_1 + x_2 + s_1 & & = 40 \\
x_1 + 2x_2 & + s_2 & = 50
\end{array}
$$

$$
\begin{array}{rl}
x_1 & \geq 0 \\
x_2 & \geq 0 \\
s_1 & \leq 0 \\
s_2 & \leq 0
\end{array}
$$

21.3.3   Generalized Standard Form

A generalized standard form of a linear optimization problem is:

Maximize (or minimize) $f = c_1 x_1 + c_2 x_2 + \ldots + c_n x_n$

Subject to the following constraints:

$$
\begin{array}{llll}
a_{11}x_1 & + a_{12}x_2 & + \ldots & + a_{1n}x_n & = b_1 \\
a_{21}x_1 & + a_{22}x_2 & + \ldots & + a_{2n}x_n & = b_2 \\
\vdots & \vdots & & \vdots & \vdots \\
a_{m1}x_1 & + a_{m2}x_2 & + \ldots & + a_{mn}x_n & = b_m
\end{array}
\tag{21.8}
$$

$$
x_i \geq 0, \quad i = 1, 2, \ldots, n.
$$

The constraints can be written in matrix form as follows:

$$
\begin{bmatrix}
a_{11} & a_{12} & \cdots & a_{1n} \\
a_{21} & a_{22} & \cdots & a_{2n} \\
\vdots & \vdots & \ddots & \vdots \\
a_{m1} & a_{2m} & \cdots & a_{mn}
\end{bmatrix}
\begin{bmatrix}
x_1 \\
x_2 \\
\vdots \\
x_n
\end{bmatrix}
=
\begin{bmatrix}
b_1 \\
b_2 \\
\vdots \\
b_m
\end{bmatrix}
\tag{21.9}
$$

Equation (22.2) can also be written as $AX = B$, in which A is an $m \times n$ matrix, X is a column vector of size n, and B is a column vector of size m.

21.3.4   Additional Definitions

To derive a basic solution to Equation (22.2), a set m of variables known as the *basic variables* is used to compute a solution. These variables are the ones left after setting the *nonbasic variables*, which is the set of $n - m$ variables chosen and set to zero.

There can be several different basic solutions in a linear optimization problem. There could be one or more sets of m basic variables for which a basic solution cannot be derived.

A *basic feasible solution* to the standard formulation of a linear optimization problem is a basic solution in which the variables are non-negative.

The solution to a linear optimization problem is the best basic feasible solution to $AX = B$ (or Equation (22.2)).

21.4 Simplex Algorithm

In addition to transforming the constraints to standard form, the expression of the objective function must be changed to an equation with zero on its right-hand side. The general expression:

$$f = c_1 x_1 + c_2 x_2 + \ldots + c_n x_n$$

is changed to

$$f - c_1 x_1 - c_2 x_2 - \ldots - c_n x_n = 0$$

This equation becomes row 0 in the complete set of equations of the problem formulation. After this transformation the Simplex method can be used to solve the linear optimization problem.

21.4.1 General Description of the Simplex Algorithm

The following is a general description of the Simplex algorithm:

1. Find a basic feasible solution to the linear optimization problem; this solution becomes the initial basic feasible solution.

2. If the current basic feasible solution is the optimal solution, stop.

3. Search for an adjacent basic feasible solution that has a greater (or smaller) value of the objective function. An *adjacent basic feasible solution* has $m - 1$ variables in common with the current basic feasible solution. This becomes the current basic feasible solution, continue in step 2.

A linear optimization problem has an *unbounded solution* if the objective function can have arbitrarily large values for a maximization problem, or arbitrarily small values for a minimization problem. This occurs when a variable with a negative coefficient in the objective row (row 0) has a non-positive coefficient in every constraint.

21.4.2 Detailed Description of the Simplex Algorithm

A shorthand form of the set of equations known as the *simplex tableau* is used in the algorithm. Each tableau corresponds to a movement from one basic variable set BVS (extreme or corner point) to another, making sure that the objective function improves at each iteration until the optimal solution is reached. The following sequence of steps describes the application of the simplex solution algorithm:

1. Convert the LP to the following form:

    (a) Convert the minimization problem into a maximization one.

    (b) All variables must be non-negative.

    (c) All RHS values must be non-negative.

    (d) All constraints must be inequalities of the form $\leq$.

2. Convert all constraints to equalities by adding a slack variable for each constraint.

3. Construct the initial simplex tableau with all slack variables in the basic variable set (BVS). The row 0 in the table contains the coefficient of the objective function.

4. Determine whether the current tableau is optimal. That is: If all RHS values are non-negative (called, the feasibility condition) and if all elements of the row 0 are non-positive (called, the optimality condition). If the answers to both questions are Yes, then stop. The current tableau contains an optimal solution. Otherwise, continue.

5. If the current basic variable set (BVS) is not optimal, determine, which nonbasic variable should become a basic variable and, which basic variable should become a nonbasic variable. To find the new BVS with the better objective function value, perform the following tasks:

    (a) Identify the entering variable: The entering variable is the one with the largest positive coefficient value in row 0. (In case of a tie, the variable that corresponds to the leftmost of the columns is selected).

    (b) Identify the outgoing variable: The outgoing variable is the one with smallest non-negative column ratio (to find the column ratios, divide the RHS column by the entering variable column, wherever possible). In case of a tie the variable that corresponds to the upmost of the tied rows is selected.

    (c) Generate the new tableau: Perform the Gauss-Jordan pivoting operation to convert the entering column to an identity column vector (including the element in row 0).

6. Go to step 4.

At the start of the simplex procedure; the set of basis is constituted by the slack variables. The first BVS has only slack variables in it. The row 0 presents the increase in the value of the objective function that will result if one unit of the variable corresponding to the *j*th column was brought in the basis. This row is sometimes known as the *indicator row* because it indicates if the optimality condition is satisfied.

Criterion for entering a new variable into the BVS will cause the largest per-unit improvement of the objective function. Criterion for removing a variable from the current BVS maintains feasibility (making sure that the new RHS, after pivoting remain non-negative). Warning: Whenever during the Simplex iterations you get a negative RHS, it means you have selected a wrong outgoing variable.

Note that there is a solution corresponding to each simplex tableau. The numerical of basic variables are the RHS values, while the other variables (non-basic variables) are always equal to zero. Note also that variables can exit and enter the basis repeatedly during the simplex algorithm.

21.4.3 Degeneracy and Convergence

A linear optimization problem (LP) is *degenerate* if the algorithm loops endlessly, cycling among a set of feasible basic solutions and never gets to the optimal solution. In this case, the algorithm will not converge to an optimal solution. Most software implementations of the Simplex algorithm will check for this type of non-terminating loop.

21.4.4 Two-Phase Method

The Simplex algorithm requires a starting basic feasible solution. The two-phase method can find a starting basic feasible solution whenever it exists. The two-phase simplex method proceeds in two phases, phase I and phase II. Phase I attempts to find an initial basic feasible solution. Once an initial basic feasible solution has been found, phase II is then applied to find an optimal solution to the original objective function.

The simplex method iterates through the set of basic solutions (feasible in phase II) of the LP problem. Each basic solution is characterized by the set of m basic variables $x_{B1}, \ldots, x_{Bm}$. The other n variables are called nonbasic variables and denoted by $x_{N1}, \ldots, x_{Nn}$.

21.5 Software Implementations

This section discusses several software tools that solve linear optimization problems using the Simplex algorithm and variations of it. Two software tools are briefly discussed and problem formulation and executions are shown with several examples, these are: LP_solve and GLPK.

21.5.1 Solution with LP_solve

LP_solve is a non-commercial linear optimization software package, written in ANSI C by Michel Berkelaar. LP_solve is a free (see LGPL for the GNU lesser general public license) linear (integer) programming solver based on the revised simplex method and the Branch-and-bound method for the integers.

LP_solve was originally developed by Michel Berkelaar at Eindhoven University of Technology. Jeroen Dirks contributed the procedural interface, a built-in MPS reader, and many fixes and enhancements to the code. Starting at version 3.0, lp_solve is released under the LGPL license. Before that, the code could only be used for non-commercial purposes.

The following example is a simple linear optimization problem formulation formatted to be solved by using LP_solve; it is stored in the script file model.lp.

```
/* model.lp */

max: 143 x + 60 y;

120 x + 210 y <= 15000;
110 x + 30 y <= 4000;
x + y <= 75;
```

The following listing shows the Linux shell command to start execution of LP_solve with the file model.lp, at a terminal window.

```
$ lp_solve model.lp
Value of objective function: 6315.62
Actual values of the variables:
x                      21.875
y                      53.125
```

21.5.1.1   Example 1 with LP_solve

The following simple linear optimization problem, which was discussed previously as Example 1, is formulated as follows:

Minimize: $4x_1 + 8x_2 + 3x_3$
Subject to the following constraints:

$$
\begin{aligned}
2x_1 &+ 5x_2 &+ 3x_3 &\geq 185 \\
3x_1 &+ 2.5x_2 &+ 8x_3 &\geq 155 \\
8x_1 &+ 10x_2 &+ 4x_3 &\geq 600 \\
&& x_1 &\geq 0 \\
&& x_2 &\geq 0 \\
&& x_3 &\geq 0
\end{aligned}
$$

To solve this problem with LP_solve, it is formatted as shown in the following listing, which is stored in the text file example1.lp:

```
/* example1.lp */
Minimize: 4 x1 + 8 x2 + 3 x3;

2 x1   + 5 x2   + 3 x3 >= 185;
3 x1   + 2.5 x2 + 8 x3 >= 155;
8 x1   + 10 x2  + 4 x3 >= 600;
```

To compute the solution to this optimization problem, the following Linux shell command is used to start execution of LP-solve with the file `example1.lp`, at a terminal window.

```
$ lp_solve example1.lp

Value of objective function: 317.5
Actual values of the variables:
x1                      66.25
x2                          0
x3                       17.5
```

On Widows, using LP-solve can be performed with a command window or using the LP-solve IDE, which may be a more convenient interface to the user.

21.5.1.2 Example 2 with LP-solve

The problem formulation of Example 2, as was defined previously, is the following:

Maximize: $143x_1 + 60x_2$
Subject to the following constraints:

$$
\begin{array}{rrcl}
120x_1 & + 210x_2 & \leq & 15000 \\
110x_1 & + 30x_2 & \leq & 4000 \\
x_1 & + x_2 & \leq & 75 \\
& x_1 & \geq & 0 \\
& x_2 & \geq & 0
\end{array}
$$

To solve this problem with LP-solve, it is formatted as shown in the following listing, which is stored in the text file `example2.lp`:

```
/* Example2.lp */
Maximize: 144 x1 + 60 x2;

 120 x1  + 210 x2 <= 15000;
 110 x1  +  30 x2 <= 4000;
     x1  +     x2 <= 75;
```

To compute the solution to this optimization problem, the following Linux shell command is used to start execution of LP-solve with the file `example2.lp` using a verbose option (v4), at terminal window.

```
$ lp_solve -v4 example2.lp
Model name:   '' - run #1
Objective:   Maximize(R0)

SUBMITTED
Model size:   3 constraints,   2 variables,      6 non-zeros.
```

```
Sets:                      0 GUB,           0 SOS.

Using DUAL simplex for phase 1 and PRIMAL simplex for phase 2.
The primal and dual simplex pricing strategy set to 'Devex'.

Optimal solution              6337.5 after           2 iter.

Excellent numeric accuracy ||*|| = 3.04201e-14

Value of objective function: 6337.5

Actual values of the variables:
x1                        21.875
x2                        53.125
```

21.5.1.3   Example 3 with LP_solve

The following linear optimization problem was presented previously; its formulation is:

Minimize: $3x_1 + 4x_2 + 6x_3 + 7x_4 + x_5$
Subject to the following constraints:

$$
\begin{array}{rcrcrcrcrcl}
2x_1 & - x_2 & + x_3 & + 6x_4 & - 5x_5 & \geq 6 \\
x_1 & + x_2 & + 2x_3 & + x_4 & + 2x_5 & \geq 3 \\
& & & & x_1 & \geq 0 \\
& & & & x_2 & \geq 0 \\
& & & & x_3 & \geq 0 \\
& & & & x_4 & \geq 0 \\
& & & & x_5 & \geq 0
\end{array}
$$

To solve this problem with LP_solve, it is formatted as shown in the following listing, which is stored in the text file `example3.lp`:

```
/* Example3.lp with LP_solve */
Minimize: 3 x1 + 4 x2 + 6 x3 + 7 x4 + x5;

 2 x1  -  x2 +  x3 + 6 x4 - 5 x5  >= 6;
   x1  + x2 + 2 x3 + x4 + 2 x5  >= 3;
```

To compute the solution to this optimization problem, the following Linux shell command is used to start execution of LP_solve with the file `example3.lp` at a terminal window.

```
$ lp_solve example3.lp

Value of objective function: 9.0

Actual values of the variables:
x1                              3
x2                              0
```

```
x3                              0
x4                              0
x5                              0
```

21.5.2 Implementation with GLPK `glpsol`

The GNU Linear Programming Kit (GLPK) is a powerful, proven tool for solving linear optimization problems with multiple constraints. GNU MathProg (GMPL GNU Mathematical Programming Language) is a high-level language for formulating and solving mathematical programming models. MathProg is specific to GLPK, but resembles a subset of AMPL and the `glpsol` program is the command line solver.

The following simple linear optimization problem, which was discussed previously as Example 1, is formulated as follows:

Minimize: $4x_1 + 8x_2 + 3x_3$
Subject to the following constraints:

$$
\begin{array}{rrrcl}
2x_1 & + 5x_2 & + 3x_3 & \geq & 185 \\
3x_1 & + 2.5x_2 & + 8x_3 & \geq & 155 \\
8x_1 & + 10x_2 & + 4x_3 & \geq & 600 \\
 & & x_1 & \geq & 0 \\
 & & x_2 & \geq & 0 \\
 & & x_3 & \geq & 0
\end{array}
$$

To solve this problem with `glpsol`, it is formatted as shown in the following listing, which is stored in the text file `example1.mod`:

```
# Example1.mod
# Using GLPK glpsol

/* Decision variables */
var x1 >= 0;
var x2 >= 0;
var x3 >= 0;

/* Objective function */
minimize z: 4*x1 + 8*x2 + 3*x3;

/* Constraints */
 s.t. R1: 2*x1  + 5*x2   + 3*x3 >= 185;
 s.t. R2: 3*x1  + 2.5*x2 + 8*x3 >= 155;
 s.t. R3: 8*x1  + 10*x2  + 4*x3 >= 600;

 end;
```

To compute the solution to this optimization problem, the following Linux shell command is used to start execution of `glpsol` with the file `example1.mod` at a terminal window.

```
$ glpsol --model example1.mod -o texample1.sol
GLPSOL: GLPK LP/MIP Solver, v4.45
Parameter(s) specified in the command line:
 --model example1.mod -o example1.sol
Reading model section from example1.mod...
example1.mod:18: warning: final NL missing before end of file
18 lines were read
Generating z...
Generating R1...
Generating R2...
Generating R3...
Model has been successfully generated
GLPK Simplex Optimizer, v4.45
4 rows, 3 columns, 12 non-zeros
Preprocessing...
3 rows, 3 columns, 9 non-zeros
Scaling...
 A: min|aij| = 2.000e+000  max|aij| = 1.000e+001
   ratio = 5.000e+000
Problem data seem to be well scaled
Constructing initial basis...
Size of triangular part = 3
      0: obj =  0.000000000e+000  infeas = 9.400e+002 (0)
*     3: obj =  4.798571429e+002  infeas = 0.000e+000 (0)
*     6: obj =  3.175000000e+002  infeas = 0.000e+000 (0)
OPTIMAL SOLUTION FOUND
Time used:   0.0 secs
Memory used: 0.1 Mb (108531 bytes)
Writing basic solution to 'example1.sol'...
```

The result of the computations is stored in file `texample1.sol` and the following listing shows the data.

```
Problem:    example1
Rows:       4
Columns:    3
Non-zeros:  12
Status:     OPTIMAL
Objective:  z = 317.5 (MINimum)
```

No.	Row name	St	Acty	Lower	Upper	Marg
1	z	B	317.5			
2	R1	NL	185	185		0.5
3	R2	B	338.75	155		
4	R3	NL	600	600		0.375

No.	Col name	St	Acty	Lower	Upper	Marginal
1	x1	B	66.25	0		
2	x2	NL	0	0		1.75
3	x3	B	17.5	0		

```
Karush-Kuhn-Tucker optimality conditions:
```

```
KKT.PE: max.abs.err = 0.00e+000 on row 0
        max.rel.err = 0.00e+000 on row 0
        High quality

KKT.PB: max.abs.err = 0.00e+000 on row 0
        max.rel.err = 0.00e+000 on row 0
        High quality

KKT.DE: max.abs.err = 0.00e+000 on column 0
        max.rel.err = 0.00e+000 on column 0
        High quality

KKT.DB: max.abs.err = 0.00e+000 on row 0
        max.rel.err = 0.00e+000 on row 0
        High quality

End of output
```

Summary

The solution to linear optimization modeling is performed by software tools known as LP solvers. These software tools implement algorithms, of which the most important one for deriving a solution to these problems is the Simplex algorithm. A common implementation is the two-phase Simplex algorithm. Some solvers require that the linear optimization problem formulation be in standard form, in which all constraints are equations and all variables non-negative. The solvers discussed and used in this chapter are open source programs. There are also several commercial proprietary solvers as well, such as Lindo.

Key Terms		
Simplex algorithm	two-phase	standard form
feasible region	infeasible region	optimal solution
binding	slack variable	excess variable
basic variables	nonbasic variables	feasible solution
simplex tableau	degeneracy	convergence

Exercises

Exercise 21.1   Minimize: $2x_1 + 3x_2$
Subject to the following constraints:

$$
\begin{array}{rl}
0.5x_1 + 0.25x_2 & \leq\ 4.0 \\
x_1 + 3x_2 & \geq\ 36 \\
x_1 + x_2 & =\ 10 \\
x_1 & \geq\ 0 \\
x_2 & \geq\ 0
\end{array}
$$

Exercise 21.2   Maximize: $2x_1 + 3x_2 + x_3$
Subject to the following constraints:

$$
\begin{array}{rl}
x_1 + x_2 + x_3 & \leq\ 40 \\
2x_1 + x_2 - x_3 & \geq\ 10 \\
- x_2 + x_3 & \geq\ 10 \\
x_1 & \geq\ 0 \\
x_2 & \geq\ 0 \\
x_3 & \geq\ 0
\end{array}
$$

Exercise 21.3   A factory produces three types of beds, A, B, and C. The company that owns the factory sells beds of type A for \$250.00 each, beds of type B for \$320.00 each, and beds of type C for \$625.00 each. Management estimates that all beds produced of types A and C will be sold. The number of beds of type B that can be sold is at the most 45. The production of different types of beds requires a different amount of resources such as: basic labor hours, specialized hours, and materials. The following table provides this data.

Type A	Type B	Type C	Resource
10 ft.	60ft.	80 ft.	Material
15 h	20 h	40 h	Basic labor
5 h	15 h	20 h	Specialized labor

The amounts of resources available are: 450 ft. of material, 210 hours of basic labor, and 95 hours of specialized labor. The problem is to optimize profit. Write the formulation of the mathematical optimization problem. Use LP_solver and GLPK `glpsol` to find a numerical solution to the linear optimization problem.

Exercise 21.4 An automobile factory needs to have a different number of employees working for every day of the week. Each employee has to work 5 consecutive days a week and have two days of rest. The following are the requirements of the factory: Monday needs 330 employees, Tuesday needs 270 employees, Wednesday needs 300 employees, Thursday needs 390 employees, Friday needs 285 employees, Saturday needs 315 employees, and Sunday needs 225 employees. To improve profitability, the factory is required to optimize the number of employees. (Hint: the key decision is how many employees begin work on each day. Employees who begin work on Monday are those who do not begin on Tuesday or Wednesday). Write the formulation of the mathematical optimization problem. Use LP_solver and GLPK glpsol to find a numerical solution to the linear optimization problem.

Exercise 21.5 Consider the problem in Exercise 21.4 with the following changes: the problem is to optimize the labor costs of the auto factory based of amount of labor time in hours worked by employees. Full-time employees work 8 hours per day and cost the factory $12.50 per hour. Part-time employees work 4 hours a day and cost $8.50 per hour. The total amount of part-time labor should be at the most 30% of total labor time. Write the formulation of the mathematical optimization problem. Use LP_solver and GLPK glpsol to find a numerical solution to the linear optimization problem.

Exercise 21.6 A factory manufactures 5 different products: P_1, P_2, P_3, P_4, and P_5. The factory needs to maximize profit. Each product requires machine time (in minutes) on three different devices: A, B, and C, each of which is available 135 hours per week. The following table provides the data on machine time needed:

Product	Device A	Device B	Device C
P_1	20	15	8
P_2	13	17	12
P_3	15	7	11
P_4	16	4	7
P_5	18	14	6

The unit sale price for products P_1, P_2, and P_3 is $7.50, $7.00, and $8.25 respectively. The first 26 units of P_4 and P_5 have a sale prices $5.85 each; all excess units have a sale price of $4.50 each. The operational costs of devices A and B are $5.25 per hour, and $5.85 for device C. The cost of materials for products P_1 and P_4 are $2.75. For products P_2, P_3, and P_5 the materials cost is $2.25. Hint: assume that x_i, $i = 1...5$ are the number of units produced of the various products and use two new variables y_4 and y_5 for the units produced of P_4 and P_5 respectively in excess of 26. Write the formulation of the mathematical optimization problem. Use LP_solver and GLPK glpsol to find a numerical solution to the linear optimization problem.

Exercise 21.7 A company produces a certain number of products and distributes these to various customer distribution centers that request a specified number of units of the product. There are two production facilities P_1 and P_2. There are three storage facilities: S_1, S_2, and S_3. The company also has four distribution centers: D_1, D_2, D_3, and D_4. This problem can be represented as a network in which there is a cost per unit to ship products from a production facility to a storage facility, and from a storage facility to a distribution center. The problem must optimize the cost of distribution of the products. The following table provides the cost per unit shipped from a source node to a destination node. Use LP_solver and GLPK `glpsol` to find a numerical solution to the linear optimization problem.

Source node	S_1	S_2	S_3	D_1	D_2	D_3	D_4
P_1	4	6	0	0	0	0	0
P_2	10	4	7	0	0	0	0
S_1	0	0	0	17	20	0	0
S_2	0	0	0	20	15	25	0
S_3	0	0	0	0	30	18	11

Chapter 22

Sensitivity Analysis and Duality

22.1 Introduction

This chapter presents the general concepts and techniques of sensitivity analysis and duality. With sensitivity analysis, we can find out how relatively small changes in the parameters of a linear optimization problem can cause changes in the optimal solution computed.

The concepts of *marginal values*, which are also known as *shadow prices*, and the *reduced costs* are extremely useful in sensitivity analysis. *Duality* helps to better understand sensitivity analysis as well as the nature of linear optimization.

22.2 Sensitivity Analysis

Sensitivity analysis deals with the effect of independent, multiple changes in the values of:

- the coefficients of the objective function of linear program models having unique solution

- the right-hand side of the constraints.

Other useful points in the analysis is the study of the change in the solution to a problem that occurs when a new constraint is added, and the changes when a new variable is added to the problem.

22.2.1 Coefficients of the Objective Function

How much can the objective function coefficients change before the values of the variables change? Or when the objective function coefficient of a single variable is changed, for what range of values of this coefficient will the optimal values of the decision variables be retained?

The concept of *reduced cost* is associated with the coefficients of the variables in the objective function. It is a very useful concept and applies to a variable with value zero in the optimal value of the objective function. The reduced cost of a variable is the amount by which the objective function will decrease when the variable is forced a value of 1.

The simplest and most direct way to find the reduced costs of variables and the allowable changes to the coefficients in the objective function is to observe the output results in the computer solution to a linear optimization problem. These are briefly discussed next for following software solver tools: LP_solve and GLPK `glpsol`.

22.2.1.1   Using LP_solve Case Study 1

To find the sensitivity of a simple problem, consider the following formulation.
Maximize: $60x_1 + 30x_2 + 20x_3$
Subject to the following constraints:

$$
\begin{array}{rrrl}
8x_1 & + 6x_2 & + x_3 & \leq 48 \\
4x_1 & + 2x_2 & + 1.5x_3 & \leq 20 \\
2x_1 & + 1.5x_2 & + 0.5x_3 & \leq 8 \\
& & x_1 & \geq 0 \\
& & x_2 & \geq 0 \\
& & x_3 & \geq 0
\end{array}
$$

The formulation of the problem is formatted for solving with LP_solve and stored in file `sensit1.lp`. The solution to this problem is specified with the `-S4` option in the Linux shell command in a terminal window, and the results can be observed in the following listing.

```
$ lp_solve -S4 sensit1.lp

Value of objective function: 280.00000000

Actual values of the variables:
x1                      2
x2                      0
x3                      8

Actual values of the constraints:
R1                     24
R2                     20
R3                      8

Objective function limits:
                      From           Till        FromValue
x1                      56             80         -1e+030
x2                 -1e+030             35             1.6
x3                      15           22.5         -1e+030

Dual values with from - till limits:
                 Dual value         From            Till
R1                       0        -1e+030          1e+030
R2                      10             16              24
R3                      10       6.666667              10
```

```
x1                    0            -1e+030            1e+030
x2                   -5              -4                 1.6
x3                    0            -1e+030            1e+030
```

From the output listing produced by LP_solve, it can be observed that the first part of the output (Value of objective function) displays the value computed for the objective function. This is the optimal value of the solution to the problem. The next part of the output (Actual values of the variables) shows the values of the variables, $x1$, $x2$, and $x3$ that make the solution optimal. Note that in this problem, variable $x2$ has a value of zero. The coefficient of this variable will have a much lower effect on the objective function.

The next part of the output (Objective function limits:) has the range of values of the coefficients of variables in the objective function that retain the conditions for the optimal solution of the problem. The range of values of the coefficient of variable $x1$ is defined by the lower bound (indicated in the From column) of 56 and the upper bound (indicated in the Till column) of 80. Any values of the coefficient of $x1$ outside this range will change the conditions of the objective function to a suboptimal value.

In a similar manner, the range of values of the coefficient of variable $x2$ is indicated from an extremely low value ($-\infty$) to an upper bound of 35, and finally the range of values of the coefficient of variable $x3$ is in the range from a lower bound of 15 to an upper bound of 22.5.

For a variable with actual value zero, the column 'FromValue' shows a value that the variable takes when its coefficient value reaches the value in the 'Till' column (for a maximization problem) or the value in the 'From' column (for a minimization problem). Variable $x2$ will take the value 1.6 when its coefficient reaches the value of 35.

Variable $x2$ has a *reduced cost* of -5 and is shown in column 'Dual value'. This is the amount the objective function would change if the value of $x2$ is changed to 1. Variables $x1$ and $x3$ are basic variables and have a zero reduced cost.

22.2.1.2   Using GLPK `glpsol` Case Study 1

To find the sensitivity of the simple problem discussed previously with the following formulation.

Maximize: $60x_1 + 30x_2 + 20x_3$
Subject to the following constraints:

$$
\begin{array}{rrrll}
8x_1 & + 6x_2 & + x_3 & \leq 48 \\
4x_1 & + 2x_2 & + 1.5x_3 & \leq 20 \\
2x_1 & + 1.5x_2 & + 0.5x_3 & \leq 8 \\
& & x_1 & \geq 0 \\
& & x_2 & \geq 0 \\
& & x_3 & \geq 0
\end{array}
$$

The formulation of the problem is formatted as follows for solving with GLPK `glpsol` and stored in file `sensit1.mod`.

```
#
/* Example for testing sensitivity */
/* sensit1.mod using glpsol */

/* Decision variables */
 var x1 >= 0;
 var x2 >= 0;
 var x3 >= 0;

 /* Objective function */
 maximize z: 60*x1 + 30*x2 + 20*x3;

 /* Constraints */
s.t. R1: 8*x1  + 6*x2 +        x3  <= 48;
s.t. R2: 4*x1  + 2*x2   + 1.5*x3  <= 20;
s.t. R3: 2*x1  + 1.5*x2 + 0.5*x3  <= 8;
end;
```

The execution of the glpsol program is initiated with the following shell command in a terminal window:

```
$ glpsol --ranges msens.sol -m sensit1.mod
```

The results of execution are stored in file sensit1.sol with the following listing.

```
GLPK 4.45 - SENSITIVITY ANALYSIS REPORT              Page   2

Problem:    sensit1
Objective:  z = 280 (MAXimum)


No. Col  St      Acty      Obj coef   Lower     Activity
                           Marginal   Upper     range
--- ---- ---  ----------- ---------- --------- -----------
1    x1   BS     2.000     60.000        .        -Inf
                              .         +Inf     4.00000

2    x2   NL       .        30.000        .      -4.00000
                          -5.00000      Inf      1.60000

3    x3   BS     8.0000    20.0000        .      -16.000
                              .         +Inf     13.3333

    Col   Obj coef     Obj value    Limiting
          range        break pt     variable
    ---- ----------   -----------   ---------
    X1    56.000       272.000        x2
          80.000       320.000        C2

    X2    -Inf         300.000        x3
          35.000       272.000        x1
```

```
X3     15.000    240.000    C2
       22.500    300.000    C3
```

```
End of report
```

The range of values of the coefficients in the objective function appear in column 'Obj coef range' for the three decision variables $x1$, $x2$, and $x3$. These values correspond to those computed by LP_solve, as shown previously.

Column 'Acty' (activity) shows the actual values of variables $x1$, $x2$, and $x3$. Note that variable $x2$ has a value of zero and column 'Activity range' shows the value 1.6, which is the value that the variable will take when its coefficient reaches the value of 35.

Variable $x2$ has a *reduced cost* of -5 and is shown in column 'Marginal.' This is the amount the objective function would change if the value of $x2$ is changed to 1. Variables $x1$ and $x3$ are basic variables and have a zero reduced cost.

22.2.1.3 Using LP_solve Case Study 2

The following simple linear optimization problem, which was discussed previously as Example 1, is formulated as follows:

Minimize: $4x_1 + 8x_2 + 3x_3$

Subject to the following constraints:

$$\begin{array}{rrrr}
2x_1 & + 5x_2 & + 3x_3 & \geq 185 \\
3x_1 & + 2.5x_2 & + 8x_3 & \geq 155 \\
8x_1 & + 10x_2 & + 4x_3 & \geq 600 \\
& & x_1 & \geq 0 \\
& & x_2 & \geq 0 \\
& & x_3 & \geq 0
\end{array}$$

To solve this problem with LP_solve, it is formatted as shown in the following listing, which is stored in the text file example1.lp:

```
/* example1.lp */
Minimize: 4 x1 + 8 x2 + 3 x3;

2 x1  + 5 x2   + 3 x3 >= 185;
3 x1  + 2.5 x2 + 8 x3 >= 155;
8 x1  + 10 x2  + 4 x3 >= 600;
```

The execution of the lp_solve program to get the solution to this problem is specified with the $-S4$ option, and the results are found in the following listing.

```
$ lp_solve -S4 example1.lp

Value of objective function: 317.50000000

Actual values of the variables:
```

```
x1                   66.25
x2                   0
x3                   17.5

Actual values of the constraints:
R1                            185
R2                         338.75
R3                            600

Objective function limits:
                     From              Till          FromValue
x1                   2                 6             -1e+030
x2                   6.25              1e+030        14
x3                   2                 4.4           -1e+030

Dual values with from - till limits:
                 Dual value            From              Till
R1                   0.5               150               450
R2                   0                 -1e+030           1e+030
R3                   0.375             246.6667          740
x1                   0                 -1e+030           1e+030
x2                   1.75              -1e+030           14
x3                   0                 -1e+030           1e+030
```

From the output listing produced by LP_solve, it can be observed that the first part of the output (Value of objective function:) displays the value computed for the objective function, 317.5. This is the optimal value of the problem. The next part of the output (Actual values of the variables:) shows the values of the variables, $x1$, $x2$, and $x3$ that make the solution optimal.

The next part of the output (Objective function limits:) has the range of values of the coefficients of variables in the objective function that retain the optimal solution of the problem. The range of values of the coefficient of variable $x1$ is defined by the lower bound of 2 and the upper bound of 6. Any values of the coefficient of $x1$ outside this range will change the objective function to a suboptimal value.

In a similar manner, the range of values of the coefficient of variable $x2$ is in the range from a lower bound of 6.25 to an extremely high value (∞) , and finally the range of values of the coefficient of variable $x3$ is in the range from a lower bound of 2 to an upper bound of 4.4. Variable $x2$ will take the value 14 when its coefficient reaches the value of 6.25.

For a variable with actual value zero, the column 'FromValue' shows a value that the variable takes when its coefficient value reaches the value in the 'Till' column (for a maximization problem) or the value in the 'From' column (for a minimization problem). Variable $x2$ will take the value 14 when its coefficient reaches the value of 6.25.

Variable $x2$ has a *reduced cost* of 1.75 and is shown in column 'Dual value.' This is the amount the objective function would change if the value of $x2$ is changed to 1. Variables $x1$ and $x3$ are basic variables and have a zero reduced cost.

22.2.1.4 Using GLPK `glpsol` Case Study 2

To solve this problem with GLPK glpsol, it is formatted as shown in the following listing, which is stored in the text file `example1.mod`:

```
# Example1.mod
# Using GLPK glpsol

/* Decision variables */
var x1 >= 0;
var x2 >= 0;
var x3 >= 0;

/* Objective function */
minimize z: 4*x1 + 8*x2 + 3*x3;

/* Constraints */
 s.t. R1: 2*x1  + 5*x2   + 3*x3 >= 185;
 s.t. R2: 3*x1  + 2.5*x2 + 8*x3 >= 155;
 s.t. R3: 8*x1  + 10*x2  + 4*x3 >= 600;

 end;
```

The execution of the `glpsol` program is initiated with the following shell command in a terminal window:

```
$ glpsol --ranges examp1sens.sol -m example1.mod
```

The results of execution are stored in file `examp1sens.sol` with the following listing.

```
GLPK 4.45 - SENSITIVITY ANALYSIS REPORT                     Page   2

Problem:    example1
Objective:  z = 317.5 (MINimum)

No. Col  St  Acty        Obj coef   Lower bound    Activity
                         Marginal   Upper bound    range
--- ---  --  ----------  ---------  -------------  -------------
1   x1   BS  66.25000    4.00000         .         92.50000
                             .          +Inf       -Inf

2   x2   NL     .         8.00000         .         -Inf
                         1.75000        +Inf        14.00000

3   x3   BS  17.50000    3.00000         .         150.00000
                             .          +Inf       -7.00000

Col  Obj coef    Obj value at   Limiting
     range       break point    variable
---- ----------- -------------  ------------
 x1   2.00000     185.00000      R3
      6.00000     450.00000      R1
```

```
x2      6.25000           -Inf
        +Inf           342.00000    x3

x3      2.00000        300.00000    R1
        4.40000        342.00000    x2
```

End of report

The range of values of the coefficients in the objective function appear in column 'Obj coef range' for the three decision variables $x1$, $x2$, and $x3$. These values correspond to those computed by LP_solve.

Column 'Acty' (activity) shows the actual values of variables $x1$, $x2$, and $x3$. Note that variable $x2$ has a value of zero and column 'Activity range' shows the value 14, which is the value that the variable will take when its coefficient reaches the value of 6.25.

Variable $x2$ has a *reduced cost* of 1.75 and is shown in column 'Marginal.' This is the amount the objective function would change if the value of $x2$ is changed to 1. Variables $x1$ and $x3$ are basic variables and have a zero reduced cost.

22.2.1.5   Using LP_solve Case Study 3

The following simple linear optimization problem, which was discussed previously as Example 2, is formulated in LP_solve as follows:

```
/* Example2.lp in LP_solve */

Maximize: 144 x1 + 60 x2;

 120 x1  + 210 x2 <= 15000;
 110 x1  +  30 x2 <= 4000;
     x1  +     x2 <= 75;
```

Executing LP_solve program at the Linux shell command in a terminal window to solve this problem, yields the following output.

```
$ lp_solve -S4 example2.lp

Value of objective function: 6337.50000000

Actual values of the variables:
x1                    21.875
x2                    53.125

Actual values of the constraints:
R1                  13781.2
R2                     4000
R3                       75

Objective function limits:
```

	From	Till	FromValue
x1	60	220	-1e+030
x2	39.27273	144	-1e+030

Dual values with from – till limits:

	Dual value	From	Till
R1	0	-1e+030	1e+030
R2	1.05	2916.667	8250
R3	28.5	36.36364	80
x1	0	-1e+030	1e+030
x2	0	-1e+030	1e+030

Following the same procedure used in the previous case studies, it can be observed that the first part of the output (Value of objective function:) displays the value computed for the objective function, 6337.50. This is the optimal value of the problem. The next part of the output (Actual values of the variables:) shows the values of the variables, $x1$ and $x2$, that make the solution optimal.

The next part of the output (Objective function limits:) has the range of values of the coefficients of variables in the objective function that retain the optimal solution of the problem. The range of values of the coefficient of variable $x1$ is defined by the lower bound of 60 and the upper bound of 220. Any values of the coefficient of $x1$ outside this range will change the objective function to a suboptimal value.

In a similar manner, the range of values of the coefficient of variable $x2$ is in the range from a lower bound of 39.27273 to a value of 144.

For a variable with actual value zero, the column 'FromValue' shows a value that the variable takes when its coefficient value reaches the value in the 'Till' column (for a maximization problem) or the value in the 'From' column (for a minimization problem). Variables $x1$ and $x2$ have a value different than zero.

Variables $x1$ and $x2$ have a *reduced cost* of zero as shown in column 'Dual value.'

22.2.1.6 Using GLPK `glpsol` Case Study 3

To solve this problem with GLPK `glpsol`, it is formatted as shown in the following listing, which is stored in the text file `example2.mod`:

```
# Example2.mod in glpsol
#
/* Decision variables */
var x1 >= 0;
var x2 >= 0;

/* Objective function */
maximize z: 144*x1 + 60*x2;

/* Constraints */
   s.t. R1: 120*x1  + 210*x2 <= 15000;
   s.t. R2: 110*x1  +  30*x2 <= 4000;
   s.t. R3:   x1  +     x2 <= 75;

end;
```

The execution of the `glpsol` program is initiated with the following shell command in a terminal window:

```
$ glpsol --ranges examp2sens.sol -m example2.mod
```

The results of execution are stored in file `examp2sens.sol` with the following listing.

```
GLPK 4.45 - SENSITIVITY ANALYSIS REPORT Page    2

Problem:    example2
Objective:  z = 6337.5 (MAXimum)

No. Col   St    Acty       Obj coef    Lower        Activity
                           Marginal    Upper        range
--- ---   --  ----------- ----------- ----------  ------------
1   x1    BS   21.87500    144.00000       .        8.33333
                                .        +Inf       36.36364

2   x2    BS   53.12500     60.00000       .          -Inf
                                .        +Inf       66.66667

Col     Obj coef   Obj val        Limiting
        range      break point    variable
----    ---------- ------------   -----------
 x1     60.00000    4500.000       R2
        220.00000   8000.000       R3

 x2     39.27273    5236.36364     R3
        144.00000   10800.000      R2

End of report
```

The range of values of the coefficients in the objective function appear in column 'Obj coef range' for the two decision variables $x1$ and $x2$. These values correspond to those computed by LP_solve.

Column 'Acty' (activity) shows the actual values of variables $x1$ and $x2$. Note that both variables $x1$ and $x2$ have a non-zero value. These variables are basic variables and have a zero reduced cost.

22.2.2 Right-Hand Side of Constraints

As mentioned previously, another part of sensitivity analysis involves finding out how changes in the right-hand side of a constraint would change the basis of the optimal solution to a problem.

The *dual value* (marginal or shadow price) in a constraint is the amount by which the objective function will decrease when the right-hand side of the constraint in incremented by one. Another way way to define a dual value is the rate of change of the objective function with respect to the right-hand side of the constraint.

The right-hand side of a single constraint can vary within a specified range. The dual value of constraint will only hold while the right-hand of the constraint is in the range of values.

A constraint that is not active has a dual value of zero. This is a constraint that is not binding because its actual value is less (or greater) than the value of the right-hand side specified in the problem formulation.

This second set of data is provided by computer solvers such as LP_solver, glpsol and display output results that can be used for sensitivity analysis. This section includes discussion of the output produced by LP_solve and glpsol for the three case studies presented previously.

22.2.2.1 Using LP_solve Case Study 1

The output produced by LP_solve shows the dual value for every constraint and the range of values that the right-hand side of each constraint can change, with the current solution of the problem remaining optimal. The following partial listing is the output produced by LP_solve for the problem in case study 1.

```
$ lp_solve -S4 sensit1.lp

Value of objective function: 280.00000000

Actual values of the constraints:
R1                      24
R2                      20
R3                       8

Dual values with from - till limits:
                Dual value        From            Till
R1                    0         -1e+030         1e+030
R2                   10            16              24
R3                   10         6.666667          10
```

An important quantity associated with constraints is provided column *Dual value*. As mentioned previously, this value is also known as *marginal value* and *shadow price*. This value indicates the amount the objective function will change when the constraint value is incremented by 1. In the output listing shown, constraint R2 has a dual value of 10, and the value of the constraint can change from 16 to 24. Constraint R3 also has a dual value of 10, and the value of the constraint can change from 6.6666667 to 10. The dual value of a constraint only holds in that range of values of the constraint.

However, constraint R1 has a dual value of zero. Its actual value is 24 and the constraint in the problem was specified as ≤ 48. Constraint R1 is said to be not active. This constraint has very minimum effect on the value of the objective function. Its range of values is actually from 24 (its actual value) to ∞.

22.2.2.2 Using `glpsol` Case Study 1

The output produced by GLPK `glpsol` shows the dual prices in the Marginals column. It also shows the range of values that the right-hand side of each constraint can change, with the current solution of the problem remaining optimal. The following listing is the output produced by `glpsol`, which is similar to the output listing produced by LP_solve for the problem in Case Study 1.

```
GLPK 4.45 - SENSITIVITY ANALYSIS REPORT              Page    1

Problem:     sensit1
Objective:   z = 280 (MAXimum)
```

No.	Row	St	Activity	Slack Marginal	Lower Upper	Acty range
1	z	BS	280.000	-280.000 .	-Inf +Inf	240.000 280.000
2	R1	BS	24.000	24.000 .	-Inf 48.000	13.333 32.000
3	R2	NU	20.000	. 10.000	-Inf 20.000	16.000 24.000
4	R3	NU	8.000	. 10.00000	-Inf 8.000	6.666 10.000

Row name	Obj coef range	Obj value at break point	Limiting variable
z	-1.00000 +Inf	. +Inf	R2
R1	-1.25000 5.00000	250.00000 400.00000	R3 R2
R2	-10.00000 +Inf	240.00000 320.00000	x3 x1
R3	-10.00000 +Inf	266.66667 300.00000	x1 x3

Note that the listing shows constraint R1 with a Slack value of 24.0, which is the difference between its specified value (48.0) and its actual value, which is 24.0. This constraint is not active in the optimal solution of the problem.

22.2.2.3 Using LP_solve Case Study 2

The output produced by LP_solve shows the dual values for each constraint and the range of values that the right-hand side of each constraint can change, with the current solution of the problem remaining optimal. The following listing is the output produced by LP_solve for the problem in Case Study 2.

The execution of the LP_solve program is specified with the −S4 option, and the results are found in the following partial listing.

```
$ lp_solve -S4 example1.lp

Value of objective function: 317.50000000

Actual values of the constraints:
R1                      185
R2                   338.75
R3                      600

Dual values with from - till limits:
                Dual value              From            Till
R1                    0.5               150             450
R2                      0          -1e+030          1e+030
R3                  0.375          246.6667             740
```

The important quantity associated with constraints is provided column *Dual value* (also known as *marginal value* and *shadow price*). This value indicates the amount the objective function will change when the constraint value is incremented by 1.

In the output listing shown, constraint R1 has a dual value of 0.5, and the value of the constraint can change from 150 to 450. Constraint R3 has a dual value of 0.375, and the value of the constraint can change from 246.6667 to 740. The dual value of a constraint only holds in that range of values of the constraint.

Constraint R2 has a dual value of zero. Its actual value is 338.75 and the constraint in the problem was specified as ≥ 155. Constraint R1 is said to be not active. This constraint has very minimum effect on the value of the objective function. Its range of values is actually from $-\infty$ to 338.75 (its actual value).

22.2.2.4 Using glpsol Case Study 2

The output produced by GLPK glpsol shows the dual prices in the Marginals column. It also shows the range of values that the right-hand side of each constraint can change, with the current solution of the problem remaining optimal. The following listing is the output produced by glpsol, which is similar to the output listing produced by LP_solve for the problem in case study 2.

```
GLPK 4.45 - SENSITIVITY ANALYSIS REPORT                Page   1

Problem:    example1
Objective:  z = 317.5 (MINimum)

No. Row   St  Activity    Slack        Lower       Activity
                                       Marginal     Upper        range
--- -----  --  ----------- ----------- ----------  -----------
1   z     BS   317.500     -317.500        -Inf     342.000
```

```
                                .          +Inf        317.500

2    R1    NL   185.000         .          185.000      150.000
                              .50000        +Inf        450.000

3    R2    BS   338.750      -183.750      155.000     1200.000
                                .           +Inf        207.500

4    R3    NL   600.000         .          600.000      246.666
                              .37500        +Inf        740.000


   Row     Obj coef     Obj value at   Limiting
           range        break point    variable
   ---   ------------  ------------   ----------
    z      -1.00000         .           x2
           +Inf           +Inf

   R1      -.50000      300.00000        x3
           +Inf         450.00000        x1

   R2      -.15385      265.38462        R1
            .18667      380.73333        x2

   R3      -.37500      185.00000        x1
           +Inf         370.00000        x3

End of report
```

The listing shows constraint R1 with a Slack value of -183.75, which is the difference between its specified value (155.0) and its actual value, which is 338.75. This constraint is not active in the optimal solution of the problem.

22.2.2.5   Using LP_solve Case Study 3

The output produced by LP_solve shows the dual value for every constraint and the range of values that the right-hand side of each constraint can change, with the current solution of the problem remaining optimal. The following partial listing is the output produced by LP_solve for the problem in Case Study 3.

```
/* Example2.lp in LP_solve */

Maximize: 144 x1 + 60 x2;

 120 x1  + 210 x2 <= 15000;
 110 x1  +  30 x2 <= 4000;
     x1  +     x2 <= 75;
```

Executing the LP_solve program at the Linux shell command terminal window to solve this problem yields the following output. Only a partial listing is shown.

```
$ lp_solve -S4 example2.lp

Value of objective function: 6337.50000000

Actual values of the constraints:
R1                      13781.2
R2                         4000
R3                           75

Dual values with from - till limits:
                  Dual value        From           Till
R1                    0            -1e+030        1e+030
R2                    1.05         2916.667        8250
R3                    28.5         36.36364         80
```

The important quantity associated with constraints is provided in column *Dual value*. This value indicates the amount the objective function will change when the constraint value is incremented by 1.

In the output listing shown, constraint R2 has a dual value of 1.05, and the value of the constraint can change from 2916.667 to 8250. Constraint R3 has a dual value of 28.5, and the value of the constraint can change from 36.36364 to 80. The dual value of a constraint only holds in that range of values of the constraint.

Constraint R1 has a dual value of zero. Its actual value in the problem was specified as, $\leq$ 15000. Constraint R1 is said to be not active. This constraint has very minimum effect on the value of the objective function. Its range of values is actually from 13781.2 (its actual value) to ∞.

22.2.2.6 Using `glpsol` Case Study 3

The output produced by GLPK glpsol shows the dual prices in the Marginals column. It also shows the range of values that the right-hand side of each constraint can change, with the current solution of the problem remaining optimal. The following listing is the output produced by `glpsol`, which is similar to the output listing produced by LP-solve for the problem in case study 3.

```
GLPK 4.45 - SENSITIVITY ANALYSIS REPORT                   Page   1

Problem:    example2
Objective:  z = 6337.5 (MAXimum)

No. Row   St  Activity     Slack        Lower      Activity
                           Marginal     Upper      range
---- ----- -- ----------- ------------- --------- -----------
 1    z    BS  6337.500    -6337.50000   -Inf      5236.36364
                                          +Inf      6337.50000

 2    R1   BS  13781.250    1218.75000   -Inf      4363.63636
                                        15000.000  15750.00000

 3    R2   NU  4000.000         .         -Inf      2916.66667
                             1.05000     4000.000   8250.00000
```

```
4    R3   NU  75.000              .            -Inf       36.36364
                           28.50000       75.00000      80.00000
```

Row	Obj coef range	Obj value at break point	Limiting variable
z	-1.00000	.	R3
	+Inf	+Inf	
R1	-.11692	4726.15385	R3
	.93333	19200.00000	R2
R2	-1.05000	5200.00000	R1
	+Inf	10800.00000	x2
R3	-28.50000	5236.36364	x2
	+Inf	6480.00000	R1

```
End of report
```

The listing shows constraint R1 with a Slack value of 1218.75, which is the difference between its specified value (15000.0) and its actual value, which is 13781.250. This constraint is not active in the optimal solution of the problem.

22.3   Duality

For every linear optimization problem, known as the *primal*, there is an associated problem known as its *dual*. The relationship between these two problems helps to understand the connection between the reduced cost and the shadow price.

22.3.1   Formulating the Dual Problem

Assume that the primal linear maximization problem has the generalized standard form:

Maximize $f = c_1x_1 + c_2x_2 + \ldots + c_nx_n$
Subject to the following constraints:

$$
\begin{aligned}
a_{11}x_1 &+ a_{12}x_2 + \ldots + a_{1n}x_n \leq b_1 \\
a_{21}x_1 &+ a_{22}x_2 + \ldots + a_{2n}x_n \leq b_2 \\
&\vdots \qquad \vdots \qquad\qquad \vdots \qquad \vdots \\
a_{m1}x_1 &+ a_{m2}x_2 + \ldots + a_{mn}x_n \leq b_m
\end{aligned}
\tag{22.1}
$$

$$x_i \geq 0, \quad i = 1, 2, \ldots, n.$$

The constraints in matrix form are as follows:

$$
\begin{bmatrix}
a_{11} & a_{12} & \cdots & a_{1n} \\
a_{21} & a_{22} & \cdots & a_{2n} \\
\vdots & \vdots & \ddots & \vdots \\
a_{m1} & a_{2m} & \cdots & a_{mn}
\end{bmatrix}
\begin{bmatrix}
x_1 \\
x_2 \\
\vdots \\
x_n
\end{bmatrix}
\leq
\begin{bmatrix}
b_1 \\
b_2 \\
\vdots \\
b_m
\end{bmatrix}
\tag{22.2}
$$

This linear problem can also be written as: maximize $f = C'X$ such that $AX \leq B$, in which C is a column vector of size n, X is a column vector of size n, A is an $m \times n$ matrix, and B is a column vector of size m. The dual problem is a minimization problem formulated in standard form as follows:

Minimize $g = b_1 y_1 + b_2 y_2 + \ldots + b_m y_m$

Subject to the following constraints:

$$
\begin{array}{lllll}
a_{11} y_1 & + a_{21} y_2 & + \ldots & + a_{m1} y_m & \geq c_1 \\
a_{12} y_1 & + a_{22} y_2 & + \ldots & + a_{m2} y_m & \geq c_2 \\
\vdots & \vdots & & \vdots & \vdots \\
a_{1n} y_1 & + a_{2n} y_2 & + \ldots & + a_{mn} y_m & \geq c_n
\end{array}
\tag{22.3}
$$

$$
y_i \geq 0, \quad i = 1, 2, \ldots, m.
$$

The constraints in matrix form are as follows:

$$
\begin{bmatrix}
a_{11} & a_{21} & \cdots & a_{m1} \\
a_{12} & a_{22} & \cdots & a_{m2} \\
\vdots & \vdots & \ddots & \vdots \\
a_{1n} & a_{2n} & \cdots & a_{mn}
\end{bmatrix}
\begin{bmatrix}
y_1 \\
y_2 \\
\vdots \\
y_m
\end{bmatrix}
\geq
\begin{bmatrix}
c_1 \\
c_2 \\
\vdots \\
c_n
\end{bmatrix}
\tag{22.4}
$$

The matrix form of this dual linear problem can also be written as: minimize $g = B'Y$ such that $A'Y \geq C$, in which C is a column vector of size n, X is a column vector of size n, A is an $m \times n$ matrix, and B is a column vector of size m. In a similar manner, if the primal problem is a minimization problem, its dual is a maximization problem. For example, the following primal maximization problem is in standard form.

Maximize $f = 144x_1 + 60x_2$

Subject to the following constraints:

$$
\begin{array}{rrcl}
120x_1 & + 210x_2 & \leq & 15000 \\
110x_1 & + 30x_2 & \leq & 4000 \\
x_1 & + x_2 & \leq & 75
\end{array}
$$

$$
x_1 \geq 0, \quad x_2 \geq 0.
$$

The dual problem of the given primal is formulated as follows:

Minimize $g = 15000y_1 + 4000y_2 + 75y_3$

Subject to the following constraints:

$$
\begin{aligned}
120y_1 &+ 110y_2 &+ y_3 &\geq 144 \\
210y_1 &+ 30y_2 &+ y_3 &\geq 60
\end{aligned}
$$

$$y_1 \geq 0, \quad y_2 \geq 0.$$

22.3.2 Transforming a Problem to Standard Form

A primal linear problem has to be formulated in standard form before its dual can be formulated. A standard maximization form is also known as the **normal maximization** form that has been explained previously.

When a linear problem is not in standard (or normal) form, then a few transformations are necessary. For example, the following linear maximization problem formulation is not in standard form.

Maximize $f = 144x_1 + 60x_2$

Subject to the following constraints:

$$
\begin{aligned}
120x_1 &+ 210x_2 &\leq 15000 \\
110x_1 &+ 30x_2 &= 4000 \\
x_1 &+ x_2 &\geq 75
\end{aligned}
$$

$$x_1 \geq 0, \quad x_2 \text{ urs.}$$

This problem has a constraint with an equal sign (=), a constraint with a $\geq$ sign, and a variable with unrestricted sign. Therefore, the problem is not formulated in standard (normal) form. The following transformation steps are necessary:

- To transform a constraint with a $\geq$ to a constraint with a $\leq$ sign, the constraint must be multiplied by -1. For example, the constraint $x_1 + x_2 \geq 75$, is transformed to $-x_1 - x_2 \leq -75$.

- To transform a constraint with an equal sign, it must be replaced by two inequality constraints, one with a $\leq$ sign and another with a $\geq$ sign. For example, the second constraint in the problem ($110x_1 + 30x_2 = 4000$) is replaced by the two constraints: $110x_1 + 30x_2 \geq 4000$ and $110x_1 + 30x_2 \leq 4000$. The first of these constraints is transformed to a constraint with a $\leq$ sign by multiplying it by -1.

- When a decision variable x_i is unrestricted in sign (urs), it means that it can take positive, negative, and zero values. The equivalence $x_i = x_i' - x_i''$ is applied; therefore, variable x_i is replaced by $x_i' - x_i''$. In the problem previously discussed, variable x_2 is unrestricted in sign, so it is replaced by $x_2' - x_2''$.

The transformed linear maximization (primal) problem formulation is now expressed as follows:

Maximize $f = 144x_1 + 60x_2' - 60x_2''$
Subject to the following constraints:

$$
\begin{array}{rrrr}
120x_1 & +210x_2' & -210x_2'' & \leq 15000 \\
110x_1 & +30x_2' & -30x_2'' & \leq 4000 \\
-110x_1 & -30x_2' & +30x_2'' & \leq -4000 \\
-x_1 & -x_2' & +x_2'' & \leq -75
\end{array}
$$

$$x_1 \geq 0, \quad x_2' \geq 0, \quad x_2'' \geq 0.$$

The final step is to find the dual problem of the primal problem discussed. Because the primal problem is a maximization problem, its dual is a minimization problem. The formulation of the dual problem is as expressed as follows:

Minimize $g = 15000y_1 + 4000y_2 - 4000y_3 - 75y_4$
Subject to the following constraints:

$$
\begin{array}{rrrrr}
120y_1 & +110y_2 & -110y_3 & -y_4 & \geq 144 \\
210y_1 & +30y_2 & -30y_3 & -y_4 & \geq 60 \\
-210y_1 & -30y_2 & +30y_2 & +y_4 & \leq -60
\end{array}
$$

$$y_1 \geq 0, \quad y_2 \geq 0, \quad y_3 \geq 0, \quad y_4 \geq 0.$$

In a similar manner, a linear minimization (primal) problem that is not in standard form can be transformed to a standard form. To transform a constraint with a $\leq$ to a constraint with a $\geq$ sign, the constraint must be multiplied by -1. The other steps are the same as outlined previously.

22.3.3 Duality Discussion

The constraint values of the primal problem are related to the variables of the dual problem. These variables are known as *shadow prices*.

The *weak duality* theorem states that the objective function of value g of the dual problem at any feasible solution: $y_1, y_2, \ldots, y_m$, is always greater than or equal to the objective value z of the primal problem at any feasible solution $x_1, x_2, \ldots x_n$. This can be expressed in matrix form as follows:

$$g = B'Y \quad \geq \quad C'X = z$$

The *strong duality* theorem specifies that the primal and dual problems have equal optimal values of the objective function. This theorem can be used to solve the primal linear optimization problem.

Recall that the *shadow price* of constraint i of a linear maximization problem is the amount by which the optimal value of the objective function increases when the right-hand value of constraint i is increased by 1. The strong dual theorem can be used to calculate the shadow price of constraint i.

Summary

Sensitivity analysis and duality are important concepts and the information is valuable in addition to the data on optimality. Sensitivity analysis helps to find the effect of small changes in the parameters of the formulation of a linear optimization problem. Duality helps to improve the understanding of a linear optimization problem.

	Key Terms	
sensitivity	duality	objective coefficients
right-hand constraints	marginal values	shadow prices
reduced costs	dual value	primal problem
dual problem	normal maximization	

Exercises

Exercise 22.1 A factory manufactures 5 different products: P_1, P_2, P_3, P_4, and P_5. The factory needs to maximize profit. Each product requires machine time on three different devices: A, B, and C, each of which is available 135 hours per week. The following table provides the data:

Product	Device A	Device B	Device C
P_1	20	15	8
P_2	13	17	12
P_3	15	7	11
P_4	16	4	7
P_5	18	14	6

The unit sale price for products P_1, P_2, and P_3 is $7.50, $6.00, and $8.25 respectively. The first 26 units of P_4 and P_5 have a sale prices $5.85 each, all excess units have a sale price of $4.50 each. The operational costs of devices A and B are $5.25 per hour, and $5.85 for device C. The cost of materials for products P_1 and P_4 is $2.75. For products P_2, P_3, and P_5 the materials cost is $2.25. Hint: assume that x_i, $i = 1 \dots 5$ are the number of units produced of the various products and use two new variables y_4 and y_5 for the units produced of P_4 and P_5 respectively in excess of 26.

Write the formulation of the mathematical optimization problem, compute the optimal solution with lp_solve, and find out the range of prices that preserve the optimal solution. If the price of P_1 changes by 35%, how does the optimal solution change?

Exercise 22.2 Find the dual of Exercise 22.1. Compute the solution with lp_solve.

Chapter 23

Transportation Models

23.1 Introduction

This chapter presents the general concepts and formulation of transportation and trans-shipment problems. These are special cases of linear optimization problems.

The main goal is to formulate these problems as linear optimization problems and compute minimum cost to transport a product as the optimal solution computed.

Transportation problems deal with finding the optimal manner by which a product or commodity produced or available at different supply points can be transported to a number of destinations or demand points. Typically the objective function is the cost of transportation subject to capacity constraints at the supply points and demand constraints at the demand points.

23.2 Model of a Transportation Problem

A transportation problem is formulated as a standard linear optimization problem. The objective function is defined to minimize the cost of transportation, subject to demand and supply constraints.

Assume there are m *supply points* and n *demand points* in a problem. Let c_{ij} denote the given unit cost of transportation from supply point i to demand (destination) point j. Let x_{ij} denote the amount of product to be transported from supply point i to demand (destination) point j. The objective function can then be expressed as follows:

Minimize z,

$$z = \sum_{i=1}^{i=m} \sum_{j=1}^{j=n} x_{ij} c_{ij}$$

This equation can be expanded and written as equations in which each row of the right-hand side of equation represents the cost of transportation from a supply point. For example, row 1 of the right-hand side of the equation represents the cost of transportation from supply point 1; row 2 represents the cost of transportation from supply point 2, the last row represents the cost of transportation from supply point m.

$$z = \begin{array}{llll} c_{11}x_{11} & + c_{12}x_{12} & + \cdots & + c_{1n}x_{1n} + \\ c_{21}x_{21} & + c_{22}x_{22} & + \cdots & + c_{2n}x_{2n} + \\ \vdots & \vdots & \ddots & \vdots & + \\ c_{m1}x_{m1} & + c_{m2}x_{m2} & + \cdots & + c_{mn}x_{mn} \end{array}$$

In transportation problems there are two types of constraints: *supply constraints* and *demand constraints*. Let s_i denote the amount of product at the supply point i. Let d_j denote the amount of product at the demand point j.

The supply constraints have the right-hand side as an upper bound. There are m supply constraints, the constraint of supply point i is expressed as follows:

$$\sum_{j=1}^{j=n} x_{ij} \leq s_i$$

This equation can be expanded to show all the quantities of the product to be shipped from supply point i. There are m supply constraints, with each row representing the total quantity of product transported from an indicated supply point. For example, the first row (1) represents the quantities of product shipped from supply point 1. The second row (2) represents the quantities of product shipped from supply point 2. The last row (m) represents the quantities of product shipped from supply point m.

$$\begin{array}{lllll} x_{11} & + x_{12} & + \cdots & + x_{1n} & \leq s_1 \\ x_{21} & + x_{22} & + \cdots & + x_{2n} & \leq s_2 \\ \vdots & \vdots & \ddots & \vdots & \vdots \\ x_{m1} & + x_{m2} & + \cdots & + x_{mn} & \leq s_m \end{array}$$

The demand constraints have the right-hand side as a lower bound. There are n demand constraints, the constraint of supply point j is expressed as follows:

$$\sum_{i=1}^{i=m} x_{ij} \geq d_j$$

This equation can be expanded to show all the quantities of the product to be received at demand point j. There are n demand constraints, with each row representing the total quantity of product to be transported and received by the indicated demand point. For example, the first row (1) represents the quantities of product at demand point 1. The second row (2) represents the quantities of product at demand point 2. The last row (n) represents the quantities of product at demand point n.

$$
\begin{array}{ccccccc}
x_{11} & + x_{21} & + \cdots & + x_{m1} & \geq d_1 \\
x_{12} & + x_{22} & + \cdots & + x_{m2} & \geq d_2 \\
\vdots & \vdots & \ddots & \vdots & \vdots \\
x_{1n} & + x_{2n} & + \cdots & + x_{mn} & \geq d_m
\end{array}
$$

The decision variables x_{ij} have sign constraint: $x_{ij} \geq 0$, for $i = 1, 2, \ldots m$ and $j = 1, 2, \ldots n$.

23.3 Transportation Case Studies

The following three case studies help to illustrate the modeling of transportation problems.

23.3.1 Transport Case Study 1

The distribution manager of a company needs to minimize global transport costs between a set of three factories (supply points) S1, S2, and S3, and a set of four distributors (demand points) D1, D2, D3, and D4. The following table shows the transportation cost from each supply point to every demand point, the supply of the product at the supply points, and the demand of the product at the demand points.

	D1	D2	D3	D4	Supply
S1	20	40	70	50	400
S2	100	60	90	80	1500
S3	10	110	30	200	900
Demand	700	600	1000	500	

The transportation unit costs for every supply point are shown from columns 2 to 5. The transportation unit cost from supply point S1 to demand point D1 is 20. The transportation unit cost from supply point S1 to demand point D2 is 40. The transportation unit cost from supply point S2 to demand point D3 is 90, and so on.

The last column in the table shows the supply capacity of the supply point, in quantity of the product. The capacity of supply point S1 is 400. The summation of the values in the last column is the total supply in the system; this value is 2800. The last row of the table shows the demand of each demand point, in quantity of the product. The demand of demand point D1 is 700, of demand point D2 is 600, and so on. The summation of the values in the last row is the total demand in the system; this value is 2800.

Note that the total supply is 2800, and the total demand is also 2800. This is calculated by summing the values of the last column and the last row. Because the value for the amount of product of total supply and the amount of total demand is the same, this transportation problem is said to be *balanced*. The objective function can be expressed as follows:

Minimize z,

$$z = \sum_{i=1}^{1=m} \sum_{j=1}^{j=n} x_{ij} c_{ij}$$

This problem has $m = 3$ supply points and $n = 4$ demand points. The objective function can be completely written with the unit cost values of the product to be transported, given in the table shown previously. The objective function is expressed as follows:

$$z = \begin{array}{llll} 20x_{11} & + 40x_{12} & + 70x_{13} & + 50x_{14} + \\ 100x_{21} & + 60x_{22} & + 90x_{23} & + 80x_{24} + \\ 10x_{31} & + 110x_{32} & + 30x_{33} & + 200x_{34} \end{array}$$

The supply constraints are:

$$\begin{array}{lllll} x_{11} & + x_{12} & + x_{13} & + x_{14} & \leq 400 \\ x_{21} & + x_{22} & + x_{23} & + x_{24} & \leq 1500 \\ x_{31} & + x_{32} & + x_{33} & + x_{34} & \leq 900 \end{array}$$

The demand constraints are:

$$\begin{array}{llll} x_{11} & + x_{21} & + x_{31} & \geq 700 \\ x_{12} & + x_{22} & + x_{32} & \geq 600 \\ x_{13} & + x_{23} & + x_{33} & \geq 1000 \\ x_{14} & + x_{24} & + x_{34} & \geq 500 \end{array}$$

The formulation of the problem is formatted for solving with LP_solve and stored in file `transport1.lp`.

```
/* File: transport1.lp
Objective function:    Cost of transportation */

Minimize: 20 x11 +   40 x12 +   70 x13 +   50 x14 +
          100 x21 + 60 x22 +   90 x23 +   80 x24 +
          10 x31 + 110 x32 +   30 x33 + 200 x34;

/*  Capacity Constraints (Supply constraints) */
   x11  +  x12   +   x13  + x14   <= 400;
   x21  +  x22   +   x23  + x24   <= 1500;
   x31  +  x32   +   x33  + x34   <= 900;

/* Demand constraints */
   x11  +  x21   + x31   >= 700;
   x12  +  x22   + x32   >= 600;
   x13  +  x23   + x33   >= 1000;
   x14  + x24    + x34   >= 500;
```

The execution of the LP_solve program to solve this linear optimization problem produces the results that can be observed in the following listing.

```
Variables      result
z              141000
x11            400
x12            0
x13            0
x14            0
x21            0
x22            600
x23            400
x24            500
x31            300
x32            0
x33            600
x34            0
```

Note that the optimal total transportation cost is 141,000 and some of the values of x are zero. If $x_{ij} = 0$, then the amount of the product to be transported from supply point i to demand point j is zero. In this problem, the total demand of demand point D1 is satisfied by the amount 400 from supply point S1, and the amount 300 from supply point S3. There was no supply from supply point D2 to demand point D1, therefore $x_{21} = 0$.

23.3.2 Unbalanced Problem: Case Study 2

The transportation problem discussed in the previous section is an example of a balanced problem. In this case, the total supply is equal to the total demand and is expressed mathematically as:

$$\sum_{i=1}^{i=m} s_i = \sum_{j=1}^{j=n} d_j$$

If the transportation problem is not balanced, the total supply may be less than the total demand, or the total supply may be greater than the total demand. When total supply is greater than the total demand, the problem is *unbalanced* and its formulation must include a *dummy demand* point to balance the problem, and the transportation costs to this demand point are zero.

The following problem has a small variation to the one discussed in the previous section. The distribution manager of a company needs to minimize global transport costs between a set of three factories (supply points) S1, S2, and S3, and a set of four distributors (demand points) D1, D2, D3, and D4.

The following table shows the transportation cost from each supply point to every demand point, the supply of the product at the supply points, and the demand of the product at the demand points.

	D1	D2	D3	D4	D5	Supply
S1	20	40	70	50	0	600
S2	100	60	90	80	0	1500
S3	10	110	30	200	0	900
Demand	700	600	1000	500	200	

The last column in the table shows the supply capacity of the supply point, in quantity of the product. The capacity of supply point S1 is 600. The summation of the values in the last column is the total supply in the system, this value is 2800. The last row of the table shows the demand of each demand point, in quantity of the product. The demand of demand point D1 is 700, of demand point D2 is 600, and so on. The summation of the values in the last row is the total demand in the system, this value is 2800.

Note that the total supply is 3000, and the total demand is 2800. This is calculated by summing the values of the last column and the last row. Because the value for the amount of product of total supply is greater than the amount of total demand, this transportation problem is said to be *unbalanced*.

The transportation unit costs for every supply point are shown from columns 2 to 5. The transportation unit cost from supply point S1 to demand point D1 is 20. The transportation unit cost from supply point S1 to demand point D2 is 40. The transportation unit cost from supply point S2 to demand point D3 is 90, and so on.

Because there is an *excess supply* of 200, the formulation of the problem must include a *dummy demand* point, D5, with a demand of 200. The transportation costs to demand point D5 are zero, and this is expressed as follows:

$$c_{i5} = 0, \quad i = 1, \ldots, 3$$

This problem now has $m = 3$ supply points and $n = 5$ demand points. The objective function can be completely written with the unit cost values of the product to be transported, given in the table shown previously. The objective function is expressed as follows:

$$
\begin{aligned}
z = \quad & 20x_{11} && + 40x_{12} && + 70x_{13} && + 50x_{14} + \\
         & 100x_{21} && + 60x_{22} && + 90x_{23} && + 80x_{24} + \\
         & 10x_{31} && + 110x_{32} && + 30x_{33} && + 200x_{34}
\end{aligned}
$$

The *supply constraints* have the right-hand side as an upper bound. There are 3 supply constraints, the constraint of supply point i is expressed as follows:

$$\sum_{j=1}^{j=n} x_{ij} \leq s_i$$

There are 3 supply constraints, with each row representing the total quantity of product transported from an indicated supply point. For example, the first row (1) represents the quantities of product shipped from supply point 1. The second row (2) represents the quantities of product shipped from supply point 2. The last row (3) represents the quantities of product shipped from supply point 3.

$$
\begin{aligned}
x_{11} + x_{12} + x_{13} + x_{14} + x_{15} &\leq 600 \\
x_{21} + x_{22} + x_{23} + x_{24} + x_{25} &\leq 1500 \\
x_{31} + x_{32} + x_{33} + x_{34} + x_{35} &\leq 900
\end{aligned}
$$

The *demand constraints* have the right-hand side as a lower bound. There are 5 demand constraints, the constraint of supply point j is expressed as follows:

$$
\sum_{i=1}^{i=m} x_{ij} \geq d_j
$$

There are 5 demand constraints, with each row representing the total quantity of product to be transported and received by the indicated demand point.

$$
\begin{aligned}
x_{11} + x_{21} + x_{31} &\geq 700 \\
x_{12} + x_{22} + x_{32} &\geq 600 \\
x_{13} + x_{23} + x_{33} &\geq 1000 \\
x_{14} + x_{24} + x_{34} &\geq 500 \\
x_{15} + x_{25} + x_{35} &\geq 200
\end{aligned}
$$

The decision variables x_{ij} have the sign constraint: $x_{ij} \geq 0$, for $i = 1, 2, 3$ and $j = 1, 2, \ldots, 5$. The formulation of the problem is formatted for solving with LP_solve and stored in file `transport1u.lp`.

```
/* Transport1u.lp
Minimize global transport costs between a set of
three factories (supply points) S1, S2, and S3,
and four distributors (demand points) D1, D2, D3, and D4.
There is an excess supply of 200, so a dummy demand point point,
D5 is added to balance the problem.
*/

/* Objective function
   Cost of transportation */

Minimize: 20 x11 +   40 x12 +   70 x13 +   50 x14 +
          100 x21 + 60 x22 +   90 x23 +   80 x24 +
          10 x31 + 110 x32 +   30 x33 + 200 x34;

/*  Capacity Constraints (Supply constraints) */
   x11  +   x12  +   x13  + x14  + x15 <= 600;
   x21  +   x22  +   x23  + x24  + x25 <= 1500;
   x31  +   x32  +   x33  + x34  + x35 <= 900;

/* Demand constraints */
   x11  +   x21  + x31  >= 700;
   x12  +   x22  + x32  >= 600;
```

```
x13  +  x23   + x33   >= 1000;
x14  + x24    + x34   >= 500;
x15  + x25    + x35   >= 200;
```

The execution of the LP_solve program to solve this linear optimization problem produces the results that can be observed in the following listing.

```
Variables   result
z           131000
x11         600
x12         0
x13         0
x14         0
x21         0
x22         600
x23         200
x24         500
x31         100
x32         0
x33         800
x34         0
x15         0
x25         200
x35         0
```

23.3.3   Unbalanced Problem: Case Study 3

When the demand exceeds the supply, the problem formulation includes a *penalty* associated with the *unmet demand*. Suppose that in the original problem discussed previously, supply point S1 produces 300 units of the product (instead of 400). This problem is now unbalanced, with an unmet demand of 100.

A dummy supply point, S4 with a supply of 100, is added to the problem formulation. The penalty for unmet demand at demand point D1 is 125; at demand point D2 is 147; at demand point D3 is 95; and at demand point D4 is 255.

The following table shows the transportation cost from each supply point to every demand point, the supply of the product at the supply points, and the demand of the product at the demand points.

	D1	D2	D3	D4	Supply
S1	20	40	70	50	300
S2	100	60	90	80	1500
S3	10	110	30	200	900
S4	125	147	95	255	100
Demand	700	600	1000	500	

Note that the dummy supply point S4 has been included in the table. The penalty amounts have also been included for this supply point. The total supply and demand is now 2800.

This problem has $m = 4$ supply points and $n = 4$ demand points. The objective function can be completely written with the unit cost values of the product to be transported, given in the table shown previously. The objective function is expressed as follows:

$$z = \begin{array}{llll} 20x_{11} & + 40x_{12} & + 70x_{13} & + 50x_{14} + \\ 100x_{21} & + 60x_{22} & + 90x_{23} & + 80x_{24} + \\ 10x_{31} & + 110x_{32} & + 30x_{33} & + 200x_{34} + \\ 125x_{41} & + 147x_{42} & + 95x_{43} & + 255x_{44} \end{array}$$

There are 4 supply constraints, with each row representing the total quantity of product transported from an indicated supply point. For example, the first row (1) represents the quantities of product shipped from supply point 1. The second row (2) represents the quantities of product shipped from supply point 2. The last row (4) represents the quantities of product shipped from supply point 4.

$$\begin{array}{llll} x_{11} & + x_{12} & + x_{13} & + x_{14} \le 300 \\ x_{21} & + x_{22} & + x_{23} & + x_{24} \le 1500 \\ x_{31} & + x_{32} & + x_{33} & + x_{34} \le 900 \\ x_{41} & + x_{42} & + x_{43} & + x_{44} \le 100 \end{array}$$

The demand constraints have the right-hand side as a lower bound. There are 4 demand constraints, with each row representing the total quantity of product to be transported and received by the indicated demand point.

$$\begin{array}{llll} x_{11} & + x_{21} & + x_{31} + x_{41} & \ge 700 \\ x_{12} & + x_{22} & + x_{32} + x_{42} & \ge 600 \\ x_{13} & + x_{23} & + x_{33} + x_{43} & \ge 1000 \\ x_{14} & + x_{24} & + x_{34} + x_{44} & \ge 500 \end{array}$$

The decision variables x_{ij} have the sign constraint: $x_{ij} \ge 0$, for $i = 1,\ldots,4$ and $j = 1,\ldots,4$. The formulation of the problem is formatted for solving with LP_solve and stored in file `transport1ub.lp`.

```
/* Transport1ub.lp
Minimize global transport costs between a set of
three factories (supply points) S1, S2, and S3,
and four distributors (demand points) D1, D2, D3, and D4.
There is an excess demand of 100, so a dummy supply point, S4
 is added to balance the problem.
Penalty costs are also included.
*/
/* Objective function -- Cost of transportation */
Minimize: 20 x11 +   40 x12 +   70 x13 +   50 x14 +
          100 x21 +   60 x22 +   90 x23 +   80 x24 +
```

```
         10 x31 + 110 x32 +  30 x33 + 200 x34 +
        125 x41 + 147 x42 +  95 x43 + 255 x44; /* penalty */

/* Capacity Constraints (Supply constraints) */
  x11  +  x12  +  x13  + x14 <= 300;
  x21  +  x22  +  x23  + x24 <= 1500;
  x31  +  x32  +  x33  + x34 <= 900;
  x41  +  x42  +  x43  + x44 <= 100; /* Dummy supply point */

/* Demand constraints */
  x11  +  x21  + x31  + x41 >= 700;
  x12  +  x22  + x32  + x42 >= 600;
  x13  +  x23  + x33  + x43 >= 1000;
  x14  +  x24  + x34  + x44 >= 500;
```

The execution of the LP_solve program to solve this linear optimization problem produces the results that can be observed in the following listing.

```
Variables     result
z             146500
x11           300
x12           0
x13           0
x14           0
x21           0
x22           600
x23           400
x24           500
x31           400
x32           0
x33           500
x34           0
x41           0
x42           0
x43           100
x44           0
```

23.4   Transshipment Models

A transshipment model includes *intermediate* or *transshipment points* in a trans-portation model. A transshipment point is an intermediate point between one or more supply points and one or more demand points. Quantities of a product can be sent from a supply point directly to a demand point or via a transshipment point.

23.4.1   General Form of Model

A transshipment point can be considered both, a supply point and a demand point. At a transshipment point, k, the total quantity of product shipped to this point must equal the total quantity of the product shipped from this intermediate point and can be expressed as follows:

$$\sum_{i=1}^{i=m} x_{i,k} = \sum_{j=1}^{j=n} x_{k,j}$$

Where m is the number of supply points, n is the number of demand points, and x_{ij} is the amount of the product shipped from supply point i to demand point j. From the previous equation, the *transshipment constraint* of a point, k, is expressed as follows:

$$\sum_{i=1}^{i=m} x_{i,k} - \sum_{j=1}^{j=n} x_{k,j} = 0 \tag{23.1}$$

23.4.2   Transshipment Problem: Case Study 4

Various quantities of a product are shipped from two cities (supply points), S1 and S2, to three destinations (demand points), D1, D2, and D3. The products are first shipped to three warehouses (transshipment points), T1, T2, and T3, then shipped to their final destinations.

The following table shows the transportation cost from each supply point to the intermediate points and to every demand point. The table also includes the supply of the product at the supply points, and the demand of the product at the demand points.

	T1	T2	T3	D1	D2	D3	Supply
S1	16	10	12	0	0	0	300
S2	15	14	17	0	0	0	300
T1	0	0	0	6	8	10	0
T2	0	0	0	7	11	11	0
T3	0	0	0	4	5	12	0
Demand	0	0	0	200	100	300	

Note that the total supply of the product is 600 and the total demand is also 600. Therefore, this a balanced problem.

The problem can be formulated directly as a standard transportation problem, using the data in the previous table. The conventional notation for the quantity of product is used and x_{ij} denotes the quantity of product shipped from point i to point j. For example: x_{23} denotes the quantity of product shipped from supply point S2 to transshipment point T3.

This problem has $m = 2$ supply points, $n = 3$ demand points, and 3 transshipment points. Because a transshipment point can be a supply point and a demand point, the problem can be formulated with a total of 5 supply points and 6 demand points.

The objective function can be completely written with the unit cost values of the product to be transported given in the table. The objective function is expressed as follows:

$$z = \begin{array}{llll} 16x_{11} & + 10x_{12} & + 12x_{13} & + \\ 15x_{21} & + 14x_{22} & + 17x_{23} & + \\ 6x_{34} & + 8x_{35} & + 10x_{36} & + \\ 7x_{44} & + 11x_{45} & + 11x_{46} & + \\ 4x_{54} & + 5x_{55} & + 12x_{56} \end{array}$$

There are three transshipment constraints, one for each transshipment point. These constraints are:

$$\begin{array}{llllll} x_{11} & + x_{21} & - x_{34} & - x_{35} & - x_{36} & = 0 \quad (T1) \\ x_{12} & + x_{22} & - x_{44} & - x_{45} & - x_{46} & = 0 \quad (T2) \\ x_{13} & + x_{23} & - x_{54} & - x_{55} & - x_{56} & = 0 \quad (T3) \end{array}$$

There are two supply constraints, each representing the total quantity of product transported from an indicated supply point.

$$\begin{array}{llll} x_{11} & + x_{12} & + x_{13} & \leq 300 \\ x_{21} & + x_{22} & + x_{23} & \leq 300 \end{array}$$

There are three demand constraints, each representing the total quantity of product transported to the indicated demand point.

$$\begin{array}{llll} x_{36} & + x_{46} & + x_{56} & \geq 300 \\ x_{35} & + x_{45} & + x_{55} & \geq 100 \\ x_{34} & + x_{44} & + x_{54} & \geq 200 \end{array}$$

There are three transshipment constraints applying the balance equations (Equation 23.1) at each one.

$$\begin{array}{llllll} x_{11} & + x_{21} & - x_{34} & - x_{35} & - x_{36} & = 0 \\ x_{12} & + x_{22} & - x_{44} & - x_{45} & - x_{46} & = 0 \\ x_{13} & + x_{23} & - x_{54} & - x_{55} & - x_{56} & = 0 \end{array}$$

The decision variables x_{ij} have the sign constraint: $x_{ij} \geq 0$, for $i = 1, \ldots, 5$ and $j = 1, \ldots, 6$. The formulation of the problem is formatted for solving with LP_solve and stored in file `transship1.lp`.

```
/* Transship1.lp
The distribution manager needs to minimize global transport
costs. There are two supply points: S1 and S2; three
transshipments points: T1, T2, and T3; three demand points:
 D1, D2, D3
*/
/* Objective function -- Cost of transportation */
Minimize: 16 x11 +  10 x12 +  12 x13 +
          15 x21 +  14 x22 +  17 x23 +
           6 x34 +   8 x35 +  10 x36 +
           7 x44 +  11 x45 +  11 x46 +
           4 x54 +   5 x55 +  12 x56;
/*  Capacity Constraints (Supply constraints) */
  x11  +  x12   +  x13  <= 300;  /* S1 */
  x21  +  x22   +  x23  <= 300;  /* S2 */

 /* Demand constraints */
  x34  +  x44   + x54  >= 200;   /* D1 */
  x35  +  x45   + x55  >= 100;   /* D2 */
  x36  +  x46   + x56  >= 300;   /* D3 */

 /* Transshipment constraints */
  x11 + x21 - x34 - x35 - x36 = 0; /* T1 */
  x12 + x22 - x44 - x45 - x46 = 0;  /* T2 */
  x13 + x23 - x54 - x55 - x56 = 0;  /* T3 */
```

The execution of the LP_solve program to solve this linear optimization problem produces the results that can be observed in the following listing.

```
Variables    result
z            12400
x11          0
x12          0
x13          300
x21          0
x22          300
x23          0
x34          0
x35          0
x36          0
x44          0
x45          0
x46          300
x54          200
x55          100
x56          0
```

23.5   Assignment Problems

An *assignment problem* is a special case of a transportation problem that is formulated as a linear optimization model. The basic goal is to find an optimal assignment of *resources* to tasks. The typical objective is to minimize the total time to complete a task or to minimize the cost of the assignment.

23.5.1   General Form of Model

A simple description of the general assignment problem is: minimize the total cost of a set of workers assigned to a set of tasks. The constraints are: each worker is assigned no more that a specified number of jobs, each job requires no more than a specified number of workers.

An assignment of a resource i to a job j is denoted by x_{ij}. Let m be the number of resources and n the number of jobs. The decision variables x_{ij} have the constraint: $x_{ij} = 1$ or $x_{ij} = 0$, for $i = 1 \ldots m$ and $j = 1 \ldots n$. The cost of resource i to complete the job j is denoted by c_{ij}. The total cost of the resource allocation is:

$$\sum_{i=1}^{m} \sum_{j=1}^{n} c_{ij} x_{ij}$$

Let P_j denote the maximum number of resources to be assigned to job j. Similarly, let Q_i denote the maximum number of jobs that resource i can be assigned to. In simple problems, all these parameters are equal to 1. The two types of constraints are m *resource constraints* and n *job constraints*. The job constraints are expressed as:

$$\sum_{i=1}^{m} x_{ij} \leq P_j, \quad j = 1 \ldots n$$

The resource constraints are expressed as:

$$\sum_{j=1}^{n} x_{ij} \leq Q_i, \quad i = 1 \ldots m$$

23.5.2   Assignment Problem: Case Study 5

A factory has 3 machines: M1, M2, and M3. These are to be assigned to four jobs: T1, T2, T3, and T4. The following table is the cost matrix that gives the expected costs when a specific machine is assigned a specific job. The goal of the problem is to optimize the assignment of machines to jobs. In this problem, each machine can only be assigned to one job, and each job can only receive one machine.

	T1	T2	T3	T4
M1	13	16	12	11
M2	15	2	13	20
M3	5	7	10	6

Note that the total number of machines is 3 and the total number of jobs is 4, so this is an unbalanced problem. A dummy machine, M4, is included with zero cost when assigned to a job.

The problem can be formulated directly using the data in the previous table. The conventional notation for an assignment of resource i to job j is used and denoted by x_{ij}. For example: x_{23} denotes the assignment of machine M2 to job T3.

The goal of the problem is to minimize the objective function, which is the cost of the assignment of the various machines to the jobs and expressed as follows:

$$
\begin{aligned}
z = \quad &13x_{11} &+\ 16x_{12} &+\ 12x_{13} &+\ 11x_{14} &+ \\
&15x_{21} &+\ 2x_{22} &+\ 13x_{23} &+\ 20x_{24} &+ \\
&5x_{31} &+\ 7x_{32} &+\ 10x_{33} &+\ 6x_{34} &+
\end{aligned}
$$

The dummy machine assignments are not included in the objective function because their costs are zero. There are four resource constraints, each representing the possible assignments of a machine.

$$
\begin{aligned}
x_{11} + x_{12} + x_{13} + x_{14} &= 1 \quad (\text{M1}) \\
x_{21} + x_{22} + x_{23} + x_{24} &= 1 \quad (\text{M2}) \\
x_{31} + x_{32} + x_{33} + x_{34} &= 1 \quad (\text{M3}) \\
x_{41} + x_{42} + x_{43} + x_{44} &= 1 \quad (\text{M4})
\end{aligned}
$$

There are four job constraints, each representing the possible assignments of the machines to it.

$$
\begin{aligned}
x_{11} + x_{21} + x_{31} + x_{41} &= 1 \quad (\text{T1}) \\
x_{12} + x_{22} + x_{32} + x_{42} &= 1 \quad (\text{T2}) \\
x_{13} + x_{23} + x_{33} + x_{43} &= 1 \quad (\text{T3}) \\
x_{14} + x_{24} + x_{34} + x_{44} &= 1 \quad (\text{T4})
\end{aligned}
$$

The decision variables x_{ij} have the constraint: $x_{ij} = 0$ or $x_{ij} = 1$, for $i = 1, \ldots, 4$ and $j = 1, \ldots, 4$. The formulation of the problem is formatted for solving with LP_solve and stored in file assign1.lp.

```
/* assignment1.lp
A factory has 3 machines: M1, M2, and M3. These are to be
 assigned to four jobs: T1, T2, , T3, and T4.
The cost matrix gives the expected costs when a specific machine
 is assigned a specific job.
*/
/* Objective function -- cost assignments */
Minimize: 13 x11 +   16 x12 +   12 x13 + 11 x14 +
```

```
            15 x21 +   2 x22 +   13 x23 +   20 x24 +
             5 x31 +   7 x32 +   10 x33 +    6 x34;

/*  Job Constraints  */
   x11   +   x21   +   x31   + x41 = 1;   /* T1 */
   x12   +   x22   +   x32   + x42 = 1;   /* T2 */
   x13   +   x23   +   x33   + x43 = 1;   /* T3 */
   x14   +   x24   +   x34   + x44 = 1;   /* T4 */

  /* Machine constraints */
   x11 + x12 + x13 + x14  = 1; /* M1 */
   x21 + x22 + x23 + x24  = 1; /* M2 */
   x31 + x32 + x33 + x34  = 1; /* M3 */
   x41 + x42 + x43 + x44  = 1; /* Dummy M4 */
```

The execution of the LP_solve program to solve this linear optimization problem produces the results that can be observed in the following listing. Note that M4 is assigned to T3, but because M4 is a dummy machine, no machine is assigned to T3.

```
Variables   result
z           18
x11         0
x12         0
x13         0
x14         1
x21         0
x22         1
x23         0
x24         0
x31         1
x32         0
x33         0
x34         0
x41         0
x42         0
x43         1
x44         0
```

Summary

Two important application areas of linear optimization are transportation and trans-shipment problems. The goal of the first type of problem is finding the minimum cost to transport a product. The goal of the second type of problem is the finding the optimal manner to transport products to destination or demand points.

Key Terms		
supply points	demand points	destination points
transportation cost	demand constraints	supply constraints
balanced problem	unbalanced problem	transshipment points
transshipment constraints	job constraints	resource constraints
dummy assignments		

Exercises

Exercise 23.1 A factory of automobile spare parts has two supply locations: S_1 and S_2. There are three demand points that require 35, 42, and 50 parts respectively. Supply point S_1 has a capacity of 85 parts and S_2 a capacity of 65 parts. The following table shows transportation costs and the costs (penalty) for excess demand of each supply point. Formulate and solve the optimization problem with lp_solve.

	D1	D2	D3	Supply
S1	25	45	42	85
S2	20	60	50	65
Demand	35	42	50	
Penalty	100	85	150	

Exercise 23.2 In Exercise 23.1, management of the factory decided to meet excess demand by purchasing additional parts and sending these to the demand points. The cost of purchasing and sending each of these parts is $140.00. Formulate and solve the optimization problem with lp_solve.

Exercise 23.3 A computer manufacturer produces three types of computer laptops: T1, T2, and T3. These are built in three facilities: F1, F2, and F3. In a week of 44 hours, the demand is for a total of 10 computers that must be produced. Formulate and solve the minimum cost of producing the computers. The following table includes the time (in hours) and the cost of producing every type of computer.

Facility	Time	T1	T2	T3
F1	5	$180	$120	$84
F2	4	$150	$90	$90
F3	3.5	$129	$60	$60

Exercise 23.4 In Exercise 23.3, each computer type has a different time (in hours) that it takes to manufacture it depending on the facility where it is built. The following table includes this data. Formulate and solve the minimum cost of producing the computers.

Facility	T1	T2	T3
F1	4.5	4	4.5
F2	4.5	4.3	5
F3	3.5	3.5	4

Chapter 24

Network Models

24.1 Introduction

This chapter presents the general concepts, formulation, and solution of problems that can be described with network models. These are special cases of linear optimization problems. The main goal is to formulate these problems as linear optimization models and compute the minimum cost or the maximum flow from a source point to a destination point in the network.

Examples of these types of problems are: shortest path problems, maximum flow problems, and minimum spanning tree problems.

24.2 Graphs

A *graph* is used as a visual representation of a network. A graph consists of a finite set of *nodes* (also known as vertices) and a finite set of *arcs* that connect pairs of nodes. A directed arc connects an ordered pair of vertices; if an arc starts at node P (head) and ends at node Q (tail), it is denoted as (P, Q). An arc will typically have an associated weight or length.

A *path* is a sequence of arcs with the property that for every arc its tail vertex is the head vertex of the next arc. The length (or cost) of the path is the summation of the lengths of the individual arcs in it.

24.3 Shortest Path Problem

A shortest path problem consists of finding the path from an initial node in the graph to a final node with the shortest length. Several algorithms have been developed to solve this general problem, the most important of which is Dijkstra's algorithm.

This section discusses the formulation of shortest path problems as transshipment problems. For this, the techniques of the previous chapters are applied. As discussed previously, a transshipment model includes *intermediate* or transshipment points in a transportation model. A transshipment point is an intermediate point between one or more supply points and one or more demand points.

The general problem consists of transporting one unit of a product from a source point P to a destination point Q. The intermediate nodes are the transshipment points. Units of the product can be sent from the source point to the destination point using one of several possible paths.

When in a graph there is no arc between two points, an *artificial arc* is included and its length is given a relatively large value, H. An arc from a node to the same node will have zero length.

The *cost* of sending 1 unit of product from node i to node j is the length of the arc and denoted by c_{ij}. For example: c_{23} denotes the cost of sending 1 unit of product shipped from node 2 to node 3. The shipment of 1 unit of the product from node i to node j is denoted by x_{ij}. Therefore, the value of x_{ij} is 1 or zero.

The objective function indicates the total cost of transporting 1 unit of the product from the source node to the destination node and can be expressed as follows:

Minimize: z,

$$z = \sum_{i=1}^{1=n} \sum_{j=1}^{j=n} c_{ij} x_{ij}$$

The number of nodes is n and there is one supply point and one demand point. A transshipment point can be considered both a supply point and a demand point. At a transshipment point, k, the total inputs to this point must equal the total outputs from this intermediate point and can be expressed as follows:

$$\sum_{i=1}^{i=p} x_{i,k} = \sum_{j=1}^{j=q} x_{k,j}$$

Where p is the number of supply points, q is the number of demand points, and x_{ij} denotes a unit of (1) product or no product (0) shipped from point i to point j. From the previous equation, the transshipment constraint of a point, k, is expressed as follows:

$$\sum_{i=1}^{i=p} x_{i,k} - \sum_{j=1}^{j=q} x_{k,j} = 0$$

24.4   Shortest Path Problem: Case Study 1

A product is shipped from a supply city (supply point), represented by node 1, to a destination city (demand point), represented by node 6. The product is first shipped to one or more cities that are represented by intermediate nodes 2–5. The goal of

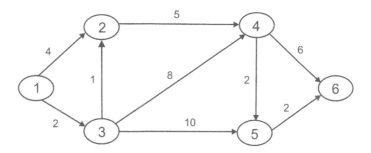

FIGURE 24.1: Graph of shortest path problem.

the problem is to find the shortest path from the supply city to the destination city. The distance between the cities is shown by the arcs between nodes in the graph of Figure 24.1.

The problem can be formulated directly as a standard transportation problem, using the data in the previous figure. This problem has one supply point and one demand point. The objective function can be completely written with the unit cost values of the product to be transported. The objective function is to minimize the following expression:

$$
z = \begin{array}{llllllll}
0x_{11} & + 4x_{12} & + 2x_{13} & + Hx_{14} & + Hx_{15} & + Hx_{16} & + \\
Hx_{21} & + 0x_{22} & + Hx_{23} & + 5x_{24} & + Hx_{25} & + Hx_{26} & + \\
Hx_{31} & + 1x_{32} & + 0x_{33} & + 8x_{34} & + 10x_{35} & + Hx_{36} & + \\
Hx_{41} & + Hx_{42} & + Hx_{43} & + 0x_{44} & + 2x_{45} & + 6x_{46} & + \\
Hx_{51} & + Hx_{52} & + Hx_{53} & + Hx_{54} & + 0x_{55} & + 2x_{56} & + \\
Hx_{61} & + Hx_{62} & + Hx_{63} & + Hx_{64} & + Hx_{65} & + 0x_{66} &
\end{array}
$$

There are two source-destination constraints, each representing 1 unit of product transported from the indicated source node to the indicated destination node.

$$
\begin{array}{llll}
x_{12} & + x_{13} & = 1 & \text{(Source node)} \\
x_{46} & + x_{56} & = 1 & \text{(Destination node)}
\end{array}
$$

There are four intermediate nodes, so there are four transshipment constraints, one for each intermediate node. These constraints are:

$$
\begin{array}{llllll}
x_{12} & + x_{32} & - x_{42} & & = 0 & \text{(Node 2)} \\
x_{13} & - x_{32} & - x_{34} & - x_{35} & = 0 & \text{(Node 3)} \\
x_{24} & + x_{34} & - x_{45} & - x_{46} & = 0 & \text{(Node 4)} \\
x_{35} & + x_{45} & - x_{56} & & = 0 & \text{(Node 5)}
\end{array}
$$

The decision variables x_{ij} have the sign constraint: $x_{ij} = 0$ or $x_{ij} = 0$, for $i =$

1,...,6 and $j = 1,...,6$. The formulation of the problem is formatted for solving with LP_solve and stored in file network1.lp.

```
/* network1.lp
Given the graph, find the shortest path from node 1 to node 6.
*/
/* Objective function -- cost assignments */
Minimize:   z =
      0 x11 +    4 x12 +     2 x13 + 100 x14 + 100 x15 + 100 x16 +
    100 x21 +    0 x22 +   100 x23 +   5 x24 + 100 x25 + 100 x26 +
    100 x31 +    1 x32 +     0 x33 +   8 x34 +  10 x35 + 100 x36 +
    100 x41 +  100 x42 +   100 x43 +   0 x44 +   2 x45 +   6 x46 +
    100 x51 +  100 x52 +   100 x53 + 100 x54 +   0 x55 +   2 x56 +
    100 x61 +  100 x62 +   100 x63 + 100 x64 + 100 x65 + 0 x66;

/*   Source and destination Constraints   */
     x12 +   x13 = 1;   /* Source */
     x46 +   x56 = 1;   /* Destination */

  /* Intermediate nodes constraints */
   x12 + x32 - x24 = 0;            /* Node 2 */
   x13 - x32 - x34 - x35 = 0;      /* Node 3 */
   x24 + x34 - x45 - x46 = 0;      /* Node 4 */
   x35 + x45 - x56 = 0;            /* Node 5 */
```

The execution of the LP_solve program to solve this linear optimization problem produces the results that can be observed in the following listing.

```
Variables   result
z           12
x11         0
x12         0
x13         1
x14         0
x15         0
x16         0
x21         0
x22         0
x23         0
x24         1
x25         0
x26         0
x31         0
x32         1
x33         0
x34         0
x35         0
x36         0
x41         0
x42         0
x43         0
x44         0
x45         1
```

```
x46     0
x51     0
x52     0
x53     0
x54     0
x55     0
x56     1
x61     0
x62     0
x63     0
x64     0
x65     0
x66     0
```

The results show that the minimum path has length $= 12$. The path selected consists of the following arcs: x13, x32, x24, x45, and x56.

24.5 Maximum Flow Problems

The are many problems in which the goal is to send the *maximum quantity* of a product from a source node to a destination node of a network. The main limitation is the *capacity* of each segment of the network represented by the arcs. For these problems, the Ford-Fulkerson algorithm was developed. In this section, the linear optimization formulation is discussed.

The following maximum flow problem illustrates the basic approach to formulate a linear optimization problem. An airline company needs to plan and setup an optimal number of flights from Chicago to Rio de Janeiro. The intermediate stops that need to be included are: first Atlanta, then Bogota and/or Caracas. Table 24.1 shows the routes, the corresponding arcs in Figure 24.2, and the capacity (maximum number of flights allowed) of each route.

TABLE 24.1: Airline routes from Chicago to Rio.

Route	Arc	Max number of flights
Chicago to Atlanta	x_{12}	7
Atlanta to Bogota	x_{23}	4
Atlanta to Caracas	x_{24}	3
Bogota to Caracas	x_{34}	6
Bogota to Rio	x_{35}	3
Caracas to Rio	x_{45}	4

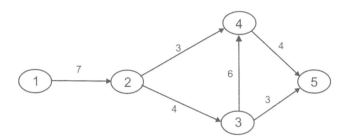

FIGURE 24.2: Graph of airline routes to Rio.

The problem can be formulated directly as a standard transportation problem, using the data in Table 24.1 and Figure 24.2.

Let x_{ij} denote the number of flights from node i to node j, and c_{ij} the capacity between nodes i and j. This problem has one supply point (source) and one demand point (destination). The objective function can be written by observing that the maximum flow from node 1 to node 5 is actually the flow from node 1 to node 2, which is represented by the variable x_{12}. The objective function is then to maximize the expression: $z = x12$.

To simplify the formulation of the problem, three types of constraints are considered:

- Source-destination constraints

- Transshipment constraints

- Capacity constraints

There is one source-destination constraint, which represents the balance in the flow from the source node and the flow into the destination node in Figure 24.2. This constraint is expressed as follows:

$$x_{12} = x_{35} + x_{45}$$

There are three intermediate nodes (2, 3, and 4), so there are three transshipment constraints, one for each intermediate node. These constraints are:

$$
\begin{array}{llll}
x_{12} & - \; x_{23} & - \; x_{24} & = 0 \quad \text{(Node 2)} \\
x_{23} & - \; x_{34} & - \; x_{35} & = 0 \quad \text{(Node 3)} \\
x_{24} & + \; x_{34} & - \; x_{45} & = 0 \quad \text{(Node 4)}
\end{array}
$$

The capacity constraints represent the maximum number of flights possible on the indicated route:

$$x_{ij} \le c_{ij}$$

These are shown as the value of the arcs in Figure 24.2. There are six arcs in the graph, therefore there are six capacity constraints.

$$x_{12} \leq 7$$
$$x_{23} \leq 4$$
$$x_{24} \leq 3$$
$$x_{34} \leq 6$$
$$x_{35} \leq 3$$
$$x_{45} \leq 4$$

The decision variables x_{ij} have the sign constraint: $x_{ij} \geq 0$, for $i = 1, \ldots, 5$ and $j = 1, \ldots, 5$. The formulation of the problem is formatted for solving with LP_solve and stored in file maxflow1.lp.

```
/* maxflow1.lp
An airline company needs to find the max number of flighs
that connect Chicago and Rio de Janeiro.
Find the maximum flow from node 1 (source)
to node 5 (destination).
*/
/* Objective function -- cost assignments */
Maximize: x12;

/*  Source and destination Constraints  */
    x12 = x45 + x35;   /* Source to destination*/

/* Arc capacity constraints */
    x12 <= 7;
    x23 <= 4;
    x24 <= 3;
    x34 <= 6;
    x35 <= 3;
    x45 <= 4;

 /* Intermediate nodes constraints */
  x12 -   x23 - x24 = 0; /* Node 2 */
  x23 -   x34 - x35 = 0; /* Node 3 */
  x24 + x34 - x45 = 0; /* Node 4 */
```

The execution of the LP_solve program to solve this linear optimization problem produces the results that can be observed in the following listing.

```
Variables    result
z            7
x12          7
x45          4
x35          3
x23          4
x24          3
x34          1
```

24.6   Critical Path Method

The critical path method (CPM) is a network model that can help in the *scheduling* of large projects. The important computations are the total time to complete the project and the interval that represents how long an activity of the project can be delayed without causing delays to the project. This method calculates the minimum completion time for a project and the possible start and finish times for the project activities.

A network model of a project typically consists of a sequence of the various activities that need to be performed, and the duration of each activity. A directed arc represents an *activity*, a node represents a *start* or *finish event* of an activity. A special initial node represents the start of the project, and a special end node represents the completion of the project.

In any project there are cases in which more than one activity needs to be completed before the next activity can start. Figure 24.3 illustrates this situation. Activity *Aj* and activity *Ak* have to be completed before activity *Al* can start. Another situation is an activity that needs to be completed before two or more activities can start. Figure 24.4 illustrates this by showing activity *Al* that needs to be completed before activities *Am*, *Ak*, and *Ap* can start.

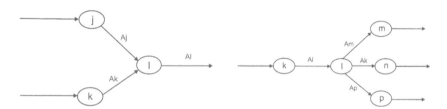

FIGURE 24.3: Several activities completed before next activity.

FIGURE 24.4: Sequencing of activities.

The following terms are used in project scheduling using the duration of the activities in the project. The *early event* time of node i, denoted by e_i, is the earliest time at which the event can occur. The *late event* time of node i, denoted by l_i, is the latest time at which the event can occur.

The *total float*, denoted by f_{ij} of an activity A_{ij} is the time interval by which the starting time of the activity can be delayed and not cause delay in the completion time of the project.

An activity that has a total float equal to zero is known as a *critical activity*. A critical path consists of a sequence of critical activities. Delays in the activities in a critical path will delay the completion of the project.

24.6.1 Using Linear Optimization

Recall that for any activity A_{ij}, the start time of the activity is the event represented by node i and the completion time of the activity is the event represented by node j. Let x_k denote the time occurrence of event k and the duration of activity A_{ij} is denoted by Δ_{ij}. For every activity A_{ij}, the completion time of the activity is given by the expression $x_j \geq x_i + \Delta_{ij}$.

Let f denote the finish node of the project, the event time of the completion of the project is denoted by x_f. Similarly, node 1 is the start node of the project and the event time of the start of the project is denoted by x_1. The total time interval or duration of the entire project is given by the expression $x_f - x_1$. Let n denote the total number of nodes in the project network, this implies that $x_f = x_n$. The formulation of the linear optimization problem that finds the critical path of a project is given by the following expressions:

Minimize: $z = x_f - x_1$
Subject to:

$$x_j \geq x_i + \Delta_{ij}, \quad i = 1, \ldots, n-1, \quad j = 2, \ldots, n$$

The variables $x_i, i = 1, \ldots, n$ are unrestricted in sign.

24.6.2 Critical Path Method: Case Study 2

A project has been defined with the activities and their duration given in Table 24.2. The goal of the problem is to find the critical path of the project. The various activities and their predecessors are shown by the arcs between nodes in the graph of Figure 24.5.

TABLE 24.2: Project data.

Activity	Predecessor	Duration
A_{12}	-	6
A_{23}	A_{12}	9
A_{35}	A_{23}	13
A_{36}	A_{23}	6
A_{34}	A_{23}	5
A_{45}	A_{34}	7
A_{56}	A_{45}, A_{35}	4

The problem can be formulated directly as a standard transportation problem, using the data in Table 24.2 and in Figure 24.5.

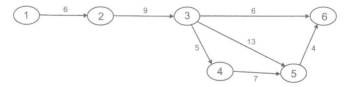

FIGURE 24.5: Graph of project activities.

This problem has node 6 as the final node of the project. The objective function is then to minimize the expression: $z = x_6 - x_1$. The constraints are expressed as follows:

$$
\begin{aligned}
x_2 &\geq x_1 + 6 & \text{Activity } A_{12} \\
x_3 &\geq x_2 + 9 & \text{Activity } A_{23} \\
x_4 &\geq x_3 + 5 & \text{Activity } A_{34} \\
x_5 &\geq x_4 + 7 & \text{Activity } A_{45} \\
x_5 &\geq x_3 + 13 & \text{Activity } A_{35} \\
x_6 &\geq x_3 + 6 & \text{Activity } A_{36} \\
x_6 &\geq x_5 + 4 & \text{Activity } A_{56}
\end{aligned}
$$

The decision variables x_i have unrestricted sign constraint for $i = 1, \ldots, 6$. The formulation of the problem is formatted for solving with LP_solve and stored in file `cpm1.lp`.

```
/* cpm1.lp
Find the minimum path for the project described in
Critical path method: case study 1 */
/* Objective function -- Total project time interval */
Minimize: x6 - x1;

/*   Activity Constraints   */
    x2 >= x1 + 6;    /* Activity A13 */
    x3 >= x2 + 9;    /* Activity A12 */
    x4 >= x3 + 5;    /* Activity A35 */
    x5 >= x4 + 7;    /* Activity A34 */
    x5 >= x3 + 13;   /* Activity A45 */
    x6 >= x3 + 6;    /* Activity A56 */
    x6 >= x5 + 4;
```

The execution of the LP_solve program to solve this linear optimization problem produces the results that can be observed in the following listing. The total time interval for completion of the project is 32 days and the start time of every activity is shown in the listing.

```
Variables       result
z               32
x6              32
x1               0
```

x2	6
x3	15
x4	20
x5	28

24.6.3   Reducing the Time to Complete a Project

When a decision is taken to *reduce* the total time to complete a project, additional resources must be allocated to the various activities. Linear optimization is used to minimize the total cost of allocating the additional resources to the project activities.

Let r_{ij} denote the number of days that the duration of activity A_{ij} is reduced, and c_{ij} denote the cost per day of allocating additional resources to activity A_{ij}. Let R denote the time (in days) that an activity can be reduced, and let T denote the new total time (in days) of the project completion. The objective function to minimize is:

$$z = \sum_{i=1}^{n-1} \sum_{j=2}^{n} c_{ij} r_{ij}, \quad \text{for all activities } A_{ij}$$

The time-reduction constraints are expressed as:

$$r_{ij} \leq R, \quad \text{for all activities } A_{ij}$$

The activity constraints are expressed as:

$$x_j = x_i + \Delta_{ij} - r_{ij}, \quad \text{for all activities } A_{ij}$$

The total time constraint is:

$$x_n - x_1 \leq T$$

24.6.4   Case Study 3

Consider a reduction of 4 days in the total time to complete the project described previously as Case Study 1. The completion time is now 28 days. The activity completion time can be reduced up to 2 days. The following table shows the cost per day of reducing each activity of the project.

The various activities and their predecessors are shown by the arcs between nodes in the graph of Figure 24.5. This problem has node 6 as the final node of the project.

TABLE 24.3: Project additional cost.

Activity	Cost
A_{12}	22.50
A_{23}	15.75
A_{35}	13.25
A_{36}	16.50
A_{34}	25.30
A_{45}	17.50
A_{56}	14.75

The goal of the problem is to find the minimum cost of reducing the total completion time of the project. The objective function is then to minimize the expression:

$$z = \quad 22.50r_{12} + 15.75r_{23} + 13.25r_{35} + 16.50r_{36} + 25.30r_{34} + \\ 17.50r_{45} + 14.75r_{56}$$

.

The time-reduction constraints are expressed as follows:

$$r_{12} \leq 2, \quad \text{Activity } A_{12}$$
$$r_{23} \leq 2, \quad \text{Activity } A_{23}$$
$$r_{34} \leq 2, \quad \text{Activity } A_{34}$$
$$r_{45} \leq 2, \quad \text{Activity } A_{45}$$
$$r_{35} \leq 2, \quad \text{Activity } A_{35}$$
$$r_{36} \leq 2, \quad \text{Activity } A_{36}$$
$$r_{56} \leq 2, \quad \text{Activity } A_{56}$$

The activity constraints are expressed as follows:

$$x_2 \geq x_1 + 6 - r_{12}, \quad \text{Activity } A_{12}$$
$$x_3 \geq x_2 + 9 - r_{23}, \quad \text{Activity } A_{23}$$
$$x_4 \geq x_3 + 5 - r_{34}, \quad \text{Activity } A_{34}$$
$$x_5 \geq x_4 + 7 - r_{45}, \quad \text{Activity } A_{45}$$
$$x_5 \geq x_3 + 13 - r_{35}, \quad \text{Activity } A_{35}$$
$$x_6 \geq x_3 + 6 - r_{36}, \quad \text{Activity } A_{36}$$
$$x_6 \geq x_5 + 4 - r_{56}, \quad \text{Activity } A_{56}$$

The decision variables x_i have unrestricted sign constraint for $i = 1,\dots,6$. The formulation of the problem is formatted for solving with LP_solve and stored in file `cpm1b.lp`.

```
/* cpm1b.lp
Find the minimum path for the project described in
CPM Case study 2. */
/* Objective function -- cost assignments */
```

```
Minimize: 22.5 r12 + 15.75 r23 + 13.25 r35 + 16.5 r36 +
          25.3 r34 + 17.5 r45 + 14.75 r56;

/* Time-reduction constraints
   An activity duration can be reduced up to 2 days
   */
   r12 <= 2; /* Activity A12 */
   r23 <= 2; /* Activity A23 */
   r34 <= 2; /* Activity A34 */
   r45 <= 2; /* Activity A45 */
   r35 <= 2; /* Activity A35 */
   r36 <= 2; /* Activity A36 */
   r56 <= 2; /* Activity A56 */

/* Activity Constraints  */
   x2 >= x1 + 6 - r12;   /* Activity A12 */
   x3 >= x2 + 9 - r23;   /* Activity A23 */
   x4 >= x3 + 5 - r34;   /* Activity A34 */
   x5 >= x4 + 7 - r45;   /* Activity A45 */
   x5 >= x3 + 13 - r35;  /* Activity A35 */
   x6 >= x3 + 6 - r36;   /* Activity A36 */
   x6 >= x5 + 4 - r56;   /* Activity A56 */

/* Total project time constraint */
   x6 - x1 <= 28;
```

The execution of the LP_solve program that solves this linear optimization problem produces the results observed in the following listing. The total cost of reducing time interval for completion of the project is \$58.50. It can be noted that the completion time of activities A_{12}, A_{36}, A_{34}, and A_{45}, were not reduced. However, the completion time of activity A_{56} was reduced in 2 days.

```
Variables    result
z            58.5
r12          0
r23          1
r35          0.999999
r36          0
r34          0
r45          0
r56          2
x2           6
x1           0
x3           14
x4           19
x5           26
x6           28
```

Summary

Linear optimization modeling can be used to study and calculate various problems that are represented by networks. The typical problems are shortest path problems, maximum flow problems and critical path method for project management. A dummy activity has zero duration and is necessary to prevent two activities with the same start node and the same end node.

Key Terms		
shortest path	routes	traffic
maximum flow	critical path	minimum spanning tree
graph	nodes	arcs
path	vertex	activity
event	early event time	late event time
total float	critical activity	transshipment point
transporting cost	artificial node	arc capacity

Exercises

Exercise 24.1 A company manufactures bicycles in two facilities: F1 and F2. Facility F1 can build a maximum of 400 bicycles per year at a cost of $850 per unit. Facility F2 can build a maximum of 300 bicycles per year at a cost of $950. There are two main destinations D1 and D2. Destination D1 demands 400 bicycles per year and D2 demands 300 bicycles per year. The following table includes the transportation costs from a facility F1, F2, or the intermediate point D3 to destination points D1, D2, and D3. The bicycles may be sent to an intermediate location D3. Compute the minimum total cost that meets the demand.

Transportation Costs			
From	D1	D2	D3
F1	$250	$200	$75
F2	$155	$130	$95
D3	$45	$35	0

Exercise 24.2 Repeat the previous problem with the additional condition that the maximum number of bicycles transported to the intermediate point (D3) is 185.

Exercise 24.3 A low-end computer laptop can be purchased for about $400 and can be used for five years with no salvage value. The maintenance of the computer is estimated to cost $85 for year 1, $140 for year 2, $210 for year 3, $250 for year 4, and $270 for year 5. Compute the minimum total cost of purchasing and using the computer for six years.

Exercise 24.4 Traffic engineers are studying traffic patterns in part of a city. The immediate problem is to find the maximum flow of vehicles from a source point, *S*, to a destination point, *D*. Figure 25.1 and the following table show the flow capacity of the various roads (between nodes). Note that the direction of the traffic is important. Formulate and solve a linear optimization problem that computes the maximum traffic flow from point *S* to point *D*.

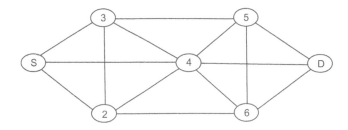

FIGURE 24.6: Graph of road capacity in city.

Arc	Capacity	Arc	Capacity
S-3	6	4-5	5
3-S	1	5-4	1
S-4	7	4-6	7
4-S	1	6-4	1
S-2	5	4-D	2
2-S	0	D-4	0
2-4	1	5-D	8
4-2	4	D-5	0
2-3	2	5-6	2
3-2	1	6-5	2
3-4	2	6-D	8
4-3	1	D-6	2
3-5	4	5-3	3

Chapter 25

Integer Linear Optimization Models

25.1 Introduction

This chapter presents the general concepts and formulation of problems that can be solved with modes of integer linear optimization. These are models with more constraint than the standard linear optimization problems.

An integer linear optimization problem in which all variables are required to be integer is called a **pure integer linear** problem. If some variables are restricted to be integer and others are not, the problem is a **mixed integer linear** problem. The special case of integer variables that are restricted to be 0 or 1 is very useful and are known as pure (mixed) 0–1 linear problems or pure (mixed) **binary** integer linear problems.

25.2 Modeling with Integer Variables

An integer linear optimization problem is a conventional linear optimization problem with the additional constraints that the decision variables be integer variables. This implies that for a maximization integer linear problem, the optimal value of the objective function is less or equal to the optimal value of the linear optimization problem.

Removing the constraints that the variables must be integer variables results in a linear problem similar to the ones discussed in previous chapters. This linear problem is known as the linear problem **relaxation** of the integer linear problem.

In most situations, computing the optimal value of the two linear problems will produce very different results. The following pure integer linear problem illustrates this. The objective function is to maximize the following expression:

$$z = 20x_1 + 20x_2 + 10x_3$$

Subject to the following restrictions:

$$
\begin{aligned}
2x_1 &+ 20x_2 &+ 4x_3 &\leq 15 \\
6x_1 &+ 20x_2 &+ 4x_3 &= 20
\end{aligned}
$$

$$x_1, x_2, x_3 \geq 0, \quad \text{integer}$$

The formulation of the problem is formatted for solving with LP_solve and is stored in file `intprob1.lp`.

```
/* intprob1.lp
variables x1, x2, and x3 are integer variables
Objective function */
Maximize: 20x1 + 10x2 + 10x3;

/* Constraints */
    2x1 + 20x2 + 4x3 <= 15;
    6x1 + 20x2 + 4x3 = 20;
/* Integer variables */
int x1, x2, x3;
```

The execution of the LP_solve program to solve this linear optimization problem produces the results that can be observed in the following listing.

```
Variables   result
z           60
x1          2
x2          0
x3          2
```

The result of solving this integer linear problem is $z = 60$, and the values of the integer variables are $x_1 = 2$, $x_2 = 0$, and $x_3 = 2$.

Relaxing the constraints that the variables $x_i, i = 1 \ldots 3$ be integer variables, results in a different numerical solution and can be observed in the following listing.

```
Variables    result
z            66.6666666666667
x1           3.33333333333333
x2           0
x3           0
```

25.3 Applications of Integer Linear Optimization

A brief description of some typical problems that can be formulated as integer optimization problems follows.

- Knapsack Problem. Given a knapsack with fixed capacity and a collection of items, each with a weight and value, find the number of items to put in the knapsack that maximizes the total value carried subject to the requirement that that weight limitation not be exceeded.

- The Transportation Problem. Given a finite number of suppliers, each with fixed capacity, a finite number of demand centers, each with a given demand, and costs of transporting a unit from a supplier to a demand center, find the minimum cost method of meeting all of the demands without exceeding supplies.

- Assignment Problem. Given equal numbers of people and jobs and the value of assigning any given person to any given job, find the job assignment (each person is assigned to a different job) that maximizes the total value.

- Shortest Route Problem. Given a collection of locations and the distance between each pair of locations, find the cheapest way to get from one location to another.

- Maximum Flow Problem. Given a series of locations connected by pipelines of fixed capacity and two special locations (an initial location or source and a final location or sink), find the way to send the maximum amount from source to sink without violating capacity constraints.

The techniques used to solve integer linear problems are Branch-and-Bound and Branch-and-Cut algorithms. They are both implicit enumeration techniques, implicit meaning that (hopefully) many solution will be skipped during enumeration as they are known to be non optimal.

25.3.1 Branch and Bound

The the most widely used method for solving integer linear optimization models is branch and bound. Subproblems are created by restricting the range of the integer variables. For binary variables, there are only two possible restrictions: setting the variable to 0, or setting the variable to 1. More generally, a variable with lower bound l and upper bound u will be divided into two problems with ranges l to q and q+1 to u respectively. Lower bounds are provided by the linear optimization relaxation to the problem. If the optimal solution to a relaxed problem is (coincidentally) integral, it is an optimal solution to the subproblem, and the value can be used to terminate searches of subproblems whose lower bound is higher.

25.3.2 Branch and Cut

For branch and cut, the lower bound is again provided by the linear optimization relaxation of the integer program. The optimal solution to this linear program is at a corner of the feasible region (the set of all variable settings which satisfy the constraints). If the optimal solution to the problem is not integral, this algorithm searches for a constraint which is violated by this solution, but is not violated by any optimal integer solutions. This constraint is called a cutting plane. When this constraint is added to the model, the old optimal solution is no longer valid, and so the new optimal will be different, potentially providing a better lower bound. Cutting planes are iteratively until either an integral solution is found or it becomes impossible or too expensive to find another cutting plane. In the latter case, a traditional branch operation is performed and the search for cutting planes continues on the subproblems.

Almost all the sample problems described in this chapter are computationally solved with lp_solve, which is a linear (integer) optimization solver based on the revised simplex method and the Branch-and-bound method for the integer variables.

25.4   Integer Linear Optimization: Case Study 1

A hiker needs to take as many items as possible in his knapsack for the next hike. The knapsack has a capacity of 25 pounds. Each items has a priority from 1 to 10 that indicates the relative importance of the item, and a weight. This data is included in Table 25.1.

TABLE 25.1: Items for Knapsack.

Item	Priority	Weight
1	7	5.5
2	5	9.5
3	10	13.5
4	6	6.5
5	8	6

The decision variables, $x_i, i = 1, \ldots, 5$, for this integer linear problem can have only two possible values: $(0, 1)$. $x_i = 1$ indicates that item i is put in the knapsack and $x_i = 0$ indicates that it is not put in the knapsack. These variables are also known as **binary** variables.

The goal of this problem is to find the best way to pack the items in the knapsack by priority, given the constraint of the weight capacity of the knapsack. The objective function is to maximize the following expression:

$$z = 7x_1 + 5x_2 + 10x_3 6x_4 + 8x_5$$

Subject to the following restrictions:

$$5.5x_1 + 9.5x_2 + 13.5x_3 + 6.5x_4 + 6x_5 \le 25$$

$$x_i \in \{0,1\}, i = 1,\ldots,5$$

The formulation of the problem formatted for solving with LP_solve is stored in file `knapsack.lp`.

```
/* knapsack.lp
The Knapsack problem. There 5 items that a hiker needs to take.
Each item has a priority and a weight.
variables x1, x2, x3, x4, and x5 are integer variables with values
0 or 1. */
/* Objective function */
Maximize: 7x1 + 5x2 + 10x3 + 6x4 + 8x5;

/* Constraints */
     5.5x1 + 9.5x2 + 13.5x3 + 6.5x4 + 6x5 <= 25;

/* Integer binary variables */
bin x1, x2, x3, x4, x5;
```

The execution of the LP_solve program to solve this linear optimization problem produces the results that can be observed in the following listing.

```
Variables   result
z              25
x1             1
x2             0
x3             1
x4             0
x5             1
```

The results listing shows that only items 1, 3, and 5 are put in the knapsack because of the weight constraint.

25.5   Integer Linear Optimization: Case Study 2

A factory manufactures three types of automobile parts. The following table has the data on the unit requirements of materials (pounds) and labor (hours), as well as the unit sales price and unit variable cost. The total available labor per week is 200 hours and the total of 170 pounds of material available per week. There are three types of machines that need to be rented, one for each type of part. The weekly costs for renting these machines are: $150.00 for machine 1, $100.00 for machine 2, and $85.00 for machine 3. The goal of the problem is to maximize the profit of producing the three types of automobile parts.

Part type	Material	Labor	Sales price	Var cost
1	7	5	34.00	18.00
2	5	4	22.00	12.00
3	7	12	40.00	22.00

In formulating this linear optimization model, let x_1 denote the number of parts manufactured of type 1, x_2 denote the number of parts manufactured of type 2, and x_3 denote the number of parts manufactured of type 3. These integer variables are used to formulate the total sales, S, with the following expression:

$$S = 34x_1 + 22x_2 + 40x_3$$

The total costs variable, V, is formulated with the following expression:

$$V = 18x_1 + 12x_2 + 22x_3$$

The cost of renting the machines (the fixed cost) depends on whether the parts of a specific type are produced. Let y_1 denote whether the parts of type 1 are produced, y_2 denote whether the parts of type 2 are produced, and y_3 denote whether the parts of type 3 are produced. These binary variables are used to formulate the total fixed cost, F, with the following expression:

$$F = 150y_1 + 100y_2 + 85y_3$$

The weekly profit is the objective function of the problem and can be expressed as:

$$P = S - V - F$$

The objective function can then be formulated with the following expression:

$$z = \begin{array}{lll} 34x_1 & + 22x_2 & + 40x_3 \\ -18x_1 & - 12x_2 & - 22x_3 \\ -150y_1 & - 100y_2 & - 85y_3 \end{array}$$

There two types of problem constraints: the first type of constraint derives from the total available labor and material per week. The second type of constraint associates the type of part produced with the corresponding machine that needs to be rented.

$$\begin{array}{llll} 7x_1 & + 5x_2 & + 7x_3 & \leq 170 \quad \text{(Material available)} \\ 5x_1 & + 4x_2 & + 12x_3 & \leq 200 \quad \text{(Labor available)} \end{array}$$

In this problem, given the total available material and labor per week, the maximum possible number of parts of type 1 that can be produced is 24. In a similar manner, the maximum possible number of parts of type 2 that can be produced is 34, and maximum possible number of parts of type 3 that can be produced is 16. These machine constraints are:

$$x_1 \leq 24y_1$$
$$x_2 \leq 34y_2$$
$$x_3 \leq 16y_3$$

The formulation of the problem is formatted for solving with LP_solve and stored in file `intprob2.lp`.

```
/* intprob2.lp
A factory manufactures three types of
automobile parts.
Case study 2.
*/
/* Objective function total profits */

max: 16x1 + 10x2 + 18x3 - 150y1 -100y2 -85y3;
/* Constraints */
7x1 + 5x2 + 7x3 <= 170;   /* Material available */
5x1 + 4x2 + 12x3 <= 200; /* labor available */

x1 <= 24y1;
x2 <= 34y2;
x3 <= 16y3;
x1 >= y1;
x2 >= y2;
x3 >= y3;

  /* y1+y2+y3 >= 2; */    /* At least 2 machines must be rented */
  /* y1 - y3 <= 1;  */    /* If machine  1 is rented,
                                machine 2 must also */
  /* y1 + y2 + y3 = 3; */   /* All three machines must be rented */

int x1, x2, x3;
bin y1, y2, y3;
```

The execution of the LP_solve program to solve this linear optimization problem produces the results that can be observed in the following listing, using the command window. The formulation of the problem is formatted for solving with LP_solve and is stored in file `intprob2b.lp`.

```
C:\computational_mod\LP_prog\LP_solve>lp_solve intprob2.lp

Value of objective function: 240.00000000

Actual values of the variables:
```

```
x1                              0
x2                              34
x3                              0
y1                              0
y2                              1
y3                              0
```

The results show that the optimal value of profit is $240.00 and only 34 automobile parts of type 2 are to be produced.

An additional constraint on the problem is: at least two machines must be rented. This constraint is formulated with the following expression:

$$y_1 + y_2 + y_3 \geq 2$$

The execution of the LP_solve program to solve the linear optimization problem with this additional constraint is stored in file `intprob2d.lp` and produces the results that can be observed in the following listing, using the command window.

```
C:\computational_mod\LP_prog\LP_solve>LP_solve intprob2b.lp

Value of objective function: 195.00000000
Actual values of the variables:
x1                              0
x2                              20
x3                              10
y1                              0
y2                              1
y3                              1
```

The results show that the optimal value of profit is now $195.00 and 20 automobile parts of type 2 and 10 automobile parts of type 3 are to be produced.

Instead of the previous constraint, another constraint on the problem is: if machine 1 is rented then machine 3 should also be rented. This constraint is formulated with the following expression:

$$y_1 - y_3 \leq 1$$

The execution of the LP_solve program to solve the linear optimization problem with this constraint produces the results that can be observed in the following listing, using the command window. The formulation of the problem is formatted for solving with **LP_solve** and stored in file `intprob2c.lp`.

```
C:\computational_mod\LP_prog\LP_solve>LP_solve intprob2c.lp

Value of objective function: 240.00000000
Actual values of the variables:
x1                              0
```

x2	34
x3	0
y1	0
y2	1
y3	0

The results are the same as the first solution to this problem. The results show that the optimal value of profit is $240.00 and only 34 automobile parts of type 2 are to be produced.

Instead of the previous constraint, another constraint on the problem is: all three machines should be rented. This constraint is formulated with the following expression:

$$y_1 + y_2 + y_3 = 3$$

The execution of the LP_solve program to solve the linear optimization problem with this constraint produces the results that can be observed in the following listing, using the command window.

```
C:\computational_mod\LP_prog\LP_solve>LP_solve intprob2d.lp

Value of objective function: 69.00000000

Actual values of the variables:
x1                       11
x2                        3
x3                       11
y1                        1
y2                        1
y3                        1
```

The results show that the optimal value of profit is $69.00 and 11 automobile parts of type 1, 3 parts of type 2, and 11 parts of type 3 are to be produced.

Summary

Integer linear optimization modeling can be used to study and calculate various problems that are represented by variables that have only integer values. The additional restrictions on decision variables, is that the values be binary, are useful on some types of problems. If not all variables have integer values then it is a mixed integer linear problem. Typical applications of problems that can be solved with integer linear optimization are: knapsack problems, transportation problems, assignment problems, shortest route problems, and maximum flow problems.

Key Terms

integer variables	binary variables	integer relaxation
shortest route	knapsack	transportation
assignment	maximum flow	branch & bound
branch & cut	mixed integer	

Exercises

Exercise 25.1   A company manufactures two types of duffel traveling bags. Three units of material are used to manufacture bags of type 1, and six units of material to manufacture bags of type 2. The total units available of material is 150. The initial setup cost to manufacture bags is $15 for type 1 and $30 for type 2. The company can sell bags of type 1 with a profit of $2 per bag and type 2 with a profit of $5 per bag. Compute the number of bags of each type to produce and the maximum profits possible.

Exercise 25.2   A computer distributer has four warehouses (W1, W2, W3, and W4) and each can ship 120 units per week. The operational costs of the warehouses are: $385 for W1, $480 for W2, $290 for W3, and $145 for W4. There are three demand points (D1, D2, and D3) that have the following weekly demands: 75 for D1, 68 for D2, and 35 for D3. The distributer has the following operational restrictions: at most, two warehouses can be operating; if warehouse W1 is operating then W2 must also be operating; either warehouse W4 or warehouse W2 must be operating. Compute the minimum weekly cost while meeting demand. The following table includes data on the cost of transporting a unit from a warehouse to a destination point.

From	D1	D2	D3
W1	$18	$36	$47
W2	$45	$12	$22
W3	$22	$34	$15
W4	$20	$46	$32

Exercise 25.3   Traffic engineers are studying traffic patterns in part of a city. The immediate problem is to find the maximum flow of vehicles from a source point, S, to a destination point, D. Figure 25.1 and the following table show the flow capacity of the various roads (between nodes). Note that the direction of the traffic is important. Formulate and solve a linear optimization problem that computes the maximum traffic flow from point S to point D.

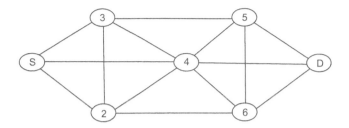

FIGURE 25.1: Graph of road capacity in city.

Arc	Capacity	Arc	Capacity
S-3	6	4-5	5
3-S	1	5-4	1
S-4	7	4-6	7
4-S	1	6-4	1
S-2	5	4-D	2
2-S	0	D-4	0
2-4	1	5-D	8
4-2	4	D-5	0
2-3	2	5-6	2
3-2	1	6-5	2
3-4	2	6-D	8
4-3	1	D-6	2
3-5	4	5-3	3

Appendix A

GNU C Compiler Tool

A.1   Command Line Documentation

On a terminal window, the following command can be typed and the short documentation on the usage of the GNU C compiler will display.

```
$ gcc --help
Usage: gcc [options] file...
Options:
```

–pass-exit-codes
    Exit with highest error code from a phase

–help
    Display this information

–target-help
    Display target specific command line options

–help=
    {common—optimizers—params—target—warnings—[8]{joined —separate—undocumented}}[,...]

    Display specific types of command line options. (Use '-v –help' to display command line options of sub-processes)

–version
    Display compiler version information

–dumpspecs
    Display all of the built in spec strings

–dumpversion
    Display the version of the compiler

–dumpmachine
    Display the compiler's target processor

–print-search-dirs
    Display the directories in the compiler's search path

–print-libgcc-file-name
> Display the name of the compiler's companion library

–print-file-name=⟨lib⟩
> Display the full path to library `<lib>`

–print-prog-name=⟨prog⟩
> Display the full path to compiler component `<prog>`

–print-multiarch
> Display the target's normalized GNU triplet, used as a component in the library path

–print-multi-directory
> Display the root directory for versions of `libgcc`

–print-multi-lib
> Display the mapping between command line options and multiple library search directories

–print-multi-os-directory
> Display the relative path to OS libraries

–print-sysroot
> Display the target libraries directory

–print-sysroot-headers-suffix
> Display the `sysroot` suffix used to find headers

–Wa,⟨options⟩
> Pass comma-separated `<options>` on to the assembler

–Wp,⟨options⟩
> Pass comma-separated `<options>` on to the preprocessor

–Wl,⟨options⟩
> Pass comma-separated `<options>` on to the linker

–Xassembler ⟨arg⟩
> Pass `<arg>` on to the assembler

–Xpreprocessor ⟨arg⟩
> Pass `<arg>` on to the preprocessor

–Xlinker ⟨arg⟩
> Pass `<arg>` on to the linker

–save-temps
> Do not delete intermediate files

–save-temps=⟨arg⟩
> Do not delete intermediate files

–no-canonical-prefixes
> Do not canonicalize paths when building relative prefixes to other gcc components

–pipe
> Use pipes rather than intermediate files

–time
> Time the execution of each subprocess

–specs=⟨file⟩
> Override built-in specs with the contents of `<file>`

–std=⟨standard⟩
> Assume that the input sources are for `<standard>`

–sysroot=⟨directory⟩
> Use `<directory>` as the root directory for headers and libraries

–B ⟨directory⟩ Add `<directory>` to the compiler's search paths

–v Display the programs invoked by the compiler

–###
> Like `-v` but options quoted and commands not executed

–E
> Preprocess only; do not compile, assemble or link

–S
> Compile only; do not assemble or link

–c
> Compile and assemble, but do not link

–o ⟨file⟩
> Place the output into `<file>`

–pie
> Create a position independent executable

–shared Create a shared library

–x ⟨language⟩
> Specify the language of the following input files. Permissible languages include: c c++, `assembler`, or `none`. 'none' means revert to the default behavior of guessing the language based on the file's extension

Options starting with `--g`, `--f`, `--m`, `--O`, `--W`, or `--param` are automatically passed on to the various sub-processes invoked by gcc. In order to pass other options on to these processes the `-W<letter>` options must be used.

A.2   GCC Command Options

The GCC tool normally performs preprocessing, compilation, assembly and linking of the source C program. See `http://gcc.gnu.org`.

Most of the command-line options that you can use with GCC are useful for C programs; when an option is only useful with another language (usually C++), the explanation says so explicitly. If the description for a particular option does not mention a source language, you can use that option with all supported languages.

The gcc program accepts options and file names as operands. Many options have multi-letter names; therefore multiple single-letter options may not be grouped: `-dv` is very different from '`-d -v`'.

You can mix options and other arguments. For the most part, the order you use doesn't matter. Order does matter when you use several options of the same kind; for example, if you specify `-L` more than once, the directories are searched in the order specified. Also, the placement of the `-l` option is significant.

Many options have long names starting with `-f` or with `-W` for example, `-fmove-loop-invariants`, `-Wformat` and so on. Most of these have both positive and negative forms; the negative form of `-ffoo` would be `-fno-foo`. This manual documents only one of these two forms, whichever one is not the default.

A.3   Options Controlling the Kind of Output

Compilation can involve up to four stages: preprocessing, compilation proper, assembly and linking, always in that order. GCC is capable of preprocessing and compiling several files either into several assembler input files, or into one assembler input file; then each assembler input file produces an object file, and linking combines all the object files (those newly compiled, and those specified as input) into an executable file.

For any given input file, the file name suffix determines what kind of compilation is performed:

file.c
> C source code that must be preprocessed.

file.i
> C source code that should not be preprocessed.

file.h C, header file to be turned into a precompiled header (default), or C, C++ header file to be turned into an Ada spec (via the `-fdump-ada-spec` switch).

Other header files: `file.hh`, `file.H`, `file.hp`, `file.hxx`, `file.hpp`, `file.HPP`, `file.h++`.

file.s

Assembler code. Other files: `file.S`, `file.sx`.Assembler code that must be preprocessed.

other

An object file to be fed straight into linking. Any file name with no recognized suffix is treated this way.

−x none

Turn off any specification of a language, so that subsequent files are handled according to their file name suffixes (as they are if −x has not been used at all).

−pass-exit-codes

Normally the gcc program will exit with the code of 1 if any phase of the compiler returns a non-success return code. If you specify -pass-exit-codes, the gcc program will instead return with numerically highest error produced by any phase that returned an error indication. The C, C++, and Fortran frontends return 4, if an internal compiler error is encountered.

−c

Compile or assemble the source files, but do not link. The linking stage simply is not done. The ultimate output is in the form of an object file for each source file. By default, the object file name for a source file is made by replacing the suffix `.c`, `.i`, `.s`, etc., with `.o`. Unrecognized input files, not requiring compilation or assembly, are ignored.

−S

Stop after the stage of compilation proper; do not assemble. The output is in the form of an assembler code file for each non-assembler input file specified. By default, the assembler file name for a source file is made by replacing the suffix `.c`, `.i`, etc., with `.s`. Input files that don't require compilation are ignored.

−E

Stop after the preprocessing stage; do not run the compiler proper. The output is in the form of preprocessed source code, which is sent to the standard output. Input files that don't require preprocessing are ignored.

−o file

Place output in file `file`. This applies regardless to whatever sort of output is being produced, whether it be an executable file, an object file, an assembler file or preprocessed C code. If −o is not specified, the default is to put an executable file in a.out, the object file for source.suffix in source.o, its assembler file in source.s, a precompiled header file in source.suffix.gch, and all preprocessed C source on standard output.

–v

Print (on standard error output) the commands executed to run the stages of compilation. Also print the version number of the compiler driver program and of the preprocessor and the compiler proper.

–###

Like −v except the commands are not executed and arguments are quoted unless they contain only alphanumeric characters or ./ − _. This is useful for shell scripts to capture the driver-generated command lines.

–pipe

Use pipes rather than temporary files for communication between the various stages of compilation. This fails to work on some systems, where the assembler is unable to read from a pipe; but the GNU assembler has no trouble.

–help

Print (on the standard output) a description of the command-line options understood by gcc. If the -v option is also specified then –help will also be passed on to the various processes invoked by gcc, so that they can display the command-line options they accept. If the −Wextra option has also been specified (prior to the –help option), then command-line options that have no documentation associated with them will also be displayed.

–target-help

Print (on the standard output) a description of target-specific command-line options for each tool. For some targets extra target-specific information may also be printed.

–help= {class | [^]qualifier}[,...]

Print (on the standard output) a description of the command-line options understood by the compiler that fit into all specified classes and qualifiers. These are the supported classes:

optimizers

This will display all of the optimization options supported by the compiler.

warnings

This will display all of the options controlling warning messages produced by the compiler.

target

This will display target-specific options. Target-specific options of the linker and assembler will not be displayed. This is because those tools do not currently support the extended --help= syntax.

params

This will display the values recognized by the --param option.

language

> This will display the options supported for language, where language is the name of one of the languages supported in this version of GCC.

common

> This will display the options that are common to all languages.

These are the supported qualifiers:

undocumented

> Display only those options that are undocumented.

joined

> Display options taking an argument that appears after an equal sign in the same continuous piece of text, such as: `--help=target`.

separate

> Display options taking an argument that appears as a separate word following the original option, such as: `-o output-file`.

Thus for example to display all the undocumented target-specific switches supported by the compiler the following can be used:

```
--help=target,undocumented
```

The sense of a qualifier can be inverted by prefixing it with the ^ character, so for example to display all binary warning options (i.e., ones that are either on or off and that do not take an argument) that have a description, use:

```
--help=warnings,^joined,^undocumented
```

The argument to `--help=` should not consist solely of inverted qualifiers. Combining several classes is possible, although this usually restricts the output by so much that there is nothing to display. One case where it does work however, is when one of the classes is target. So for example to display all the target-specific optimization options the following can be used:

```
--help=target,optimizers
```

The `--help=` option can be repeated on the command line. Each successive use will display its requested class of options, skipping those that have already been displayed. If the `-Q` option appears on the command line before the `--help=` option, then the descriptive text displayed by `--help=` is changed. Instead of describing the displayed options, an indication is given as to whether the option is enabled, disabled or set to a specific value (assuming that the compiler knows this at the point where the `--help=` option is used). The output is sensitive to the effects of previous command-line options, so for example it is possible to find out which optimizations are enabled at −O2 by using: `-Q -O2 --help=optimizers`

Alternatively you can discover which binary optimizations are enabled by -O3 by using:

```
gcc -c -Q -O3 --help=optimizers > /tmp/O3-opts
gcc -c -Q -O2 --help=optimizers > /tmp/O2-opts
diff /tmp/O2-opts /tmp/O3-opts | grep enabled
```

–no-canonical-prefixes

Do not expand any symbolic links, resolve references to /../ or /./, or make the path absolute when generating a relative prefix.

–version

Display the version number and copyrights of the invoked GCC.

–wrapper

Invoke all subcommands under a wrapper program. The name of the wrapper program and its parameters are passed as a comma separated list.

```
gcc -c t.c -wrapper gdb,--args
```

This will invoke all subprograms of gcc under gdb --args, thus the invocation of cc1 will be gdb --args cc1 ....

–fplugin=name.so

Load the plugin code in file name.so, assumed to be a shared object to be dlopen'd by the compiler. The base name of the shared object file is used to identify the plugin for the purposes of argument parsing (See -fplugin-arg-name-key=value below). Each plugin should define the callback functions specified in the Plugins API.

–fplugin-arg-name-key=value

Define an argument called key with a value of value for the plugin called name.

–fdump-ada-spec[-slim]

For C and C++ source and include files, generate corresponding Ada specs. See Generating Ada Bindings for C and C++ headers, which provides detailed documentation on this feature.

–fdump-go-spec=file

For input files in any language, generate corresponding Go declarations in file. This generates Go const, type, var, and func declarations which may be a useful way to start writing a Go interface to code written in some other language.

@file

Read command-line options from file. The options read are inserted in place of the original @file option. If file does not exist, or cannot be read, then the option will be treated literally, and not removed.

Options in file are separated by whitespace. A whitespace character may be included in an option by surrounding the entire option in either single or double quotes. Any character (including a backslash) may be included by prefixing the character to be included with a backslash. The file may itself contain additional @file options; any such options will be processed recursively.

A.4   Options to Request or Suppress Warnings

Warnings are diagnostic messages that report constructions that are not inherently erroneous but that are risky or suggest there may have been an error.

The following language-independent options do not enable specific warnings but control the kinds of diagnostics produced by GCC.

–fsyntax-only

Check the code for syntax errors, but don't do anything beyond that.

–fmax-errors=n

Limits the maximum number of error messages to n, at which point GCC bails out rather than attempting to continue processing the source code. If n is 0 (the default), there is no limit on the number of error messages produced. If --Wfatal-errors is also specified, then --Wfatal-errors takes precedence over this option.

–w

Inhibit all warning messages.

–Werror

Make all warnings into errors.

–Werror=

Make the specified warning into an error. The specifier for a warning is appended, for example -Werror=switch turns the warnings controlled by --Wswitch into errors. This switch takes a negative form, to be used to negate --Werror for specific warnings, for example --Wno-error=switch makes --Wswitch warnings not be errors, even when --Werror is in effect.

The warning message for each controllable warning includes the option that controls the warning. That option can then be used with --Werror= and --Wno-error= as described above. (Printing of the option in the warning message can be disabled using the --fno-diagnostics-show-option flag.)

Note that specifying --Werror=foo automatically implies --Wfoo. However, --Wno-error=foo does not imply anything.

–Wfatal-errors

This option causes the compiler to abort compilation on the first error occurred rather than trying to keep going and printing further error messages. You can request many specific warnings with options beginning '--W', for example --Wimplicit to request warnings on implicit declarations. Each of these specific warning options also has a negative form beginning '--Wno-' to turn off warnings; for example, --Wno-implicit. This manual lists only one of the two forms, whichever is not the default. For further, language-specific options also refer to C++ Dialect Options and Objective-C and Objective-C++ Dialect Options.

When an unrecognized warning option is requested (e.g., -Wunknown-warning), GCC will emit a diagnostic stating that the option is not recognized. However, if the -Wno- form is used, the behavior is slightly different: No diagnostic will be produced for -Wno-unknown-warning unless other diagnostics are being produced. This allows the use of new -Wno- options with old compilers, but if something goes wrong, the compiler will warn that an unrecognized option was used.

–pedantic

Issue all the warnings demanded by strict ISO C and ISO C++; reject all programs that use forbidden extensions, and some other programs that do not follow ISO C and ISO C++. For ISO C, follows the version of the ISO C standard specified by any --std option used.

Valid ISO C and ISO C++ programs should compile properly with or without this option (though a rare few will require -ansi or a -std option specifying the required version of ISO C). However, without this option, certain GNU extensions and traditional C and C++ features are supported as well. With this option, they are rejected.

--pedantic does not cause warning messages for use of the alternate keywords whose names begin and end with __. Pedantic warnings are also disabled in the expression that follows __extension__. However, only system header files should use these escape routes; application programs should avoid them. See Alternate Keywords.

Some users try to use -pedantic to check programs for strict ISO C conformance. They soon find that it does not do quite what they want: it finds some non-ISO practices, but not all—only those for which ISO C requires a diagnostic, and some others for which diagnostics have been added. A feature to report any failure to conform to ISO C might be useful in some instances, but would require considerable additional work and would be quite different from -pedantic. We don't have plans to support such a feature in the near future. Where the standard specified with -std represents a GNU extended dialect of C, such as 'gnu90' or 'gnu99', there is a corresponding base standard, the version of ISO C on which the GNU extended dialect is based. Warnings from -pedantic are given where they are required by the base standard. (It would not make sense for such warnings to be given only for features not in the specified

GNU C dialect, since by definition the GNU dialects of C include all features the compiler supports with the given option, and there would be nothing to warn about.)

Appendix B

Introduction to Linux

B.1   Introduction

Linux is a multi-user, multi-tasking operating system that is Unix-like and performs many of the same functions as UNIX. Linux is freely available; to acquire Linux is to purchase a Linux distribution.

A Linux distribution is an organized bundle that includes the kernel (core of the operating system), a large set of utility programs, which includes an installation utility. Some of the most widely-used Linux distributions are: Red Hat, Debian, Ubuntu, CentOS, Slackware, and others.

B.2   Operating System Interfaces

Users and application programmers can communicate with an operating system through its interfaces. There are three general levels of interfaces provided by an operating system and they are:

1. Graphical user interface (GUI), for example, Gnome, KDE, the Windows desktop and others.

2. Command line interpreter, also known as the shell.

3. System-call interface.

The highest level is the graphical user interface (GUI) because the user is presented with intuitive icons, menus and other graphical objects. With these objects, the user interacts with the operating system in a relatively easy and convenient manner, for example using the click of a mouse to select an option or command.

The user at this level is completely separated from any intrinsic detail about the underlying system. This level of operating system interface is not considered an essential part of the operating system; the interface is rather an add-on system software component.

397

The second level of interface, the shell (or command line interpreter), is a text-oriented interface. The advanced users and the application programmers will normally directly communicate with the operating system at this level. In general, the graphical user interface and the shell are at the same level in the structure of an operating system.

The third level, the system call interface, is used by the application programs to request the various services provided by the operating system.

A user interacts with Linux using one of several available interfaces that are grouped in two categories:

- A text-based command line interface

- A graphical user interface (GUI).

Linux has several graphical user interfaces available, such as GNU Network Object Model Environment (GNOME) and the K Desktop Environment (KDE). The graphical interface or a variation of it is typically used with Linux installed in your computer.

The most common interface when connected to a remote Linux server is the command line. The Command Line Interpreter, called the *shell*, accepts commands in text form and executes them one after the other. This interface is also used from a Linux installed in your computer and activating the Terminal program.

When you connect to a Linux system via a network, you will need to type in your user account name and your password. After you have logged in, a command prompt will appear as a $ symbol. This prompt indicates that the shell is ready to accept a command; the line with the prompt displays:

```
[Home directory/current directory] $
```

A similar login procedure is normally necessary when starting a session on Linux installed in a standalone computer.

B.3 Command Line Interface

This section explains the basic use of the command line interface and the most common commands used in Linux. Each command has a name, followed by a set of parameters. In general, commands are entered using the keyboard (Standard Input) and the system response appears on the screen (Standard Output). The shell accepts and executes the commands as they are entered by the user.

The shell is the command interpreter. Linux has many different Shells available and these include:

- Bourne

- Korn

- C shell

- Bash (default on Linux)

- tcsh

- dash

While these shells vary in some of the more complex features that they provide, they all have the same basic usage. This section presents a quick introduction to using the most basic Linux commands.

B.4 Files and Directories

In Linux, there are three types of files: ordinary files, directories, and device files. A file is simply a named organized collection of data. A file can contain data or a program (in binary or in source form). The directory structure of Linux allows a convenient way to organize files and subdirectories. The top directory is the system root, denoted by / (forward slash). All other directories and files are located under the system root.

When you are working with a *Linux Shell*, you navigate the system's directory structure to perform the desired operations. The Shell has the concept of the *Current Directory*, which is the directory that you are currently working in. When you login, the Shell will initially position you at your *Home Directory*, which is the unique directory provided for each user of the system. For example: `/home/jgarrido` is a home directory.

In addition to your Home Directory, the system usually has other directories with various system files that you may use. Some common system directories are shown in Table B.1.

`/bin`	Contains binary or executable files
`/sbin`	Contains system binaries
`/usr`	Contains user libraries and applications
`/etc`	Typically contains configuration information
`/tmp`	Directory for temporary files
`/dev`	Contains device drivers for the hardware components

TABLE B.1: System directories.

B.4.1 Specifying Paths

A directory or file is specified by typing the *path* to the desired directory or file. A path consists of the names of directories in the directory tree structure, with a "/" between directory names. (Note that Windows uses "\" to separate directory names). If the path begins with a directory (or file) name, then the path begins at the *current directory*. If the path begins with a (/), then the path begins at the root of the entire directory structure on this system.

An absolute path includes the complete and hierarchical structure. To access a file with the indicated absolute path, Linux starts at the top and searches downward every directory specified. For example, an absolute path of file `batch.cpp` is: `/home/jgarrido/psim3/batch.cpp`. A relative path specifies the location of a file in relation to the current working directory.

Unlike MS Windows, Linux pathnames are case-sensitive. Thus, `MyFile`, `myfile`, and `MYFILE` all specify different files.

There are two special directory names (a single period, and 2 consecutive periods):

- A reference to the current directory is specified as '.' In the command line.

- A reference to the parent directory one level up in the tree, is specified with '..' in the command line.

Thus, the command: `ls ..` will display the parent directory.

B.4.2 Wildcards

When using the Bash shell, it is possible to select several files names that have some sequence of characters in common. Some commands allow you to specify more than one file or directory. The special characters that are used are wildcard characters. There are three wildcards:

1. If you specify a '?' in a directory or file name, it will match any character in that position. Thus `h?` will match `he` or `hf`, but not `her`.

2. Similarly, if you specify '*' in a directory or file name, it will match any sequence of characters in that position. Thus `h*` will match `he`, `hf` or `her`.

3. A pair of square brackets '[]' enclosing a list of characters allow the shell to match individual characters of the list.

The following examples show the convenience of using wildcards with the `ls` command:

```
ls *.dat
ls re*asd
ls *.?
ls bat[edfg].cpp
ls [a-j]*.dat
```

B.5 Basic Linux Commands

B.5.1 The `passwd` Command

This command is used to change your password. It will prompt you to enter your current password and then ask you to enter your new password twice. For example:

```
passwd
Changing password for user jgarrido.
Changing password for jgarrido
(current) UNIX password:
New UNIX password:
Retype new UNIX password:
passwd: all authentication tokens updated successfully.
```

B.5.2 The `man` Command

To request help, if you cannot remember what the parameters to a command are, or the specific use of a command, you can type the `man` command. If you type `man commandname`, it will display manual page for that command including the parameters for that command. It accepts wildcards, so if you cannot remember exactly how the command is spelled, you can use a wildcard to have it try to match the name of the desired command. For example, the following command typed at the the shell prompt, will display the manual command page shown in next listing:

```
man passwd
```

```
PASSWD(1)              User Commands              PASSWD(1)
```

NAME
> passwd - change user password

SYNOPSIS
> passwd [options] [LOGIN]

DESCRIPTION
> The `passwd` command changes passwords for user accounts. A normal user may only change the password for his/her own account, while the superuser may change the password for any account. The `passwd` command also changes the account or associated password validity period.

Password Changes
> The user is first prompted for his/her old password, if one is present. This password is then encrypted and compared against the stored password. The user has only one chance to enter the correct password. The superuser is permitted to bypass this step so that forgotten passwords may be changed.

After the password has been entered, password aging information is checked to see if the user is permitted to change the password at this time. If not, `passwd` refuses to change the password and exits.

The user is then prompted twice for a replacement password. The second entry is compared against the first and both are required to match in order for the password to be changed.

Then, the password is tested for complexity. As a general guideline, passwords should consist of 6 to 8 characters including one or more characters from each of the following sets:

- lower case alphabetics
- digits 0 thru 9
- punctuation marks

Care must be taken not to include the system default erase or kill characters. `passwd` will reject any password which is not suitably complex.

Hints for user passwords

The security of a password depends upon the strength of the encryption algorithm and the size of the key space. The legacy UNIX System encryption method is based on the NBS DES algorithm. More recent methods are now recommended (see ENCRYPT_METHOD).

The size of the key space depends upon the randomness of the password which is selected.

Compromises in password security normally result from careless password selection or handling. For this reason, you should not select a password which appears in a dictionary or which must be written down. The password should also not be a proper name, your license number, birth date, or street address. Any of these may be used as guesses to violate system security.

You can find advice on how to choose a strong password on
`http://en.wikipedia.org/wiki/Password_strength`

OPTIONS

The options which apply to the `passwd` command are:

-a, –all

> This option can be used only with `-S` and causes show status for all users.

-d, –delete

> Delete a user's password (make it empty). This is a quick way to disable a password for an account. It will set the named account passwordless.

-e, –expire

> Immediately expire an account's password. This in effect can force a user to change his/her password at the user's next login.

-h, –help

> Display help message and exit.

-i, –inactive INACTIVE

This option is used to disable an account after the password has been expired for a number of days. After a user account has had an expired password for INACTIVE days, the user may no longer sign on to the account.

-k, –keep-tokens Indicate password change should be performed only for expired authentication tokens (passwords). The user wishes to keep their non-expired tokens as before.

-l, –lock

Lock the password of the named account. This option disables a password by changing it to a value which matches no possible encrypted value (it adds a '!' at the beginning of the password).

Note that this does not disable the account. The user may still be able to login using another authentication token (e.g. an SSH key). To disable the account, administrators should use `usermod --expiredate 1` (this set the account's expire date to Jan 2, 1970).

Users with a locked password are not allowed to change their password.

-n, –mindays MIN_DAYS

Set the minimum number of days between password changes to MIN_DAYS. A value of zero for this field indicates that the user may change his/her password at any time.

-q, –quiet

Quiet mode.

-r, –repository REPOSITORY

change password in REPOSITORY repository

-S, –status

Display account status information. The status information consists of 7 fields. The first field is the user's login name. The second field indicates if the user account has a locked password (L), has no password (NP), or has a usable password (P). The third field gives the date of the last password change. The next four fields are the minimum age, maximum age, warning period, and inactivity period for the password. These ages are expressed in days.

-u, –unlock

Unlock the password of the named account. This option re-enables a password by changing the password back to its previous value (to the value before using the -l option).

-w, –warndays WARN_DAYS

Set the number of days of warning before a password change is required. The WARN_DAYS option is the number of days prior to the password expiring that a user will be warned that his/her password is about to expire.

-x, −maxdays MAX_DAYS

Set the maximum number of days a password remains valid. After MAX_DAYS, the password is required to be changed.

CAVEATS

Password complexity checking may vary from site to site. The user is urged to select a password as complex as he or she feels comfortable with.

Users may not be able to change their password on a system if NIS is enabled and they are not logged into the NIS server.

passwd uses PAM to authenticate users and to change their passwords.

FILES

/etc/passwd
User account information.

/etc/shadow
Secure user account information.

/etc/pam.d/passwd
PAM configuration for passwd.

EXIT VALUES

The passwd command exits with the following values:

0
success

1
permission denied

2
invalid combination of options

3
unexpected failure, nothing done

4
unexpected failure, `passwd` file missing

5
passwd file busy, try again

6
invalid argument to option

SEE ALSO passwd(5), shadow(5), usermod(8).

B.5.3 The `ls` Command

The `ls` (list directory) command is used to provide a directory listing. It has the following format:

```
ls [options] directory-name
```

This command will provide a listing of the specified directory. If directory-name is omitted, it will provide a listing of the *Current Directory*. The following listing shows the result of using the ls without including a directory name.

```
addrptr.c
APP_C.C
APR91.DAT
areacir.c
arrayex4.cpp
arrayex5.cpp
arraymulb.c
arraymult.c
arrmax.c
arrmult.c
AUG91.DAT
averagef.c
basic_lib.c
basic_lib.h
basic_lib.o
bool.h
```

The `ls` command provides several options; the most important ones are: long listing (-l), list all files (including the system files) (-a), list files by order of date and time starting with the most recent (-t). To use the *long* listing option, type the `ls` command with a hyphen and an letter l, -l. For example:

```
ls -l
```

```
total 1952
-rw-r--r-- 1 jgarrido jgarrido    389 Dec 31 12:07 addrptr.c
-rw-r--r-- 1 jgarrido jgarrido    889 Sep  5 10:22 APP_C.C
-rw-r--r-- 1 jgarrido jgarrido   3180 Mar 22  1993 APR91.DAT
-rw-r--r-- 1 jgarrido jgarrido    686 Dec 27 22:41 areacir.c
-rw-r--r-- 1 jgarrido jgarrido   1480 Sep  5 12:32 arrayex4.cpp
-rw-r--r-- 1 jgarrido jgarrido   1467 Sep  5 12:36 arrayex5.cpp
-rw-r--r-- 1 jgarrido jgarrido    856 Dec 31 19:26 arraymulb.c
-rw-r--r-- 1 jgarrido jgarrido    596 Dec 31 15:00 arraymult.c
-rw-r--r-- 1 jgarrido jgarrido    404 Dec 30 19:59 arrmax.c
-rw-r--r-- 1 jgarrido jgarrido    428 Dec 31 14:46 arrmult.c
-rw-r--r-- 1 jgarrido jgarrido   3071 Mar 22  1993 AUG91.DAT
-rw-r--r-- 1 jgarrido jgarrido    426 Dec 30 19:42 averagef.c
-rw-r--r-- 1 jgarrido jgarrido   6581 Oct 19 09:41 basic_lib.c
-rw-r--r-- 1 jgarrido jgarrido    550 Sep 14 15:18 basic_lib.h
-rw-r--r-- 1 jgarrido jgarrido   2902 Oct 19 10:06 basic_lib.o
-rw-r--r-- 1 jgarrido jgarrido     55 Dec 20 12:22 bool.h
```

The command will list all the files and folders in the current directory with the long listing option. With the *long* option, detailed information about the files and subdirectories (subfolders) is shown. The listing shows the `ls` command with the long option, the additional information displayed includes the file type, the permissions mode, the owner of the file, the size, the date and time of last update, and other data about the file.

Typing the `ls` command and a name of a directory displays all the files and subdirectories in the specified directory. By typing `ls os` in the command line, the `ls` command is used with the `os` directory.

B.5.4   The `cp` Command

The `cp` command will copy one or more files from one directory to another. The original files are not changed. The command has the following format:

```
cp [options] source destination
```

This command will copy a source file on some directory to a destination directory. You can use wildcards to specify more than one file or directory to be copied. For example:

```
cp psim2/rrsched.cpp psim3
cp data/*.dat mydir/data
```

The first line copies the file `rrsched.cpp` from the `psim2` subdirectory to directory `psim3`. The second line copies all the files with an `.dat` extension to subdirectory `mydir/data`.

The most common options of the `cp` command are the following. Interactive option `-i`, which prompts the user when a file will be overwritten. Recursive option `-r`, which can copy directories and their files. Verbose `-v`, which displays the name of each file copied.

B.5.5   The `mv` Command

The `mv` command can be used to change the name of a file (or directory). This can also be done with the `cp` command. The `mv` command has the following format:

```
mv [options] source target
```

Some of the options for this command (interactive and verbose options) are similar to the ones for command `cp`. The following example illustrates the use of the `mv` command.

```
mv rrsched.cpp timesh.cpp
mv my.dat mydir/data
```

The first line of the example changed the name of file `rrsched.cpp` to `timesh.cpp`. The second line moves file `my.dat` to directory `mydir/data`.

B.5.6   The `rm` Command

The `rm` command will remove (delete) a file from the directory where it is currently located. The command has the following format:

```
rm [options] filename
```

This command will remove (delete) the specified file from the current directory. The following example shows the use of the `rm` command.

```
rm rrsched.cpp
rm mydat/*.dat
```

The first line in the example deletes file `rrsched.cpp` from the current working directory. The second line removes all the files whose names end with `.dat`. The most common options of the `rm` command are the following. Interactive option `-i`, which prompts the user when a file will be removed. Recursive option `-r`, which will remove the entire directories (all their files). Verbose `-v`, which displays the name of each file before removing it.

B.5.7   The `cd` Command

The `cd` command (Change Directory) changes the current directory to the one specified. The command has the following format:

```
cd directory-path
```

This command will change the Current Directory to the specified directory. Thus, the command line `cd ..` will move to the parent directory. When specifying a directory, be careful to include (or not include) the beginning '/' depending on whether you are intending to use an absolute or relative path. If a directory name is not given, the `cd` command changes to the user's home directory. For example:

```
cd
cd psim3/data
cd /usr/misc
```

The first line of the example changes to the user's home directory. The second line changes to directory `psim3/data`, which is a subdirectory under the current directory. The third line changes to directory `/usr/misc`, for which an absolute path is given.

B.5.8   The `mkdir` Command

The `mkdir` command (Make Directory) creates a new directory with the specified name. The newly created directory is empty (it has no files). The command has the following format:

```
mkdir directory-name
```

This command will create a new directory with the specified name. One usually uses this with a relative path to create a sub-directory, but it is possible to specify an absolute path. For example:

```
mkdir data/models
```

The mkdir command in the example creates a new subdirectory called models located on the subdirectory data.

B.5.9   The rmdir Command

The rmdir (Remove Directory) removes a directory with the specified name. The command has the following format:

```
rmdir directory-name
```

This command will remove (delete) the specified subdirectory from the current directory. The following example removes directory psim3/data from the current working directory.

```
rmdir psim3/data
```

B.5.10   I/O Re-Direction and Pipe Operators

In Linux, three default files are available for most commands, standard input, standard output, and standard error. Using direction and piping the shell user can change these defaults. With *I/O Re-Direction*, the Shell can:

- Take Input from another file, called input direction.

- Send Output to another file, called output redirection.

Input redirection is carried out by using the less-than symbol ($<$). The command will take the file specified after the ($<$) as the input. For example, the executable file a.out will execute with data from the file myfile.dat:

```
a.out < myfile.dat
```

Output redirection is carried out by using the greater-than symbol ($>$). The command will take the file specified after the ($>$) as the output. For example, the executable file a.out will execute and send the results to the file myresult.dat:

```
a.out > myresult.dat
```

With pipes, the standard output of a command can be connected to the standard input of another command. Several commands can be connected with pipes, which use the pipe character (|) between commands. For example to pipe the ls command with the more command:

Command	Action
ls > tmp.txt	Take the Standard Output from the ls command and write it to the file tmp.txt.
xyz < tmp.txt	The xyz application will use the file tmp.txt as its Standard Input.
ls | xyz	The Standard Output of the ls command will be used as the Standard Input for the xyz application.

TABLE B.2: I/O redirection.

```
ls -l | more
```

The pipe between these two commands is useful because there will be too much data to be displayed by the ls command. The output of this command is sent as input to the more command and the data is displayed by the more command screen by screen. Table B.2 shows several simple examples of I/O redirection and piping.

The tee command sends the output of a command to a file and pipes the output to another command. For example:

```
./a.out | tee myout.txt | more
```

In this example, the first command executes the file a.out and sends the output to file myout.txt and is also displayed on the screen by the more command.

B.6   Shell Variables

In Linux, there are two types of variables that are used with the shell:

- System variables are created and maintained by Linux itself. This type of variable defined in CAPITAL LETTERS.

- User defined variables are created and maintained by the user. The name of this type of variable is defined in lowercase letters.

The following are examples of system variables.

```
HOME=/home/vivek
PATH=/usr/bin:/sbin:/bin:/usr/sbin
SHELL=/bin/bash
USERNAME=vivek
```

```
echo      $USERNAME
echo      $HOME
echo      $SHELL
```

To print the content of a system variable, you must use the character $ followed by the variable name. For example, you can print the value of the variables with the `echo` command, as follows:

The following example defines a user variable called `vech` having value *Bus*, and a variable called n having value 10.

```
vech=Bus
no=10       # this is ok
10=no       # Error,Value must be on right side of = sign.
```

To print the content of variable `vech`, you must use the character ($) followed by the variable name. For example:

```
echo $vech # will print 'Bus'
echo $n    # will print value of 'n'
echo vech  # prints 'vech' instead its value 'Bus'
echo n     # prints 'n' instead its value '10'
```

Associated with each user login are a set of Environment Variables. The `set` command is used to change the value of an environment variable. The set command has the following format:

```
set variable=value
```

On Ubuntu, the `env` command will display the values of all environment variables. The result of executing this command will produce the following listing:

```
SSH_AGENT_PID=1765
GLADE_PIXMAP_PATH=:/usr/lib/glade3/modules
TERM=xterm
SHELL=/bin/bash
XDG_MENU_PREFIX=xfce-
DESKTOP_STARTUP_ID=
XDG_SESSION_COOKIE=7e5a45522af0aa74e4541e7250c77a70-1357570206.
    725759-1268713858
WINDOWID=41943044
GNOME_KEYRING_CONTROL=/run/user/jgarrido/keyring-SBNRn1
GTK_MODULES=overlay-scrollbar
USER=jgarrido
LS_COLORS=rs=0:di=01;34:ln=01;36:mh=00:pi=40;33:so=01;35:do=01;
    35:bd=40;33;01:
    cd=40;33;01:or=40;31;01:su=37;41:sg=30;43:ca=30;41:tw=30;
        42:ow=34;42:st=37;44:
    ex=01;32:*.tar=01;31:*.tgz=01;31:*.arj=01;31:*.taz=01;
        31:*.lzh=01;31:*.lzma=01;31:
    *.tlz=01;31:*.txz=01;31:*.zip=01;31:*.z=01;31:*.Z=01;
```

```
      31:*.dz=01;31:*.gz=01;31:
*.lz=01;31:*.xz=01;31:*.bz2=01;31:*.bz=01;31:*.tbz=01;
      31:*.tbz2=01;31:*.tz=01;31:
*.deb=01;31:*.rpm=01;31:*.jar=01;31:*.war=01;31:*.ear=01;
      31:*.sar=01;31:*.rar=01;31:*.ace=01;31:*.zoo=01;
      31:*.cpio=01;31:*.7z=01;31:*.rz=01;31:
*.jpg=01;35:*.jpeg=01;35:*.gif=01;35:*.bmp=01;35:*.pbm=01;
      35:*.pgm=01;35:
*.ppm=01;35:*.tga=01;35:*.xbm=01;35:*.xpm=01;35:*.tif=01;
      35:*.tiff=01;35:
*.png=01;35:*.svg=01;35:*.svgz=01;35:*.mng=01;35:*.pcx=01;
      35:*.mov=01;35:*.mpg=01;35:*.mpeg=01;35:*.m2v=01;
      35:*.mkv=01;35:*.webm=01;
      35:*.ogm=01;35:*.mp4=01;35:*.m4v=01;35:*.mp4v=01;
      35:*.vob=01;35:*.qt=01;35:
*.nuv=01;35:*.wmv=01;35:*.asf=01;35:*.rm=01;35:*.rmvb=01;
      35:*.flc=01;35:
*.avi=01;35:*.fli=01;35:*.flv=01;35:*.gl=01;35:*.dl=01;
      35:*.xcf=01;35:*.xwd=01;35:
*.yuv=01;35:*.cgm=01;35:*.emf=01;35:*.axv=01;35:*.anx=01;
      35:*.ogv=01;35:
*.ogx=01;35:*.aac=00;36:*.au=00;36:*.flac=00;36:*.mid=00;
      36:*.midi=00;36:
*.mka=00;36:*.mp3=00;36:*.mpc=00;36:*.ogg=00;36:*.ra=00;
      36:*.wav=00;36:
*.axa=00;36:*.oga=00;36:*.spx=00;36:*.xspf=00;36:
XDG_SESSION_PATH=/org/freedesktop/DisplayManager/Session0
GLADE_MODULE_PATH=:/usr/share/glade3/pixmaps
XDG_SEAT_PATH=/org/freedesktop/DisplayManager/Seat0
SSH_AUTH_SOCK=/tmp/ssh-cHQs41SbLuUg/agent.1729
SESSION_MANAGER=local/jgarrido-OptiPlex-755:@/tmp/.ICE-unix/
      1776;unix/jgarrido-OptiPlex-755:/tmp/.ICE-unix/1776
DEFAULTS_PATH=/usr/share/gconf/xfce.default.path
XDG_CONFIG_DIRS=/etc/xdg/xdg-xfce:/etc/xdg:/etc/xdg
PATH=/usr/lib/lightdm/lightdm:/usr/local/sbin:/usr/local/bin:
      /usr/sbin:/usr/bin:/sbin:/bin:/usr/games:/usr/local/games
DESKTOP_SESSION=xfce
PWD=/home/jgarrido
GNOME_KEYRING_PID=1718
LANG=en_US.UTF-8
MANDATORY_PATH=/usr/share/gconf/xfce.mandatory.path
UBUNTU_MENUPROXY=libappmenu.so
PS1=$
GDMSESSION=xfce
SHLVL=1
HOME=/home/jgarrido
LOGNAME=jgarrido
XDG_DATA_DIRS=/usr/share/xfce:/usr/local/share/:/usr/share/:
      /usr/share
DBUS_SESSION_BUS_ADDRESS=unix:abstract=/tmp/dbus-D7rsvgkGfD,
      guid=34ee11f9d27c26dd593398fd50eae09f
LESSOPEN=| /usr/bin/lesspipe %s
XDG_RUNTIME_DIR=/run/user/jgarrido
DISPLAY=:0.0
GLADE_CATALOG_PATH=:/usr/share/glade3/catalogs
XDG_CURRENT_DESKTOP=XFCE
LESSCLOSE=/usr/bin/lesspipe %s %s
```

```
COLORTERM=Terminal
XAUTHORITY=/home/jgarrido/.Xauthority
_=/usr/bin/env
```

An important environment variable is PATH. This environment variable has a list of directories in which the system will look for the program file whenever you request an application be executed.

B.6.1 The pwd Command

This command will print (display) the absolute name of the current working directory. The command will display the full path name of the current directory. Note that the dot (.) references the current working directory and is used in other commands that manipulate files and directories (ls, cp, mv, mkdir, rmdir among others).

B.6.2 The more Command

The more command will display the contents of a file one page at a time. In using this command, Linux assumes that the specified file is a text file (or a file that can be viewed on the screen). This command can also be used with a pipe to display the output of another command.

```
more mydata.dat
ls | more
```

The first line of the example lists the contents of file mydata.dat one page at a time. The second line displays a listing of the current directory one page at a time.

B.6.3 The script Command

The script command is useful when record a complete (or partial) Linux session. For example, the following command line starts recording or logging a session and all commands used in the session including the system responses are stored in the specified text file *mysession1.txt*.

```
script mysession1.txt
```

When you are done recording a session, the exit command can be used.

B.6.4 The exit Command

The exit command can be used to logout of Linux. When you are finished using Linux, type the exit command. This will cause the *shell* to terminate and you will be logged off the system. This command is normally used in a script file (a command file) to exit with an optional integer error code.

B.7 Text Editing

There are several text editors available for Linux. With the command level interface, the three most popular are *vi*, *nano*, and *emacs*. When using the graphical interfaces, *gedit* and *kedit* may also be used. Figure B.1 shows the screen that appears when using *nano*. To start *nano* to edit a file, one simply types the following command:

```
nano filename
```

In the command line, `filename` is the name of the file to be edited. The top part of the screen in Figure B.1 shows part of the file being edited. The last 2 lines of the screen show the editing commands that one can use. The (ˆ) character refers to the *Ctrl* key. Thus, *Ctrl-G* is the editing command that will open the help pages and *Ctrl-X* will exit the editor. Refer to the Help screen (by using *Ctrl-G*) for a complete discussion of the available editing commands.

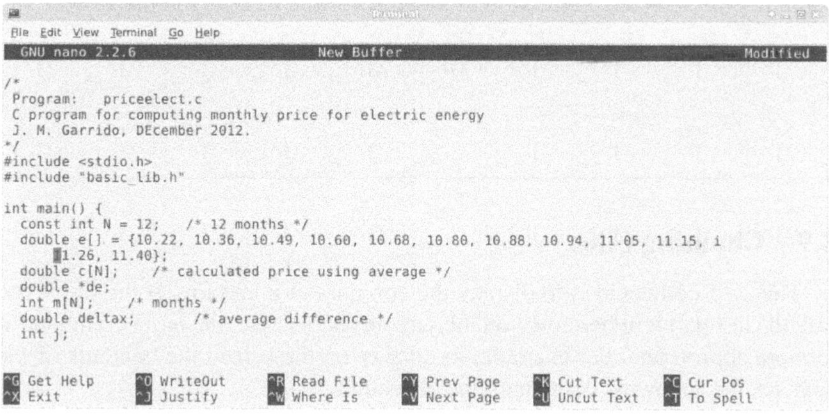

FIGURE B.1: Nano text editor screen.

B.8 File Access Permissions

A user may deny or allow access to files and directories; these are normally called permission modes. The following types of permission modes are allowed for files and directories:

- Read (r)

- Write (w)

- Execute (x)

In Linux, there are three types of users for file access:

1. Owner of the file or directory (user)

2. Group, users in the same group as the owner

3. Others, all other users

A user can change the file permissions on one or more files and directories by using the *chmod* command. The following example changes the permission mode to give read access permission to all users for file batch.cpp and changes the permission mode to remove write permission to other users.

```
chmod a+r batch.cpp
chmod o-w batch.cpp
```

When writing a shell script, it is necessary to change permission and enable execute permission of the script file to the user or owner of the file. The following example gives execute permission to the user for the (script) file myscript.

```
chmod u+x myscript
```

B.9 Chaining Files

The cat command is to display the contents of a text file. If the file is too big, only the last part will be shown on the screen; in this case, the *more* command would be more appropriate. For example, to display on the screen the contents of file *mytasks.txt*, the following command line is needed.

```
cat mytasks.txt
```

An important purpose of the cat command is to join two or more files one after the other. In the following example, three files are chained together, *alpha*, beta, and gamma. The redirection operator is required to create the output file

```
cat alpha beta gamma > myfile
```

The cat command can also be used to append data to a file using the append ($\gg$) operator. The following example illustrates the use of the append operator ($\gg$) to append data to the end of an existing file. The example shows that data from file *res.h* is appended to the end of file *batch.txt*.

```
cat res.h >> batch.txt
more batch.txt
```

B.10   Commands for Process Control

A shell command can be internal (built-in) or external. Internal commands are part of the Linux system. The external commands are the utilities in Linux, application programs, or script files. Linux creates a new process for every external command that is started and assigns a Process ID number (PID) to every process created.

The ps command lists the active processes of the user. With the l option, this command displays a long listing of the data about the processes.

```
/home/jgarrido ~ $ps
PID TTY          TIME CMD
4618 pts/3    00:00:00 bash
4644 pts/3    00:00:00 ps
/home/jgarrido ~ $

/home/jgarrido ~ $ps -l
F S   UID   PID  PPID  C PRI  NI ADDR SZ WCHAN   TTY
    TIME CMD
0 S  1374  5279  5277  0  75   0 - 13493 wait    pts/3
00:00:00 bash
0 R  1374  5318  5279  0  76   0 -  1093 -       pts/3
00:00:00 ps
/home/jgarrido ~ $ps -lj
F S   UID   PID  PPID  PGID   SID  C PRI  NI ADDR SZ WCHAN
TTY           TIME CMD
0 S  1374  5279  5277  5279  5279  0  76   0 - 13493 wait
pts/3    00:00:00 bash
 0 R  1374  5319  5279  5319  5279  0  76   0 -  1093 -
pts/3    00:00:00 ps
/home/jgarrido ~ $
```

The ps commands with the l option display the attribute values of the processes; the most important of these are the PID and the process state (S).

When a user needs to force the termination of a process, he or she can use the kill command. The command sends a signal to terminate the specified process. The user can terminate only processes that he or she owns. The general syntax for this command is:

```
kill [options] pid
```

For example, in the first of the following lines the shell will terminate process 15097. The second line will terminate process 15083 by sending it signal with number 9.

```
kill 15097
kill -9 15063
```

B.11 Foreground and Background Processes

When a process is started in a Linux shell, the user waits until the process terminates and Linux displays the shell prompt. Processes that execute in this manner, run in the *foreground*. For long-running processes, Linux allows the user to start a process so it will execute in the *background*. Processes that run in the background have a lower priority compared to the foreground processes.

For processes to start execution in the background, the user needs to type the name of the command followed by the ampersand (&) symbol. To start a process called *myjob* as a background process: `myjob &`. The following example shows the start of a script file `psim3` in the background.

```
./psim3 batch.cpp &
[1] 5325
```

The Linux shell returns two numbers when a background process is started by the user; the job number (in brackets) for the process, and its PID. In the second command line of the previous example, 1 is the job number and 5325 is its PID.

A background job (a process running in the background) can be brought to run in the foreground with the `fg` command. The job number has to be specified as an argument to the command.

```
nanoo tarea.kpl &
[1] 15165
fg %1
nano tarea.kpl
```

In this example, the *nano* editor is started as a background job with job number 1. The next command makes the shell bring the background job to the foreground. The *nano* screen will be shown with the file `tarea.kpl` and editing can proceed normally.

Another use of the `fg` command is to resume a suspended process. A process running in the foreground can be suspended by the Ctrl-Z keys combination. It can later be resumed by using the `fg` command and the job number of the process. A suspended process can be resumed to run in the background using the `bg` command and the job number.

B.12 Script Commands

A shell script is basically a text file with shell commands. The shell executes the script by reading and executing the commands line by line. The syntax of scripts is relatively simple, similar to that of invoking and chaining together utilities at the command line.

A shell script is a *quick and dirty* method of prototyping a complex application. Achieving a limited subset of the functionality to work in a shell script is often a useful first stage in project development. Shell scripting provides the user with the ability to decompose complex projects into simpler subtasks, of chaining together components and utilities.

A shell script file must be readable and executable to the user of the script. The chmod command must be used to change these permissions modes of the script file.

B.12.1   Comments in Scripts

Comments lines are lines that begin with a # symbol. Comments may also occur following the end of a command. Comments may also follow a white space at the beginning of a line. For example:

```
# This line is a comment.
 echo "A comment will follow."  # Comment here.
     #        ^ Note whitespace before #
     # A tab precedes this comment.
```

B.12.2   Positional Parameters

The Shell uses these parameters to store the values of command-line arguments, and are used as read-only variables in the script. A $ symbol precedes the parameter number. The parameters in the script are numbered 0–9. In addition to positional parameters, a script file normally uses system and user variables. The following two command lines are included in a script file.

```
rm $1   # remove file $1
echo $1 $2
```

B.12.3   Command Substitution

Command substitution evaluates a command and makes available the output of the command for assignment to a variable. The format of this is notation `command`; this notation is known as backquotes or backticks. An alternative notation is $(command). For example:

```
echo Current date is:  `date`
echo Today is:  $(date)
mydate=`date`
```

B.12.4   The test Command

Use the test command is used to test or evaluate the conditions you formulate in loops and branches. The command is used to:

- Compare numbers

- Compare strings

- Check files

The `test` command has two syntaxes:

- Uses the keyword *test* before the expression to be evaluated

- Uses brackets to enclose the expression to be evaluated

The test command supports several operators Testing files and integers Testing and comparing strings Logically connecting two or more expressions to form complex expressions

Table B.3 shows several operators used in the `test` command that evaluate expression with numbers. Table B.4 shows several operators used in the `test` command that evaluate expression with strings. Table B.5 shows several operators used in the `test` command that evaluate expression with files.

Operator	Action
n1 -eq n2	Evaluates to True if *n*1 and *n*2 are equal
n1 -le n2	Evaluates to True if *n*1 is less than or equal to *n*2
n1 ge n2	Evaluates to True if *n*1 is greater than or equal to *n*2
n1 gt n2	Evaluates to True if *n*1 is greater then *n*2
n1 ne n2	Evaluates to True if *n*1 and *n*2 are not equal

TABLE B.3: Test command operators for expressions with numbers.

B.12.5 The `if` with the `test` Commands

The basic form of the `if` command is used for two-way branching in the flow of control of the execution of the script. The expression in the `if` statement can

Operator	Action
str1	Evaluates to True if *str1* is not an empty string
str1 = str2	Evaluates to True if *str1* and *str2* are equal
str1 != str2	Evaluates to True if *str1* and *str2* are not equal
n str1	Evaluates to True if the length of *str1* is greater than zero
z str1	Evaluates to True if the length of *str1* is zero

TABLE B.4: Test command operators for expressions with strings.

Operator	Action
f file	Evaluates to True if *file* is an ordinary file
d file	Evaluates to True if *file* is a directory
e file	Evaluates to True if *file* exists
s file	Evaluates to True if *file* is non-empty
r file	Evaluates to True if *file* is readable
w file	Evaluates to True if *file* is writable
x file	Evaluates to True if *file* is executable

TABLE B.5: Test command operators for expressions with files.

be evaluated with the `test` command, which returns true or false after evaluating the expression. The following example illustrates the use of the `if` and `test` commands.

```
if test $# -ne 1   # number of arguments is not 1
    then
        echo Error in number of arguments
        exit 1
fi
#
f [ -f $1 ]
    then
        echo File is:   $1
    else
        echo File $1 is not an ordinary file
        exit 1
fi
```

The following shell script file, called *psim3*, uses most of the commands discussed previously. This shell script file is used to compile and link C programs with the GSL.

```
#!/bin/bash
echo "script for compiling and linking a C program with the GSL"
if [4# == o ]
  then
    echo "No source file"
    exit 1
fi
if test -e "$1"
  then
    echo "Source file: $1"
  else
    echo "File: $1 does not exist"
    exit 1
fi
if test $# -eq 2
  then
```

```
   echo "Excutable file: $2"
   objf = $2
 else
   echo "Executable file: a.out"
   objf="a.out"
fi
#
gcc $1 -o $objf basic_lib.o -lgsl -lgslcblas -lm
#
if [ $? -ne 0 ]
 then
   echo "Compiling/linking failed"
   exit 1
 else
   echo "Success . . . "
fi
exit 0
```

The `if` command can be used for multiway branching; this is considered an extension of the basic structure of the `if` command. The general syntax for this follows.

```
if expr1
    then
        commands
    elif expr2
        commands
        elif expr3
            commands
            . . .
        else
            commands
 fi
```

The following example illustrates the use of the multi-branch if command in a shell script.

```
if [ -f $1 ]
    then
        echo File $1 is an ordinary file
        exit 0
    elif [ -d $1 ]
        nfiles = $(ls $1 | wc w)
        echo The number of files in $1 is:  $nfiles
        exit 0
    else
        echo Argument $1 is not file or directory
        exit 1
 fi
```

B.12.6 The `set` Command

When used with arguments, the set command allows one to reset the positional parameters. After resetting the parameters, the old parameters are lost. The positional parameters can be saved in a variable. Command substitution can be used with the set command. The following script file includes several examples using the set command.

```
# Script name: posparam
# Example script using set command
echo "Total number of arguments: " $#
echo "All arguments: " $*
echo "1st is: " $1
echo "2nd is: " $2
prevargs=$*      # save arguments
set V W X Y Z      # resetting positional parameters
echo "New parameters are: " $*
echo "Number of new parameters: " $#
set `date`    # Reset parameters with command
                    # could also be typed as $(date)
echo $*          # display new parameters
echo "Today is: " $2 $3 $6
echo Previous parameters are: " $prevargs
set $prevargs
echo $3 $1 $2 $4
exit 0
```

The following command lines execute the script file and show the output produced by the script.

```
/home/jgarrido ~ $./posparam a b c d
Total number of arguments:   4
All arguments:   a b c d
1st is:   a
2nd is:   b
New parameters are:   V W X Y Z
Number of new parameters:   5
Fri Sep 15 13:39:06 EDT 2006
Today is:   Sep 15 2006
Old parameters are:   a b c d
c a b d
/home/jgarrido ~ $
```

B.12.7   Repetition

The for is used for repetitive execution of a block of commands in a shell script. The repetitions of a block of commands are normally known as loops.The number of times the block of commands is repeated depends on the number of words in the argument list. These words are assigned to the specified variable one by one and the commands in the command block are executed for every assignment. The general syntax of this command follows.

```
for variable [in argument list]
do
             command block
done
```

The following example shows a short script file with a for loop.

```
# File: loopex1. Example of for-loop
#     for system in  Linux Windows Solaris OS2 MacOSX
do
   echo $system
done
```

The execution of the script file is shown in the following listing:

```
./loopex1
Linux
Windows
Solaris
OS2
MacOSX
```

The while loop allows repeated execution of a block of commands based on a condition of an expression. The general syntax of the while command follows.

```
while expression
do
    command-block
done
```

The expression in the while command is evaluated and if true, the command block is executed and the expression is again evaluated. The sequence is repeated until the expression evaluates to false. The following example of a script file shows a while loop.

```
# while example
#secret code
secretcode=Jose
echo n "Enter guess: "
read yguess
while [ $secretcode != $yguess ]
do
       echo "Wrong guess"
       echo n "Enter guess again: "
       read yguess
done
echo "Good guess"
exit 0
```

The loop command is often used with the break and continue commands, which are used to interrupt the sequential execution of the loop body. The break command transfers control to the command following done, terminating the loop prematurely. The continue command transfers control to done, which results in the evaluation of the condition again.

B.13   Searching Data in Files

For searching for lines with specific data in text lines, Linux provides several utilities. These utilities are `grep`, `egrep`, and `fgrep`. For example, the following command line uses the `grep` command and will search for the lines in file `pwrap.cpp` that contain the string *include*.

```
/home/jgarrido ~ $grep -n include pwrap.cpp
42:#include "pthread.h"
43:#include <iostream.h>
165://#include <unistd.h> //included to test sleep(), usleep(),
          wait()
210:    /*this is included to adjust
/home/jgarrido ~ $
```

As the following example shows, the `grep` command also allows the search of a string in multiple files, using a wildcard.

```
/home/jgarrido ~ $grep -n include *.cpp
dscann.cpp:10:// A disk service includes the rotational delay,
          seek delay, and
dscann.cpp:18:#include "proc.h"
dscann.cpp:19:#include "queue.h"  // additional library classes
dscsstfn.cpp:10:// A disk service includes the rotational delay,
          seek delay, and
dscsstfn.cpp:19:#include "proc.h"
dscsstfn.cpp:20:#include "queue.h" // additional library classes
```

The `grep` command can also list names of files. The following uses the `grep` command with the `l` option to display the names of the files where string *include* is found.

```
home/jgarrido ~ $grep -l include *.cpp
dscann.cpp
dscsstfn.cpp
dsfcfs.cpp
dsfcfsn.cpp
dsscan.cpp
dssstf.cpp
pwrap.cpp
/home/jgarrido ~ $
```

Regular expressions can be used with the `grep` command. For example, the following command line uses the `grep` command to display lines in file *dscan.cpp* that start with letters *d* through *f*.

```
/home/jgarrido ~ $grep '^[d-f]' dsscan.cpp
double rev_time = 4;           // revolution time (msec.)
double transfer_rate = 3;      // in Mb/sec (Kb/msec)
double seek_tpc = 15.6;        // seek time per cylinder (msec.)
double simperiod;              // simulation period
```

The following command lines provide more examples using the `grep` command with regular expressions.

```
grep '[a-f]\{6\}' pwrap.cpp
grep '<cl'myfile.dat
grep 'et\> myfile.dat
egrep n "lass|object" pwrap.cpp
egrep v "class|object" pwrap.cpp
egrep "^O" pwrap.cpp
egrep "^O|^C" pwrap.cpp
```

The `egrep` command does not directly accept a string as input, it accepts files. When there is no file specified, the `egrep` command takes input from standard input. The following script file takes two input strings as parameters and checks if one string is a substring of the other. The script file converts the first string into a file with the `echo` command. The file created by executing the `echo` command is piped into the `egrep` command.

```
# An example script file that locates a substring using egrep
# Script file: testegrep
#
if [ $# != 2 ]
then
    echo "Two parameters are required"
    exit 1
fi
#
# egrep does not directly accept a string as input
if echo $1 |  egrep $2 > myfile
then
    echo " $2 is in $1"
else
    echo "Not found"
fi
exit 0
```

The command line shows the execution of the `testegrep` script file and the output produced.

```
/home/jgarrido ~ $./testegrep September emb
 emb is in September
/home/jgarrido ~ $
```

B.14   Evaluating Expressions

The expr command evaluates an expression and sends the result to standard output. The expressions can be used to manipulate numbers and strings. The following operators used with the expr command to manipulate strings.

With arg1 : arg2, the expr command searches for the pattern *arg2* (a regular expression) in *arg1*. The result is the number of characters that match. The following examples show the use of the (:) operator.

```
# number of lowercase letters at the beginning of var1
expr $var1 : '[a-z]*'
#Match the lowercase letters at the beginning of var1
expr $var1 : '\([a-z]*\)'
#Truncate var1 if it contains five or more chars
Expr $var1 : '\(..\)' \| $var1
#Rename file to its first five letters
Mv $mf `expr $mf : '\(..\)' \| $mf`
```

With the index string character-list format, the expr command searches the *string* for the first possible character in the *character list*. The following example illustrates the use of the *index* operator of the expr command.

```
/home/jgarrido ~ $expr index "jose" def
4
```

With the length string format, the expr command gets the length of the *string* variable. The following example illustrates the use of the *length* operator of the expr command.

```
/home/jgarrido ~ $expr length "kennesaw"
8
```

With the substr string start length format, the expr command searches for the portion of the *string* variable starting at the specified location. The following example illustrates the use of the *subr* operator of the expr command.

```
/home/jgarrido ~ $expr substr "kennesaw" 5 3
esa
```

Writing more powerful shell scripts is possible with additional external commands (shell utilities), such as sed and awk.

Bibliography

[1] Hossein Arsham. *Deterministic Modeling: Linear Optimization with Applications*. Web site (University of Baltimore): http://home.ubalt.edu/ntsbarsh/opre640a/partIII.htm.

[2] J. J. Banks, S. Carson, and B. Nelson. *Discrete-Event System Simulation*. Second ed. Englewood Cliffs, NJ: Prentice-Hall, 1996.

[3] John W. Eaton. *GNU Octave Manual*. Network Theory Limited, United Kingdom, 2002.

[4] Joyce Farrell. *Programming Logic and Design*. Second edition. Course Technology (Thompson) 2002.

[5] Behrouz A. Forouzan. *Foundations of Computer Science: From Data manipulation to Theory of Computation*. Brooks/Cole (Thompson), 2003.

[6] G. Fulford, P. Forrester, and A. Jones. *Modeling with Differential and Difference Equations*. Cambridge University Press, 1997.

[7] Garrido, J. M. *Object-Oriented Discrete-Event Simulation with Java*. New York: Kluwer Academic/Plenum Pub., 2001.

[8] GNU Scientific Library. Reference Manual. Free Software Foundation, 2011. www.gnu.org/software/gsl/manual.

[9] Roe Goodman. *Introduction to Stochastic Models*, Reading, MA: Addison-Wesley, 1988.

[10] Rod Haggarty. *Discrete Mathematics for Computing*. Addison Wesley (Pearson), Harlow, UK, 2002.

[11] Dan Kalman. *Elementary Mathematical Models*. The Mathematical Association of America, 1997.

[12] Jack P. C. Kleijnen. *Simulation: A Statistical Perspective*. New York: Wiley, 1992.

[13] Roland E. Larsen, Robert P. Hostteller, and Bruce H. Edwards. *Brief Calculus with Applications*. Alternate third ed., D. C. Heath and Company, 1991.

[14] Averill M. Law and W. David Kelton. *Simulation Modeling and Analysis*. Third Ed. New York: McGraw-Hill Higher Education, 2000.

[15] Charles F. Van Loan and K.-Y. Daisy Fan. *Insight Through Computing: A MATLAB Introduction to Computational Science and Engineering*. SIAM-Society for Industrial and Applied Mathematics, 2009.

[16] E. B. Magrab, S. Azarm, B. Balachandran, J. H. Duncan, K. E. Herold, and G. C. Walsh. *An Engineer's Guide to MATLAB: With Applications from Mechanical, Aerospace, Electrical, Civil, and Biological Systems Engineering*. Third Ed. Prentice Hall, Pearson, Upper Saddle River, NJ, 2011.

[17] J. Medhi. *Stochastic Processes*. 2nd ed. New York: John Wiley, 1994.

[18] I Mitrani. *Simulation Techniques for Discrete Event Systems*. Cambridge: Cambridge University Press, 1982 (Reprinted 1986).

[19] Douglas Mooney and Randall Swift. *A Course in Mathematical Modeling*. The Mathematical Association of America, 1999.

[20] Holly Moore. *MATLAB for Engineers*. Sec Ed. Prentice Hall, Pearson, Upper Saddle River, NJ, 2009.

[21] Barry L. Nelson. *Stochastic Modeling, Analysis and Simulation*. New York: McGraw-Hill, 1995.

[22] E. Part-Enander, A. Sjoberg, B. Melin, and P. Isaksson. *The MATLAB Handbook*. Addison-Wesley Longman, Harlow, UK, 1996.

[23] D. M. Etter. *Engineering Problem Solving with C*. Pearson/Prentice Hall, 2005.

[24] Harold J. Rood. *Logic and Structured Design for Computer Programmers*. Third edition. Brooks/Cole (Thompson), 2001.

[25] Rama N. Reddy and Carol A. Ziegler. *C Programming for Scientists and Engineers*. Jones and Bartlett Pub. Sudbury, Massachusetts, 2010.

[26] Angela B. Shiflet and George W. Shiflet. *Introduction to Computational Science: Modeling and Simulation for the Sciences*. Princeton University Press, Princeton, NJ, 2006.

[27] L. F. Sampine, R. C. Allen, and S. Prues. *Fundamentals of Numerical Computing*. John Wiley and Sons, New York, 1997.

[28] David M. Smith. *Engineering Computation with MATLAB*. Addison-Wesley, Pearson Education, Boston MA, 2010.

[29] Robert E. White. *Computational Mathematics: Models, Methods, and Analysis with MATLAB and MPI*. Chapman and Hall/CRC. September 17, 2003.

[30] Wayne L. Winston. *Operations Research: Applications and Algorithms*. PWS-Kent Pub. Co., 1987.

[31] Wing, Jeannette M. "Computational Thinking". *Communications of the ACM*. March 2006. Vol. 49, No. 3.

Index

Milton Keynes UK
Ingram Content Group UK Ltd.
UKHW031125141024
449569UK00006B/440